CAMBRIDGE LIBRARY COLLECTION

Books of enduring scholarly value

Technology

The focus of this series is engineering, broadly construed. It covers technological innovation from a range of periods and cultures, but centres on the technological achievements of the industrial era in the West, particularly in the nineteenth century, as understood by their contemporaries. Infrastructure is one major focus, covering the building of railways and canals, bridges and tunnels, land drainage, the laying of submarine cables, and the construction of docks and lighthouses. Other key topics include developments in industrial and manufacturing fields such as mining technology, the production of iron and steel, the use of steam power, and chemical processes such as photography and textile dyes.

Lightning Conductors and Lightning Guards

As a result of being asked to give public lectures on the subject, the eminent physicist Oliver Lodge (1851–1940) published in 1892 a pioneering study of the protection of buildings, cables and telegraphic instruments from the devastation caused by lightning strikes. This work led him almost immediately to the discovery of electromagnetic wave transmission and ultimately to the development of a version of radio telegraphy. Lodge also saw that many of the current theories about the nature of lightning were seriously in error, and his investigations led to a number of significant changes in the design of lightning conductors and lightning guards. Some of the methods and procedures that Lodge advocated have since become standard practice. They are described with Lodge's characteristic flair and accompanied by a wealth of illustrations that give a fascinating insight into how contemporary scientists and engineers tackled this significant problem.

Cambridge University Press has long been a pioneer in the reissuing of out-of-print titles from its own backlist, producing digital reprints of books that are still sought after by scholars and students but could not be reprinted economically using traditional technology. The Cambridge Library Collection extends this activity to a wider range of books which are still of importance to researchers and professionals, either for the source material they contain, or as landmarks in the history of their academic discipline.

Drawing from the world-renowned collections in the Cambridge University Library and other partner libraries, and guided by the advice of experts in each subject area, Cambridge University Press is using state-of-the-art scanning machines in its own Printing House to capture the content of each book selected for inclusion. The files are processed to give a consistently clear, crisp image, and the books finished to the high quality standard for which the Press is recognised around the world. The latest print-on-demand technology ensures that the books will remain available indefinitely, and that orders for single or multiple copies can quickly be supplied.

The Cambridge Library Collection brings back to life books of enduring scholarly value (including out-of-copyright works originally issued by other publishers) across a wide range of disciplines in the humanities and social sciences and in science and technology.

Lightning Conductors and Lightning Guards

Oliver Lodge

CAMBRIDGE UNIVERSITY PRESS

Cambridge, New York, Melbourne, Madrid, Cape Town,
Singapore, São Paolo, Delhi, Mexico City

Published in the United States of America by Cambridge University Press, New York

www.cambridge.org
Information on this title: www.cambridge.org/9781108052153

This edition first published 1892
This digitally printed version 2012

ISBN 978-1-108-05215-3 Paperback

LIGHTNING CONDUCTORS AND LIGHTNING GUARDS.

THE

SPECIALISTS'

SERIES.

LIGHTNING PHOTOGRAPH BY MR. A. W. CLAYDEN, SHOWING THE CURIOUS PHOTOGRAPHIC APPEARANCE KNOWN AS THE "DARK FLASH."

Frontispiece.

LIGHTNING CONDUCTORS

AND

LIGHTNING GUARDS.

A TREATISE ON THE PROTECTION OF BUILDINGS, OF TELEGRAPH INSTRUMENTS AND SUBMARINE CABLES, AND OF ELECTRIC INSTALLATIONS GENERALLY, FROM DAMAGE BY ATMOSPHERIC DISCHARGES.

BY

OLIVER J. LODGE, D.Sc., F.R.S., LL.D., M.I.E.E.,
PROFESSOR OF PHYSICS IN UNIVERSITY COLLEGE, LIVERPOOL.

LONDON:
WHITTAKER & CO., 2, WHITE HART STREET,
PATERNOSTER SQUARE;
GEORGE BELL & SONS, YORK STREET, COVENT GARDEN.
1892.

CHISWICK PRESS:—C. WHITTINGHAM AND CO., TOOKS COURT,
CHANCERY LANE.

PREFACE.

THIS book is the outcome of a couple of lectures which in 1888 I was asked by Sir Trueman Wood, Secretary to the Society of Arts, to deliver in memory of the late Dr. Robert Mann, an enthusiastic advocate of lightning rods in South Africa and elsewhere, and they are published with the sanction and approval of the Society.

Experiments made by me shortly before the lectures showed me that several of the current ideas on the subject were unfounded and incorrect; mainly because the momentum of an electric current and the energy of an electrostatic charge had both been more or less overlooked by those who had treated the subject. The application of the known fact of electrokinetic momentum revolutionized the treatment of certain phenomena. The old drainpipe idea of conveying electricity gently from cloud to earth was thus proved fallacious, and the problem of protection became at the same time more complex and more interesting.

An idea at one time got abroad that my experiments proved existing lightning conductors to be useless or dangerous; this is an entire misrepresentation. Almost any conductor is probably better than none, but few or no conductors are absolute and complete safeguards. Certain habits of lightning rod practice may be improved, and the curious freaks or vagaries of lightning strokes in protected buildings are intelligible without any blame attaching to the conductor; but this is very different from the contention that lightning rods are unnecessary and useless. They are essential to anything like security. A summary of the changed views will be found in chapter xxvi., p. 366. A still fuller summary was written a year or two ago for the American monthly "The Forum," but, so far as I know, this latter article has not yet appeared, or I would have appended it to this book.

The lightning photographs, which have lost terribly in reproduction and diminution of size, are mostly the property of the Royal Meteorological Society. I acknowledge indebtedness to many writers on the subject; but principally perhaps to the author and reviser of the third edition of "Lightning Conductors," by Richard Anderson, 1885, and to the "Report of the Lightning Rod Conference," 1882, both published by Spon. Remarkably extensive references to the literature of the subject are contained in these works.

To workers who had been led by experience along something of the same lines as experiment and theory have led me, I attempt to do justice in an appendix.

The protection of cables and telegraphic instruments in general from lightning constitutes another branch of the subject, and one that can be treated far more definitely and distinctly than can the protection of buildings. This branch of the subject is discussed at length in Part II.; and instruments made by Dr. Alexander Muirhead are there depicted which are now on trial in several important cable stations.

Many points of considerable theoretical interest arose in the course of the investigation, and these are several times briefly referred to; in fact, the whole matter became far more extensive and interesting than could at first sight have been expected. It rests now with architects, engineers, and practical men generally to determine the modifications which they may think desirable to introduce into existing practice.

O. J. L.

University College: Liverpool.
July, 1892.

CONTENTS.

LIST OF PLATES.

LIGHTNING CONDUCTORS AND LIGHTNING GUARDS.

CHAPTER I.

GENERAL CONSIDERATIONS CONCERNING ATMOSPHERIC ELECTRICITY AND LIGHTNING.

ONE hundred and fifty years ago the nature of lightning was unknown. Several persons surmised that it had some connection with the phenomena excited in a piece of glass tube when rubbed, phenomena which were called electric; but no proof of the connection was attempted. The proof that lightning was in fact nothing but a large electric spark was given in 1751 by that comprehensive and common-sense genius, the Philadelphian printer, Benjamin Franklin.

This great man, a statesman of the first magnitude, might have made a leading experimental philosopher had he lived in quiet times instead of in the stirring period of the Declaration of Independence and the great American War. A space of some twelve years limits his active devotion to electrical matters, but in that time he acquired a masterly grip of the subject, expressing himself very accurately and precisely concerning electrical theory, his statement of which is far superior to a great deal that has quite lately passed current in text-books; indeed, it is only now becoming capable of improvement through the labour and the inspiration of some of the still greater giants of our own day.

From the time that Franklin flew his kite at Philadelphia and ascertained beyond cavil the true nature of lightning—from that time to the present, the protection of buildings and ships from its destructive agency has been mainly a matter of detail and application of the laws of electricity so far as they were known.

For a long time the erection of lightning conductors was opposed by the religious world as heretical and impious. But, first in some Protestant provinces in Germany, and later in France and England, the use of the heretical rods gradually extended. The extension of their use in England and their application to ships, were greatly aided by the labours of that enthusiastic worker, Sir W. Snow Harris. Their extension in our South African colonies, where violent thunderstorms are frequent, is largely due to the influence of the late Dr. Mann, who contributed important papers on the subject to the Society of Arts,[1] who edited the second edition of the hitherto standard British work on the subject, Anderson's "Lightning Conductors," and in whose honour the course of lectures which constitutes the foundation of the present volume were established by the Society of Arts.

Concerning the origin of atmospheric electricity, I have not much to say. It seems to me probably due to friction, as in Armstrong's hydro-electric machine. Faraday showed that the exciting cause in that apparatus of Armstrong's was the friction of water spray driven by steam over the solid surface of the jet; so also I picture winds in the atmosphere driving the spray of mist against

[1] "Journal of the Society of Arts," April 30, 1875, and March 15, 1878.

rock and ice surfaces, and thus gradually producing a certain difference of potential between the upper layers of the atmosphere and the surface of the earth.

It has been discovered by meteorologists that thunderstorms are often associated with curious V-shaped troughs or depressions among the isobars (see for instance Mr. Abercromby's book on "Weather," in the International Scientific Series, pp. 240-259), evidencing a whirl or cyclone with its axis horizontal. Now I would suggest that a horizontal cyclone is very like a cylinder electrical machine, with the earth acting as rubber and the upper regions of air acting as prime conductor; the air which has been charged by friction being discharged as soon as it is carried up to these higher regions, and thus electrifying them continually until they locally discharge. There are, however, various types of thunderstorm, and probably there are many varieties of cause to account for the electrification. Some variety of friction seems to me, however, most probable in all cases.

I have spoken as if I thought the friction had to be between mist globules and solid matter. It seems doubtful whether friction against *air* will suffice to render water electric. If it be efficient, then the constant slipping of mist or dust-particles through the upper layers of the atmosphere is one effective cause of atmospheric electrification; whatever may be the cause, it probably acts continuously, and a rain-shower probably carries down some of the charge to earth, so that after a spell of dry weather there is liable to be an accumulation of electricity, because it has had no recent opportunity of escape.

In the polar regions electrical discharges are mainly silent, or brush-like, giving the fantastic forms of aurora. But in our latitude these silent discharges in the upper semi-conducting rarefied layers of atmosphere are seldom visible. We see the effect in another form. The electrification gets occasionally conducted down by clouds into the lower and denser layers of atmosphere where silent discharge is impossible, until, when the potential rises high enough, they flash either into each other or into the earth, and the strain is partially relieved.

It may thus be that clouds play only a secondary part in the phenomena. The upper regions of atmosphere are at a different potential from the earth—cloud or no cloud. Clouds are able to conduct it down towards the earth, and thick dense clouds are therefore the usual prelude to a thunderstorm.

So much might be said whatever the type of storm; but when it is the type associated with an isobaric "trough," or whirl round an horizontal axis, then the formation of electrified cloud at the summit of the whirl is a simple and natural consequence; the ascending air becoming chilled by expansion and condensing its vapour in the ordinary way. And this visible cloud being a semi-conductor, whereas invisible vapour is a good insulator, we have all the ingredients necessary for an accumulation of electricity by frictional electric-machine-like action without postulating an ascent into the still higher conducting layers of atmosphere. For remember the great height at which these must exist. Air can hardly be called conducting at a greater pressure than about 1 inch of mercury, or say $\frac{1}{30}$ atmosphere; now such a pressure exists at about $3\frac{1}{2}$ times "the height of the homogeneous atmosphere" appropriate to the temperature; and at the freezing point the homogeneous atmosphere is 8 kilometres high. The average temperature will be lower, but one cannot suppose the conducting layers to be much less than 20 miles high, except in excessively cold regions. But the clouds caused by a horizontal whirl may be readily found at a height of a mile or two, or even less.

It is often asserted and believed that the coalescence of small charged globules into larger globules is competent to raise their potential enormously,[1] but Dr. Everett has shown ("Nature," vol. 38, p. 343) that so long as the charge adheres to the individual globules their aggregation makes no difference to the average potential. The mistake arose by forgetting that what would be true for any one isolated globule would not be true of an assemblage; the potential of any one of a crowd being $\Sigma\left(\frac{q}{d}\right)$, which is independent of size, instead of the simple $\frac{q}{r}$ for an isolated sphere.

The converse phenomenon, or the tendency to coalescence of globules produced by feeble electrical charges in their neighbourhood, is, however, a real and important phenomenon, and is best illustrated as discovered by Lord Rayleigh, by means of a very simple experiment with a water-jet and sealing-wax.

A vertical jet about one-twentieth of an inch in diameter and three or four feet high does best, but almost any fairly vertical jet of clean water will serve. A little vertical fountain can be easily arranged at any water-tap by help of a bit of india-rubber tubing and a glass nozzle. It spreads out at the top in the well-known brush of spray, and the drops fall as a scattered shower, like fine rain, until a stick of excited sealing-wax is held a yard or two away; the jet then at once shrinks

[1] See a lecture on "Thunderstorms," by Prof. Tait, in "Nature," August, 1880.

upon itself in width, changing its appearance entirely; its drops collect into large globules and fall as a thunder-shower.

Lord Rayleigh has shown, by examining what goes on by means of intermittent illumination, and by other experiments on jets impinging on each other, that the scattering is due to collisions among the particles, and that two colliding drops or two colliding jets do not unite, but rebound from each other, so long as their surfaces are clean and so long as they are at the same electric potential. But a difference of potential of even one or two volts is sufficient to affect the joint boundary surface during a collision sufficiently to cause a puncture, as it were, or at any rate to unite the surfaces and stop any rebound. Thus it is, that under feeble electrical influences drops accidentally striking each other unite, and thus rapidly grow in size. The obvious connection between this experiment and the notably large drops of thunder-showers need not be insisted on. Every detail of the explanation may not be considered perfectly clear, but the facts are undeniable, and the same kind of explanation which serves for the small-scale experiment must serve also for the large-scale observation.

A process of electrical aggregation of a somewhat analogous kind was detected also by the writer, in conjunction with the late Mr. J. W. Clark, when examining the behaviour of dust-laden air.[1] They found that on electrifying such air by a brush discharge all the particles rushed together, and grew rapidly into flakes and streamers, and rapidly cleared away, either by reason of attraction to the walls of the vessel, or by simple gravitational subsidence. The same thing was found to occur

[1] It has since been pointed out that the first discoverer of this effect was a Mr. C. F. Guitard; in the "Mechanics' Magazine" for 1850,

with the water-dust of visible steam-like cloud or fog, as with any form of solid smoke or dust particles. A cloud exposed in this way to a non-uniform electric field can, therefore, be made to rain. For remember, that in one sense clouds are always raining: the water globules are always sinking through the air, only the rate of sinking is so slow that up-currents more than counterbalance the descent, or else the drops dry up on the way. To make them reach the earth it is only necessary to accelerate their descent by increasing their size, and this is exactly what neighbouring electrification can accomplish: first aggregating the minute globules into drops, which begin to fall as a fine shower, rebounding against each other as they fall; but being liable to further electric influence if they pass anywhere near an electrified body, whereby their collisions result in coalescence, and their rapidity of fall becomes violent.

These experiments throw light on the connection between rain and electrical states of the atmosphere, and render probable the assumption that the weather is more affected by electrical condition than may have been supposed, and that if ever the weather is to be in any sort artificially controlled it must be by the agency of extensive works for the supply of high-tension electricity of definite sign. Perhaps for neighbouring supplies of opposite sign, in order to secure the needful inequality of field. Perfect uniformity of field would tend to keep globules separate, and would result in fog.

Still more recently another effect, possibly a distinct effect, has been observed by the late Robert von Helmholtz, viz., that a cloud of visible steam was darkened and rendered much more opaque by the discharge of electricity into it from a point.

The experiment has been several times exhibited by

Mr. Shelford Bidwell; and it is certainly very striking to see the instantaneous change from a light fleecy cloud of steam, issuing from an orifice of a vessel of boiling water, into an opaque, dark, lurid cloud, the instant it is electrified. The peculiar heavy colour of a thunder-cloud is exactly reproduced, and it is impossible not to suppose the two things connected, though the precise cause of the extra opacity is not specially clear. It seems most likely that extra condensation goes on, and more vapour is condensed, by means of the presence of an extra number of nuclei, in accordance with the discoveries of Mr. Aitken. But whether these nuclei are dust-nuclei, *i.e.*, are fragments of metal torn off by the brush discharge, or whether the chemical dissociation going on in such discharge has itself a nucleus-like power, such as Mr. Aitken found dry flame to have, are matters not yet quite settled.

We have now to think of ourselves as living always between the coatings of a large condenser or Leyden jar, the upper coating of which is the sky, the under coating the earth, while the common air is the dielectric between. Ordinarily the sparking distance is far too great. Every now and then portions of the upper coating protrude down as clouds, and we are then liable to a disruptive discharge. Some square miles of cloud and some square miles of land are the two coatings, and the interval of separation need not be extremely great. If the cloud and the earth were perfect conductors, all this great area would be relieved in a single flash of awful size; but, fortunately, the conduction of cloud is a slow process, and it usually takes a good many flashes from different parts to remove its charge.

The total maximum energy of a given area of cloud at a given height from the earth is easily estimated, for it is well known that as soon as the electric tension of air reaches the limit of about half a gramme weight per square centimetre, disruption occurs. Supposing it all equally on the verge of giving way, the energy of the dielectric per cubic centimetre is therefore $\frac{981}{2}$ ergs, and per cubic mile is $\frac{4{,}110 \times 10^{12}}{2 \times 3 \times 10^{7}}$ foot-tons $= 70{,}000{,}000$ foot-tons.

Given, then, the whole area of cloud facing the earth, and its height when on the point of discharging at every point, and you have the number of cubic miles of strained dielectric, and an approximation to the energy of the storm, at the rate of 70,000,000 foot-tons per cubic mile.

The potential needed to give a spark a mile long is so enormous[1] that the quantity of electricity required to give this energy need not be very great; 217 million electrostatic units of quantity per square mile would give a bursting tension. Now, 217 million electrostatic units is just about 70 coulombs, or not enough to decompose one-hundredth of a gramme (1-7th of a grain) of water. Faraday stated this, but it is often disbelieved.

The energy of any ordinary flash can be accounted for by the discharge of a very small portion of charged cloud, for an area of 10 yards square at the height of a mile would give a discharge of over 2,000 foot-tons energy.

One is not, however, to suppose that because the total quantity is small therefore the current caused by it is weak. Two factors enter into strength of current,

[1] The difference of potential for a spark a mile long, between flat plates, is roughly sixteen million electrostatic units, each one of which is equal to 300 volts; that is, nearly 5,000 million volts.

quantity and time, and if the time of passage of a given quantity be short, as in the case of lightning it is, the current may be furiously strong.

Thinking now of a cloud and of the earth under it as forming the two coats of a Leyden jar, in the dielectric of which houses and people exist, we have now to consider what determines a discharge, and what happens when the discharge occurs. The maximum tension which air can stand is $\frac{1}{2}$ gramme weight per square centimetre. At whatever point the electric tension rises to this value, smash goes the air. The breakage need not amount to a flash, it must give way along a great length to cause a flash; if the break is only local, nothing more than a brush or fizz need be seen. But when a flash does occur it must be the weakest spot that gives way first—the place of maximum tension—and this is commonly on the smallest knob or surface which rears itself into the space between the dielectrics.

If there be a number of small knobs or points, the glows and brushes become so numerous that the tension is greatly relieved, and the whole of a moderate thunder-cloud might be discharged in this way without the least violence. This is by far the best way of protecting anything from lightning; do not let the lightning flash occur if you can possibly avoid it. But one cannot always prevent it, even by a myriad of points. Sometimes a cloud will descend so quickly, or it will have such a tremendous store of energy to get rid of, that no points are sufficiently rapid for the work, and crash it all comes at once. One specially noteworthy case when points are no protection occurs when one cloud sparks into another, and thence to the ground; or, in general, whenever electric strain is thrown quite suddenly upon a layer of air. (See Chaps. IX. and XVI. for details.)

When a flash occurs, a considerable area is relieved of strain, and the rush of electricity along the cloud and along the ground toward the line of flash sets up a state of things very encouraging to another or secondary flash or flashes, practically simultaneous with the first.

One consequence of this is known as the back stroke or return stroke. It was studied by Lord Mahon, and depicted in his work, "Principles of Electricity," published in 1780. Professor Tyndall has made it extremely well known, so that a reference to it is made in almost every elementary science school in the country.

But the popular account of the matter is, I believe, very inadequate. It says the man's electrical condition is disturbed by the inductive action of the cloud, and that on the cloud being discharged the man's original condition is restored; the passage of the induced charge from his hat to the ground being enough to kill him.

Now the shock that a man could get that way is but feeble; it is only twice what he would get if completely isolated and exposed to inductive action. The amount of charge stored up in a man's hat is not great; the electric tension is more likely to pull his hat off than its release is likely to do him any damage.

I do not deny the existence of this static return shock, but I assert it to be impotently feeble. One can feel all there is to feel by holding a hat near an electric machine till nearly bursting tension, and then discharging the machine. There is no special object in waiting for a thunderstorm.

There is more in the real "return stroke" than a mere recovery from statical disturbance of equilibrium. It is a matter to which I must return at some length, but I will just say here that it is due to electrical momentum or inertia; it is a surging and splash off of an electric charge whose equilibrium has been disturbed by a local discharge at some distance. It is a kind of echo of the main flash; the echo being caused, not by mere static induction or its release, but by the far more powerful agency of electro-magnetic momentum. (See Chapter X.)

The following figures are reduced copies from photographs of actual lightning flashes. Plates I. and II. illustrate multiple and branching flashes, and show how a building may be struck in a number of places at once. Plates III. and IV. show the curious meandering of some flashes, principally those between cloud and cloud. The ribbon effect of Plate III. is supposed to be due to the unsteadiness of the camera; and if so, demonstrates that this particular flash was of the slow or fizzing order, not the sudden and destructive kind.

I.

MULTIPLE FLASHES.

To face p. 12.

II.

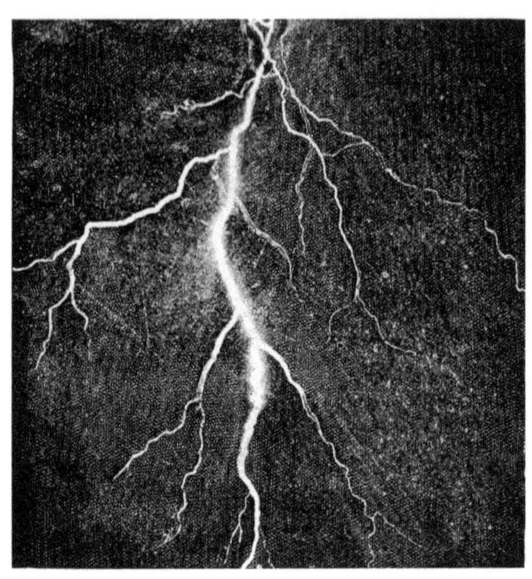

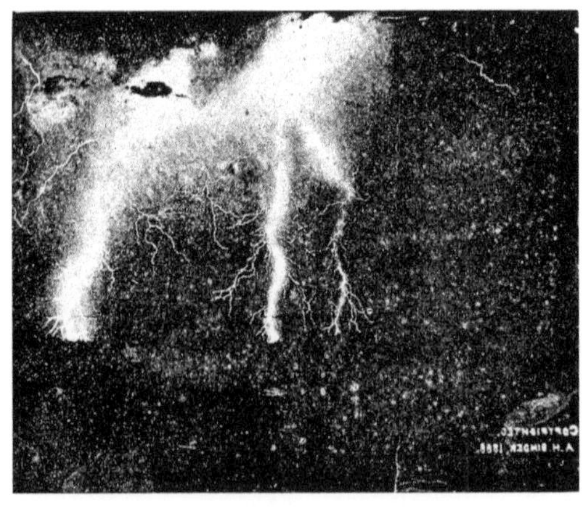

MULTIPLE AND BRANCHING FLASHES.

To face p. 12.

III.

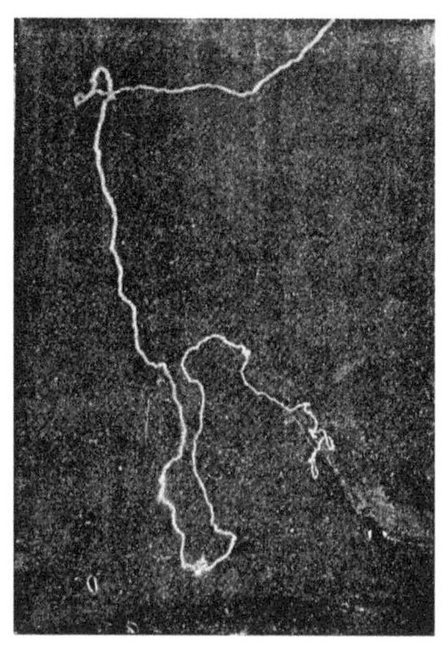

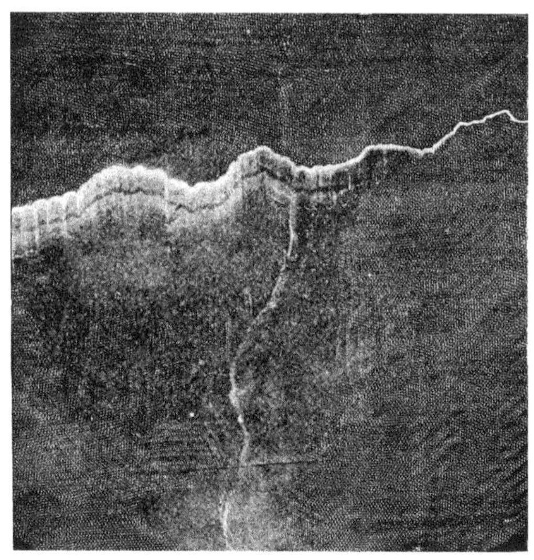

MEANDERING FLASH AND REPEATED OR INTERMITTENT FLASH IN MOVING CAMERA.

To face p. 12.

IV.

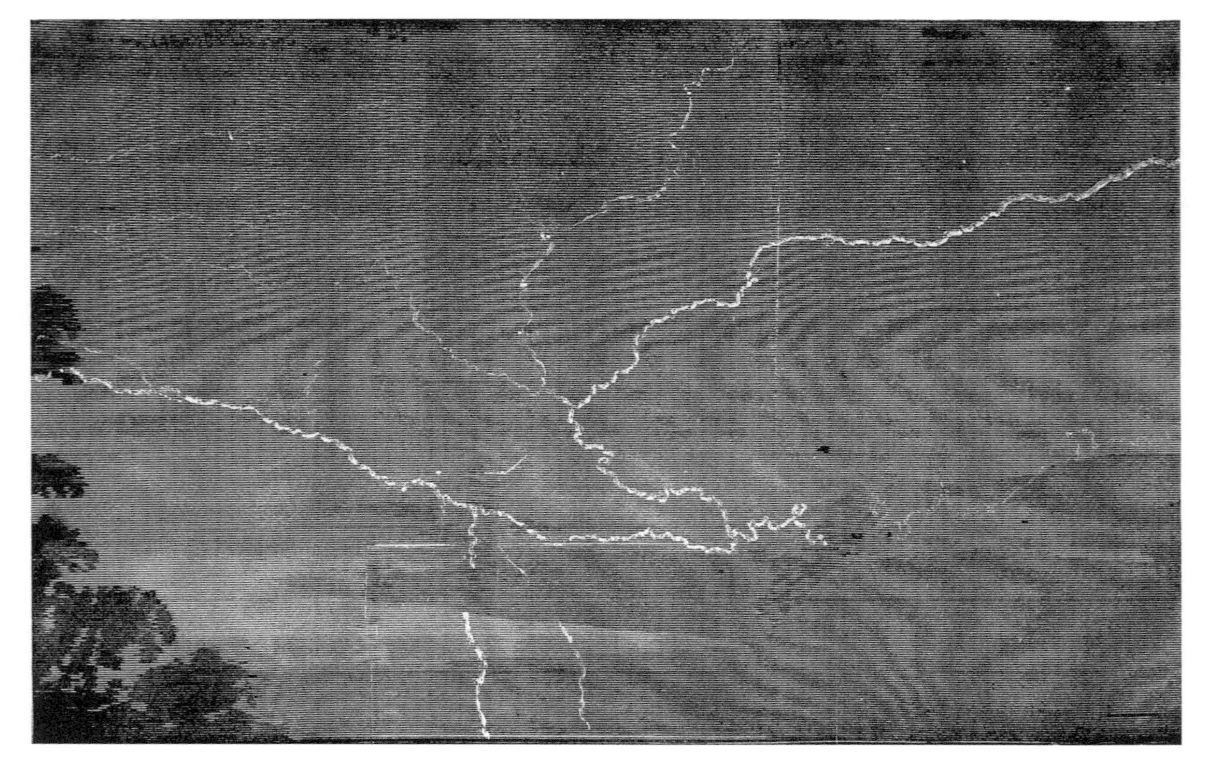

MEANDERING FLASH.

To face p. 12.

V.

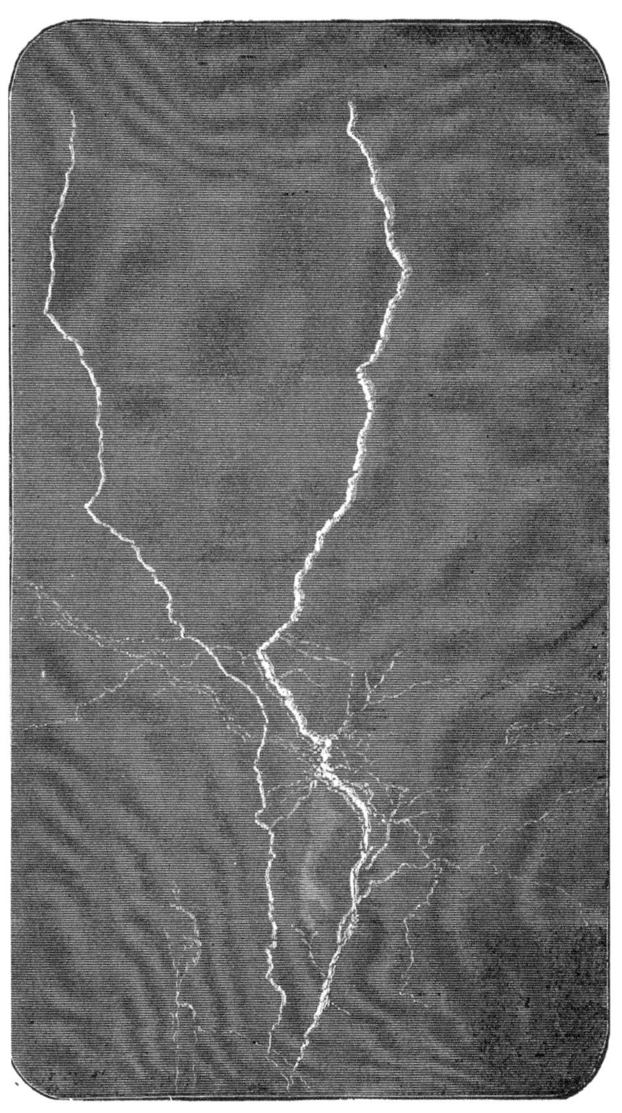

MULTIPLE FLASH.

To face p. 12.

CHAPTER II.

GENERAL CONSIDERATIONS REGARDING DAMAGE BY LIGHTNING.

Now proceed to the kind of damage done when a building is struck, and to the customary and orthodox modes of protecting them from the effects of the flash when it does occur, as well as if possible of warding off the flash altogether by silent discharge. The two main destructive aspects of a lightning flash are—(1) its disruptive, or expanding or exploding violence; (2) its heat.

The heating effect is more to be dreaded when the flash is slow and much resisted; the bursting effect when conducted well except at a few places. A noteworthy though obvious thing is, that the energy of the discharge must be got rid of somehow. The question is, how best to distribute it. It is no use trying to hocus-pocus the energy out of existence by saying you will conduct the charge to earth quite easily and quietly. Conducting the charge to earth is no secure mode of getting rid of the energy, and unless the energy is exhausted the charge will rise up again, and so may swing up and down a good many times before the store of energy is all gone; and nothing can be worse as regards disruptive effect than this repeated and violent passage of an enormous electric current.

The disruptive effect is well shown by the effect of lightning on trees. It is as if every cell were burst by the expansion in the path of the discharge. The effect on conductors is, however, just as marked. Here are five specimens of wires deflagrated on glass by a Leyden jar discharge; gold, silver, copper, iron, and platinum; each has its characteristic trace, by which it can easily be recognized. (Plates VI., VII., VIII.)

St. George's, Leicester, is a curious case of explosive damage to a building, for the rod of the vane conducted the flash half-way down the spire, where it blew a ring of stones out, and so dropped the top half of the spire neatly inside the bottom half, making a tremendous smash, carrying away all the floors of the tower, and beating in the foundation-arch. (Plate IX.)

Ships may have a mast utterly destroyed and split to pieces—thick iron hoops binding the mast being rent asunder and flung about by the force of the expansion; or on the other hand, if the discharge is slower the heat of the flash may ignite the sails, or other combustible matter.

Now take a few examples of buildings or ships more or less protected by conductors. Some of these examples are very instructive as calling attention to the vagaries and unexpected behaviour of powerful flashes. It is these vagaries which I consider have been hitherto unexplained, and it is precisely these to which I wish to direct special attention.

I shall hope to show how closely and completely they can be illustrated by laboratory experiments, and I believe that it is to a neglect and misunderstanding of these phenomena that so many of the partial failures of conductors have been due.

For that conductors often fail is undeniable. When this

VI.

EFFECT OF LIGHTNING ON TREE.

DEFLAGRATION OF GOLD LEAF.

To face p. 14.

VII.

DEFLAGRATED SILVER WIRE.

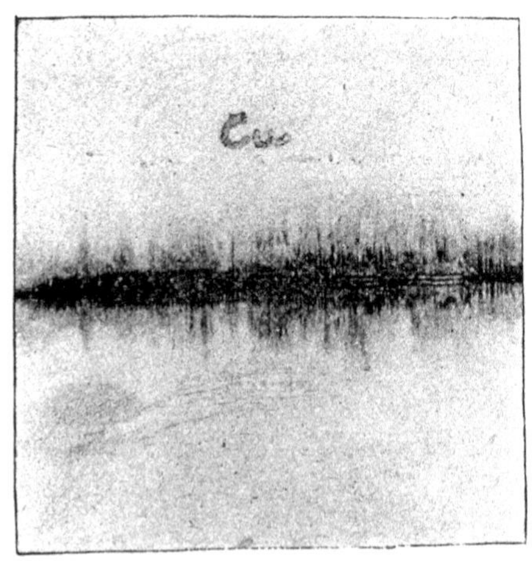

DEFLAGRATED COPPER WIRE.

To face p. 14.

VIII.

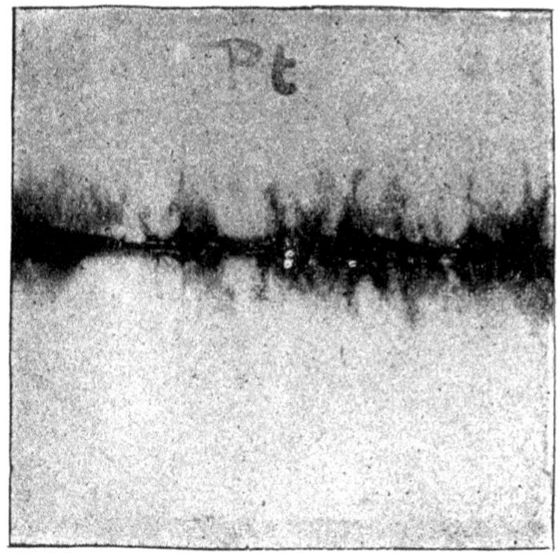

DEFLAGRATED PLATINUM WIRE.

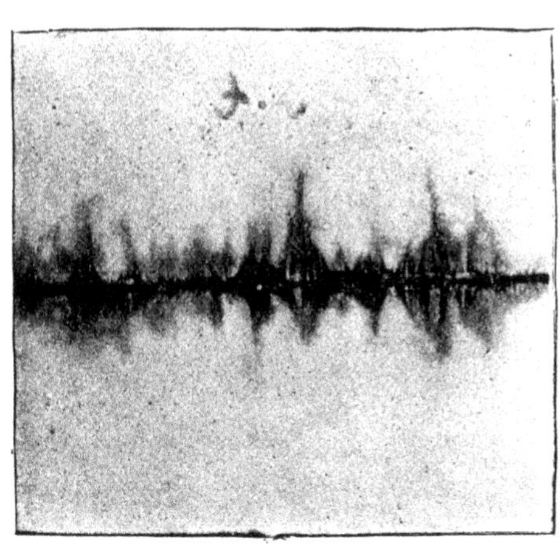

DEFLAGRATED IRON WIRE.

To face p. 14.

IX.

ST. MICHAEL'S CHURCH, CORK.

SECTION OF ST. GEORGE'S CHURCH, LEICESTER.
(*The dotted lines show the course of the discharge.*)

ST. GEORGE'S, LEICESTER.

To face p. 14.

happens it is customary to say they are not properly made, or that there was a faulty joint, or that there was a bad earth. A bad earth is the favourite excuse. A good earth is a good thing undoubtedly, and one cannot well have too much of it; but for a flash to leave a fine thick copper conductor on a tall chimney while still high up, and begin knocking holes in the brickwork in order to make use of the soot, or the smoke, or some bolts, or other miserable conductors of that sort, because it is not satisfied with the moderate allowance of earth provided for it at the bottom, is evidence either of simple perverseness or else of something more deep-seated, and not yet properly called attention to.

If the earth is bad, the flash can show its displeasure when it gets there by tossing it about, and boring holes into it, and breaking water and gas mains; but at least it might leave the top and middle of the chimney alone, it might wait till it got to the badly conducting place before doing the damage. Yet it is notorious that on high chimneys a flash often refuses to follow a thoroughly good conductor more than a quarter or half way down, but takes every opportunity of jumping out of it and doing damage.[1] Why is this? Well, that is the main question I shall attempt to answer in this book.

It may be said that the effect of the bad earth is to make the whole path so highly resisting that the discharge necessarily declines to take it. Well, if that were so, it need not have come into the conductor at all. It is supposed with one breath to strike the conductor because it affords an easy path to earth, and with the next it is

[1] For an example of the entire failure of a brass rod, 1 in. in diameter, see Anderson's book, p. 137 (reproduced at end of present chapter). For cases of capricious leaving of a conductor for such things as safes and fowling-pieces, see pp. 250 and 273 (reproduced at end of Chapter XI.).

said to leave the conductor because after all it finds it a bad one.

Besides, it need not be so very particular about a little resistance; it has already come through, say, half-a-mile of clear air, it might manage a few feet of dry soil. It strikes violently through the air, enters the conductor, and begins to go quietly. Why does it not continue to go quietly till it gets to the bottom of the good conductor, and *then* begin displaying its vigour by boring holes below, as it has done above? Why should one end have to be so persistently cockered up? Why not insist upon having not only a good "earth," but also a good "sky"?[1]

Let me repeat, a good earth is a good thing, and it is not possible to have too much of it, except on the score of expense; but even if it were so good an earth that it might be almost called a heaven, it would not stop the tendency to side-flash. One would still be liable to spittings out from conductors, especially from tall stout ones, as I shall hope to clearly show further on. However bad an earth may be, it can hardly be worse than one afforded by soot, and bricks, and yards of air.

My position at present is that a good earth is desirable chiefly to prevent damage to pipes, foundations, and other things buried therein; also secondarily to keep down the total obstruction to the discharge and lessen the tendency to side-flash as far as may be. No earth can prevent a tendency to side-flash, but a bad earth may aggravate that tendency unnecessarily. On the ordinarily received principles however, side-flash ought not to

[1] This sentence has been mistakenly read as stating that I consider two or three points towards the ground a sufficient earth. It may now be perceived without much trouble that all this portion is argumentative rather than didactic; practical recipes come later.

X.

HOUSE STRUCK BY LIGHTNING, ILLUSTRATING APPARENT SIDE-FLASH FROM THE CONDUCTOR TO A RAIN WATER SPOUT.

To face p. 16.

occur unless the bottom of the conductor is further (conductively) from the ground than are intermediate portions; on ordinarily received principles it ought to suffice if the conductor was cut off level with, or even a foot or two above, the ground. It would knock things about in jumping the rest of the way, but it should not be expected to leave the conductor until it gets to the break, any more than it should be expected to take the conductor half way down instead of at the top. It is true it sometimes *does* partially take the conductor at the middle, just as it sometimes partially leaves it at the middle, but I say that it is not to be expected to do either of these things on ordinarily accepted principles. There are some differences between the behaviour of positive and negative electricity, but none of such extent as will explain the extraordinary difference between the two or three points pointing skywards, and the extensive roots quite properly advocated at the other end.

Take the examples depicted in Plates X. and XI. A house where the lightning, having struck the conductor, flashes off from it in two places to get to a roof-gutter and a distant water-butt. In a ship the lightning often strikes two places at once, the top of the mast and the yard-arm. In Plate XI. it is striking at the top of the conductor, and also at a point on deck, and at two places on land as well. This is a most instructive example. I have little doubt that the three are echoes or reverberations of the main flash, and excited by it; not excited by induction, as in a coil, still less by mere static return, but by a surging or momentum of electricity, of which I have more to say. The training-ship "Conway," now in the Mersey, was once struck, but was perfectly protected, except as to its flag-staff. The water was seen to be luminous in conducting away this flash from the sides of the ship. Plate XI. indicates the arrangement by Snow Harris of conductors able to accomplish this protection—simple and effective. A ship is an easy thing to protect provided you realize that every mast and every spar is liable to be struck. If you protect the masts you may chance the spars if you like; but you are not to think of areas of protection; all such ideas are perfectly illusory.

Incident referred to in footnote, page 15.

"The little town of Rosstall in Franconia, Bavaria, had a church the steeple of which was 156 feet high, and this stood on the brow of a hill, overlooked the country far and wide, and was itself visible for many miles. Being necessarily much exposed to the influence of lightning clouds, it had been provided with one of the best brass-wire conductors, arranged by Professor von Yelin himself, and made of unusual thickness, being over an inch in diameter. Nevertheless, on the evening of April 30th, 1822, while a dark storm-cloud of great density was passing over Rosstall, a heavy flash of lightning was seen to fall vertically upon the church steeple, and this was followed by a terrible peal of thunder which seemed to shake the earth.

When people looked up they saw that the church clock had been thrown from its place, and that part of the lower wall of the edifice had been scattered upon the ground. It was manifest that the electric discharge from the atmosphere had been of unusual intensity, but it was equally clear that the trusted conductor had not done its work."

XI.

FLASH STRIKING MAST AND DECK AND SHORE SIMULTANEOUSLY.

SNOW HARRIS'S SYSTEM OF PROTECTION.

To face p. 18.

XII.

PROTECTED BUILDINGS.

To face p. 19.

XIII.

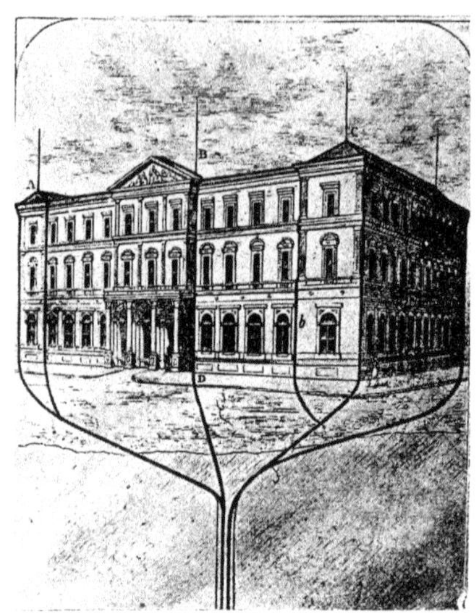

ELABORATELY PROTECTED BUILDINGS.

To face p. 19.

CHAPTER III.

GENERAL CONSIDERATIONS CONCERNING CONDUCTORS FOR HOUSE-PROTECTION.

CONSIDER first a few examples of elaborately protected dwellings and public buildings, according to present ideas. The diagrams in Plates XII. and XIII. pretty well explain themselves.

Nothing projecting upwards is left unspiked, and the earths are thorough. These appear to be examples of excellent, though certainly expensive, protection.

If these houses were powder magazines we should have to be more careful still and make a more critical examination, but as ordinary houses they are safe so long as the conductors are in decent condition.

Now rapidly run over ordinary good orthodox conductors. First, the sky end. Points are good, as explained, though their importance may easily be exaggerated. They can however do no possible harm, and there are occasions when they may do good ; it is customary to make them of platinum when expense is no object, but platinum points are liable to be melted. The best points are cones of copper, not too sharp, and thickly gilt. Gold is better than platinum, in being just as durable and much better conducting, and therefore less liable to melt. Many points are better than one, especially as some are apt to get fused and blunted by some discharge; others may still remain sharp. (Plate XIV.)

True, anything will act as a point when the tension rises high enough, but it is desirable to keep down these dangerous tensions if anyway possible. So soon as a knob begins to emit brushes, the sparking point cannot be far distant; but sharp points will glow and reduce the strain far before the sparking point.

Whether the glow of sharp points is however as effective as has been imagined, is a question for experiment to decide. So far as experiments begun by me and continued by Mr. A. P. Chattock have gone, they assert clearly that the discharging power of needle points is insignificant until they begin to audibly fizz, at which tension small knobs would do nearly equally well. It is very curious to find that hundreds of needle points are insufficient to discharge the electricity supplied by a small inductive machine, unless their tension be raised by bringing near them an earth-connected body. In the free middle of a room, their glow effects next to nothing. However, points have always been believed in since the subsidence of the famous old controversy in favour of knobs, and certainly points are good so far as they go.

Perhaps the best protected building in the world is the Hôtel de Ville at Brussels, the hobby of M. Melsens. The whole system used in this building is excellent and theoretically perfect, so far as I know, in every respect; but it is not cheap, and some people might perhaps hold that it was not artistic.

As for the main conductors, one finds rod, rope, and ribbon. The plan most approved perhaps by the Lightning Rod Conference is this copper tape, which is very nice, neat, and flexible, and free from joints. It is quite the

XIV.

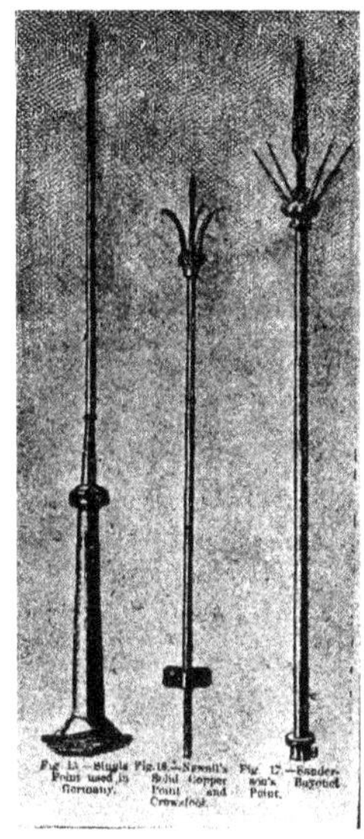

VARIOUS TERMINALS IN USE.

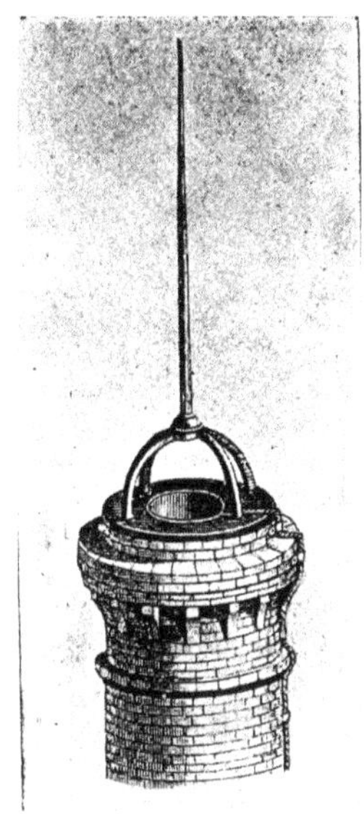

CHIMNEY TERMINALS OFTEN EMPLOYED ABROAD.

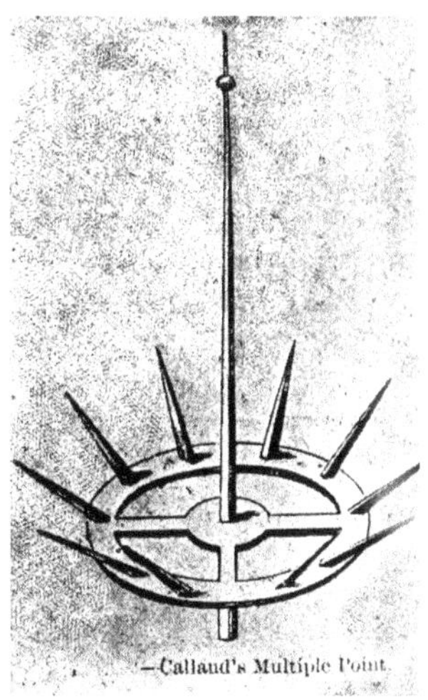

CALLAUD'S MULTIPLE POINT.

To face p. 20.

XV.

TERMINALS OF THE BRUSSELS HÔTEL DE VILLE.

To face p. 20.

XVI.

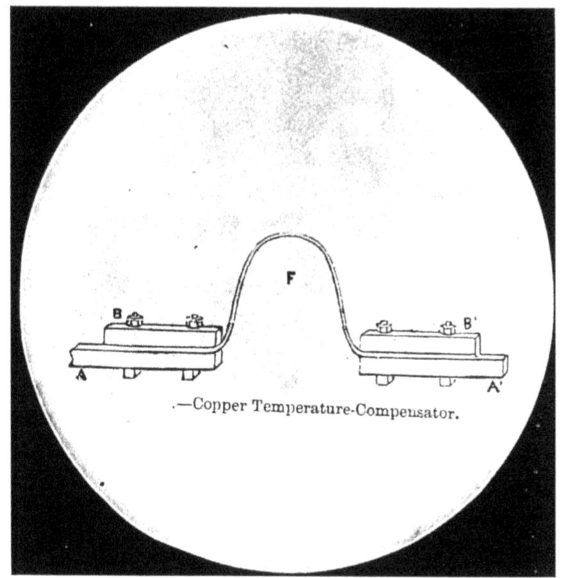

COMPENSATOR ALLOWING EXPANSION.

SANDERSON'S COPPER TAPE.

To face p. 20.

best kind of conductor from the old point of view, and is the kind supplied by the most eminent firms.

Two important matters to be thought of in connection with the conductor are—that it shall not corrode away in places, and that it shall not be liable to be stolen. Another point that must not be overlooked in fixing up any length of conductor is that it is liable to expand and contract. An allowance must be left for this. When it is remembered that it is liable to be exposed to the full glare of the sun, backed up sometimes by a kitchen chimney behind, and then again at another time exposed to the coldest nights, a range of 100° Centigrade is not excessive.

I once rigged up some copper rod battery conductors of three-eighths of an inch thickness between two walls, and one morning after a frost found one snapped clean in half by the contraction. This would be a bad thing to happen to a lightning conductor; so either bends must be left in the rope when put up, or else special compensators must be introduced at intervals.

The allowance for copper is one in 500; say an inch in every forty feet. The best place for a compensator is just above a holdfast, so that it will have to support no weight. A bight or bend in a flexible rope answers every purpose, but bends should not be made too sharp, or the discharge will jump across instead of going round. That is a thing to be remembered. Flexible conductors are very convenient, but their convenience must not be abused. Always take them as straight as possible. Lightning has no time to go round circles—it will jump across sooner. Why should it not be let to jump across? Well, because it burns the conductor. That is the real objection to bad joints—the extra heat; a sort of arc produced there, and the liability to fusing and destruction of the conductor at these parts. There is, moreover, some danger from fire.

Now about the earth end. We have had examples of good earths already. Here is a cheap one advocated by Dr. Mann: wire rope opened out into a brush, and the two ends of another short bit, similarly treated, spliced across the first. Two of the fuzzed out ends make contact with deep soil, one with the surface soil, so that one or other is pretty sure to reach moisture (Plate XVII.).

The whole conductor introduced into the Cape by Dr. Mann, is simple, cheap, and admirable for cottage purposes and for emigrants. Squatters in the States, or Canada, or at the Cape, are far more liable to thunderstorms than we are in this country, and they should certainly rig up one of these homely things. A bit of iron fencing rope will do, with both ends fuzzed out, one supported by a tube fixed to the chimney, the other sunk deep into moist ground, or swamp if available.

In towns, where there are water or gas mains very near the terminus of a lightning conductor, I surmise that they had better be connected to it; and this mainly for their own protection. For if they be not connected, the lightning will not scruple to still make use of them if it chooses, and having to jump across a yard or two of bad conductor on the way, it can easily knock a hole in them or melt them, instead of getting to them quietly.

It must be understood that what I say of the mains underground does not apply to the pipes in a house. To connect lead water-pipes with a lightning conductor might possibly lead to their being melted; but to connect the house gas-pipe with a conductor is a most dangerous proceeding. The neighbourhood of gas-pipes in a house must be scrupulously avoided. It is probably better when possible to avoid even the mains underground, but

XVII.

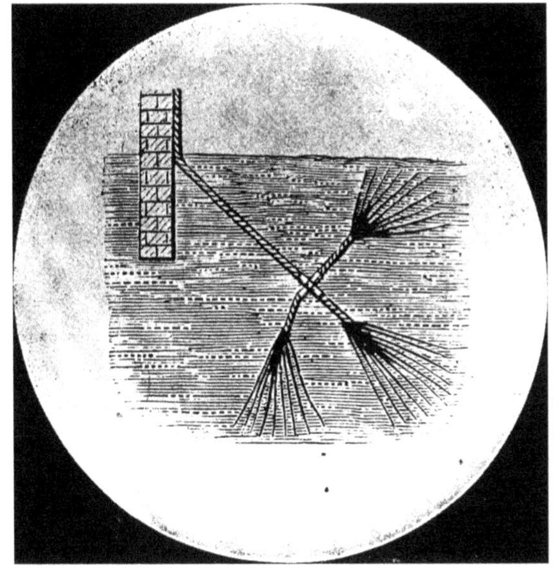

DR. MANN'S CHEAP EARTH.

MAXWELL'S ARRANGEMENT.

To face p. 22.

certainly one does not want lightning rushing along compo-pipes, picking out all the bad joints, and lighting the gas there. In so far as a house contains escaping gas, or weak gas-pipes, it must be treated like a powder magazine, and great care be taken. A ridiculously minute spark may ignite gas without being noticed; the hole in the pipe may quietly enlarge, and the house be burnt down. A considerable amount of damage has been done in this way. So soon as Swan lamps are in universal use, lightning may occasionally play havoc with their filaments and fuse a few cut-outs, but it will not find the leading-wires easily combustible or capable of burning the house down.

Whether it be gas or electric light, however, lightning should, if possible, be kept out of the house-leads—not only because of the danger it may do at joints and insulations, but because gas-brackets and chandeliers are usually conveniently suspended over desks and near arm-chairs, just where an unsuspecting person's head is likely to be; and a spark to one's head is unsafe.

Hitherto I have spoken of the orthodox system of protection, the gather up and carry away system. But, as you know, there is another system suggested by Clerk-Maxwell, the birdcage or meat-safe principle; that is to say, the protective action of a closed conducting sheath. It has long been known that inside an empty hollow conductor there is no trace of electrostatic charge (it was proved last century by Cavendish), and this same screening action may apply to violent electric discharges; at least that is evidently Maxwell's opinion. He does not enter into detail, and possibly he did not contemplate some difficulties that might be suggested on the ground that a hollow conductor is no protection to ordinary currents, and might therefore be no protection to the furious currents of lightning. The screening effect of a hollow conductor is clearer now than it was to any ordinary people when he wrote, possibly clearer than it was even to Maxwell himself, though this would be a rash thing to say. However, the fact is true that in a banker's strong room you are absolutely safe. Even if it were struck, nothing could get at you. In a birdcage, or in armour, you are moderately safe. I should not care to try armour myself, the joints might get unpleasantly hot and explosive. And even the birdcage, if struck by a big enough flash, might get melted. A melted patch on one's protective armour would be extremely disagreeable. Sometimes one is told to get thoroughly wet through instead of seeking shelter in a thunderstorm; but it is a question whether a stroke is more unpleasant than rheumatic fever.

However, a sufficiently stout and closely-meshed cage or netting all over a house will undoubtedly make all inside perfectly safe. Only, if that is all the defence, you must not step outside or touch the netting while outside, for fear of a shock. It would be unpleasant, when you reached home out of a storm, to find it so highly charged as to knock you down directly you tried to go in. An earth connection is necessary as well.

A wire netting all over the house, a good earth connection to it at several points, and all over the roof a plentiful supply of that barbed wire which serves so abominably well for fences, and you have an admirable system of defence.

CHAPTER IV.

FURTHER DETAILS REGARDING CONDUCTORS.

Now let us see how far most people agree, and where they begin to branch out and differ. The old and amusing political controversy between knobs and points has disappeared. Points to the sky are recognized as correct; only it may be better to have more of them, any number of them, rows of them, like barbed wire—not necessarily very prominent—along ridges and eaves. For a point has in no case a very great discharging capacity. It takes several points to discharge readily all the electricity set in motion by a moderately-sized Voss or Wimshurst machine, even under favourable circumstances; hence, if you want to neutralize a thunder-cloud, three points are not so effective as 3,000.

No need, however, for great spikes and ugly tridents, so painful to the architect. Let the lightning come to you, do not go to meet it. Protect all your ridges and pinnacles, not only the highest, and you will be far safer than if you built yourself a factory chimney to support your conductor upon. At present the immediate neighbourhood of a factory chimney or steeple is not a safeguard, but a source of mild danger, even when itself thoroughly protected. If it have no conductor it is, no doubt, still more dangerous.

Next, as to the conductor. Should it be iron or

should it be copper? Should it be insulated from the building, or should it be connected with all the metal it contains? These are questions at present in dispute. The lightning-rod conference approves copper, though not putting it specially and strongly before iron. Durability is its main recommendation. Under all circumstances I am not sure whether it is more durable than galvanized iron. Mr. Preece has great experience of wires in chemical and all other districts, and I believe he upholds iron. Franklin, and the Americans to this day, prefer iron. Certainly it is much cheaper, and not so easy to melt. We will consider the question, and come to a definite conclusion later.

Also the question about connecting up the conductors to all metal masses, roofs, girders, balconies, water-gutters, etc., we had better leave that open too. Nearly everyone condemns insulators, but one eminent authority, M. Callaud, advocates caution and circumspection in what things you connect and what you do not connect. He points out, for instance, that if you connect up a balcony to the conductor, a person standing thereon may become one of its striking terminals. I must say I agree with him. Some there are who advocate connecting *both* ends of a roof-gutter, or other such nearly closed contour, and not only one end. I decidedly agree with this also, for reasons which before long will become abundantly clear. On this point I have, in fact, no doubt.

As regards the shape of the conductor, whether rod or ribbon? Many experiments have been made, notably some by Mr. Preece on the discharge of Dr. De la Rue's battery through conductors of various sectional shapes, to see if extent of periphery matters. Hitherto the results have been negative. But theory clearly points to the fact that a bundle of detached wires is electrically better than a solid rod of the same weight per foot, in every respect except durability. But durability is an essential feature. No shape can be considered satisfactory which aids corrosion. One thing is obvious; plenty of surface encourages cooling, and slightly diminishes danger of melting. Its other much more important advantages we will consider later.

Lastly, the "earth" and its testing. An earth is necessary, or you will have your foundations knocked about and your garden ploughed up. A good earth is desirable. A few tons of coke with the conductor coiled up amongst it is a well known and satisfactory plan if the soil be permanently damp. A bag of salt might perhaps be buried with it to keep it damp throughout; or rain water may be led there. Often, however, the most violent thunderstorms occur after a spell of fine weather, and the soil is likely to be dry. It is best therefore to run your conductor pretty deep, and there make earth.

It is very well to connect the conductor to water-mains if near, but they may be far off or non-existent; and in the most elaborate cases they should not, in my opinion, be used as sole earths. Gas-mains at any rate, if used at all, should certainly be supplemented by a deeper reaching and more reliable earth. In dry weather gas-mains are not earthed at all well, and a strong charge may then surge up and down them and light somebody else's gas in the most surprising way. It may indeed do so even when the soil is damp if other conditions are favourable; and it is difficult to prevent accidents of this kind. The best plan is to have a good deep earth—a well if possible, a boring if not—and to lead the conductor down into it. If the flash likes to make a disturbance when it has to leave the conductor a long way down, no one need grumble. It can't do much harm there.

There is, of course, no magic in water unless it forms a large continuous sheet. An isolated puddle in a rock, such as has been used before now for a lighthouse, is no earth at all. A thoroughly good earth is really a geological question; and for an important building a geological specialist should be consulted.

An occasional test of an earth, in ordinary weather, is no real security as to what may happen after a long-continued drought. It is desirable occasionally to make some test of the underground portion of a conductor just to make sure that it is still there, and has not been accidentally or purposely removed by a workman engaged on drainage or other jobs; and there would be no difficulty in arranging a plan whereby just raising a handle shall give sufficient information as to the state of the earth, without any skilled operator. To this end, two earths should be provided, quite independent of each other (one a water-main, for instance, the other a ton of coke), and they should be connected, first to each other and then to the conductor, by a substantial copper band. Now let the band connecting the two earths pass through some covered outhouse, and have a well over-lapping junction of two flat areas pressed together by a spring, but capable of being raised on or off the other by pulling at a handle or a rope. A galvanometer indicator and Leclanché cell, permanently connected so as to send a current between the two earths directly the handle is raised, will show by its deflection the state of conductivity of the two earths. Very likely the two earths themselves will suffice to give the necessary current without an auxiliary battery.

There is this to be said, however. If a building is so situated, either on high sandy ground or on impervious rock, that a decent earth is very difficult to get, then, at least, the house is not likely to make a better earth than the conductor. That is a weak point in the excuse so often made concerning an accident to a protected building, that the earth was not sufficiently good. It can very seldom be shown that the earth apparently chosen in preference was any better; often it was obviously worse.

It is a superstition to place much reliance on the testing of conductors with a galvanometer and Wheatstone bridge. A galvanometer and Wheatstone bridge are powerless to answer many important questions. A Leclanché cell can no more point out what path lightning will take, than a trickle down a hill-side will fix you the path of an avalanche. The one is turned aside by every trivial obstacle, and really chooses the line of least resistance; the other crashes through all obstacles, and practically makes its own path. A flash strikes a house at one corner, rushes apparently part way down the conductor, then flashes off sideways to a roof-gutter, sends forks down all the spouts, and knocks a lot of bricks out. Another branch bangs through a wall in order to run aimlessly along some bell-wires. Another goes through a window, and down a spade or something propped up against the wall to earth. The lightning tester comes with his galvanometer and Leclanché cell, and reports that the earth of the conductor has 100 ohms resistance; and the accident is therefore accounted for! But how much resistance would he have found in the paths which the lightning seemed to choose in preference to the 100 ohms? Something more like 1,000,000 probably. Or, perhaps, there is a bad joint in the conductor somewhere, the parts being separated by one-sixteenth of an inch. But why should it prefer to jump several yards, and knock holes in walls and windows, rather than jump one-sixteenth of an inch? No; the galvanometer, and Wheatstone's bridge, and Ohm's law, and conductivity, are simply not in it.

Something has been left out of consideration, and something very important too; and until that something is fully taken into account, no satisfactory and really undeniable security can be guaranteed.

That something is inertia—electrical inertia.

Suppose you have a pipe or U-tube full of water, used as perpetual overflow to a cistern, and you want it to be equal to all demands. You test it, and find it perfectly easy to pour the water either way; both ends are perfectly open; the pipe is a good conductor. Then comes someone and hits the stagnant water in your pipe a tremendous blow with a hammer, bursts the pipe, and scatters the water all about. That is what lightning does to your lightning conductor and to the electricity in it. It is no gentle push, it is a terrific blow.

CHAPTER V.

EXPERIMENTS ESTABLISHING THE IMPORTANCE OF ELECTRICAL INERTIA, AND AFFORDING A MEANS OF COMPARING THE EFFECTIVENESS OF DIFFERENT CONDUCTORS.

I MADE several assertions in the last chapter which it is my business now to justify by actual experiment.

The word "inertia" one uses as conveying a correct general notion of the behaviour of an electric circuit to sudden electro-motive forces; a behaviour which is caused by the influence or induction which every portion of a circuit exerts on every other portion. Consider a conducting rod as analyzed into a bundle of parallel wires or filaments, and let a current be suddenly started in all. The rising current in any one filament exerts an opposing force on all the others; and this self-generated opposition E.M.F., due to induction between the different filaments of the conductor, exactly imitates the effects of ordinary inertia as observed in massive bodies submitted to sudden mechanical forces. (For some illustrations of these well-known mechanical effects see a letter by Mr. Maclean in "Nature," vol. 37, p. 612.)

The term commonly employed to denote the electrical inertia-like effect is "self-induction"; which is becoming gradually shortened to "inductance"; its original form when first dealt with by Sir William Thomson was the "electro-dynamic capacity" of the circuit.

Now since electric inertia is due to a mutual action between the filaments into which a conductor may be supposed to be divided, it is manifest that the closer packed they are the greater the inertia of the whole will be; and that to diminish inertia it is only necessary to separate the filaments and spread them out.

The main count of the indictment against ordinary procedure is, that too much attention has been hitherto paid to conducting power, and too little to inertia. In fact, it is not too much to say that practically nothing but conductivity has been attended to, or thought of, in the erection of lightning conductors.

I want to show that conductivity is, from many points of view, of hardly any moment, and that the circumstances of a discharge are regulated far more by inertia than by conductivity. I can even show that, under certain circumstances, high conductivity appears to be an actual objection, and that a stout rod of good conducting copper carries off a flash less well and quietly than a thin wire of badly-conducting iron.

Let us proceed to verify this paradoxical statement at once.

Experiment of the Alternative Path.

The first form of experiment I have to describe is a very simple one. I call it the experiment of the alternative path. It consists in giving a Leyden-jar discharge the choice between a certain conductor and a certain length of air, and in adjusting the length of air until it had as lief take one path as the other.

I am not aware that the particular mode of carrying

out this simple experiment has any special significance, but, to be definite, I depict, in Fig. 1, the symmetrical arrangement I have most frequently, though not exclusively, adopted.

The knobs marked *A* are the ordinary terminals of a Voss machine. The jars stand on an ordinary wood table, and their outer coats are led to the discharger, *B*, the distance of whose air-space can be varied. The alternative path, *L*, is shown by a dotted line. The discharge has to choose between *B* and *L*. Sometimes *L* is absent,

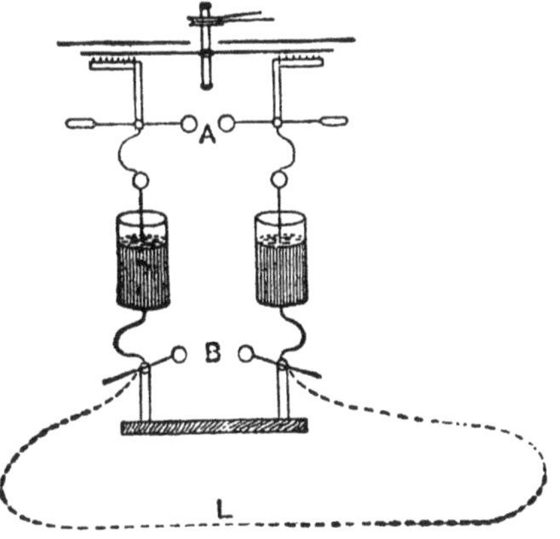

Fig. 1.

and in that case the charging of the jars is quite well effected through the wood of the table; this is the advantage of having the jars imperfectly insulated. At the same time the conducting power of the wood is too low to enable the jars to discharge themselves at all satisfactorily, unless the knobs, *B*, are within striking distance, or unless some path, *L*, is provided. The only discharge obtained at *A* when both paths, *B* and *L*, are absent, is a feeble spitting or intermittent and frequent sparking, very different from the loud report heard as soon as the knobs, *B*, are brought within striking distance. But it is not to be supposed that the *B* knobs must be as close together as the *A* knobs, in order to permit complete discharge ; on the contrary, the *B* knobs may be almost twice as far apart as the *A* knobs, and yet the discharge shall be complete and noisy.

It will be understood that the two sparks occur together ; a spark at *A* precipitates, and is the cause of, the spark at *B*. Not *vice versâ,* for until the *A* spark occurs there is not the slightest tendency for a spark at *B*. The two *B* knobs are at the same zero potential, and may be touched with impunity except at the instant when the flash occurs at *A*. Remember that the jars are standing on a common table all the time.

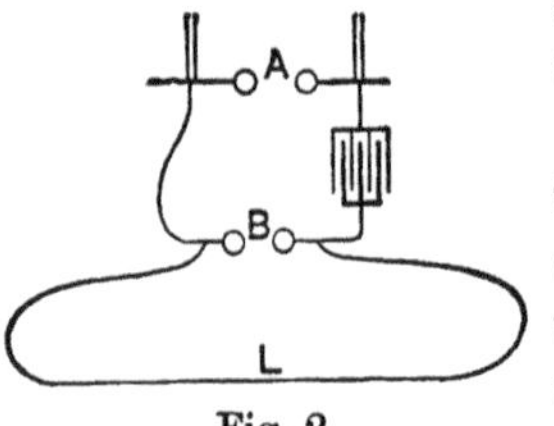

Fig. 2.

Lest it be thought that there is anything occult in this mode of obtaining the spark at *B*, let us subjoin as Fig. 2 another arrangement of connection, which does just as well, and, in fact, represents the first experiment I tried. The condenser is a very large one of tinfoil and glass plates, with carefully insulated terminals. My object in making it was to obtain a sudden rush of a considerable quantity of electricity (like lightning), and then study its behaviour under various circumstances. I found afterwards that for most experiments an ordinary gallon or even pint Leyden jar served just as well, was much quicker in use, and less dangerous. Moreover, the use of insulated terminals necessitates the continued presence of some alternative path (*L*) or other, else of course the condenser declines to charge.

It may be noted at once that with either arrangement

the spark *A* was very loud whenever the spark was allowed to occur at *B* as well; but so soon as the discharge was compelled to traverse the alternative conductor, *L*, by putting the *B* knobs too far apart, then the noise of the discharge was much diminished, not merely because there is now only one spark instead of two, but plainly because, for some reason or other, the discharge meets with considerable obstruction in the wire, whereby its duration is lengthened, and its noise therefore very greatly lessened.

The numbers given below are extracted from a page of the laboratory note-book, and refer to an experiment with two Leyden jars, of a size which I sufficiently specify by calling them "gallon" jars, arranged as in Fig. 1. The length of the *A* spark was maintained at one inch throughout the experiment; or, as it happened to be accidentally convenient to measure lengths in tenths of an inch, I will call the *A* length 10. The *B* length is variable, being altered until a *B* spark sometimes passes and sometimes misses.

The first alternative path I show is a length of about 40 feet of stout No. 1 (B.W.G.) copper wire or rod, suspended round the room by silk ribbon. Its resistance to ordinary currents is very small, being ·025 ohm. Nevertheless, we shall find that the discharge refuses to take this apparently easy path, and persists in jumping the air-gap *B* instead, until the *B* knobs are separated 14·3 tenths of an inch. This is the critical distance. If they are further apart than this, the discharge chooses the thick copper wire by preference, and its noise or suddenness is then much less.

As a contrast with this, I next try a similar length of fine iron wire (No. 27, B.W.G.), whose resistance to ordinary currents is 33·3 ohms, or 1,300 times as great as the other. We find that the discharge very distinctly prefers this wire to the other. For if the *B* knobs remain at the distance of 14·3 we never now get a *B* spark, nor do we get one until they have been brought distinctly nearer. The critical spark length is now 10·3. I confess I was surprised at this result.

Let us next try an enormous resistance—a capillary tube of liquid (very dilute acid), giving to ordinary measurements some 300,000 ohms. The critical spark-length at *B* is indeed a little increased by this great resistance, but not much above that found for the stout copper wire. I have not an exact measure of it, but 16 or 17 will not be far out. The spark at *A* now becomes very quiet indeed, pointing to the fact that what we were observing all along was, in some sense or other, the effect of true resistance; for the undeniable resistance in the capillary tube gives just the same kind of effects as does the copper or iron wire, only a little more pronounced.

I have suspended three other conductors of about the same length, which it is easy to try in the same way, keeping the *A* spark-length 10 all the time. The results are here summarized:—

Alternative Path.				Critical *B* spark-length.
Stout copper wire, No. 1	R = ·025	ohm	...	14·3
Ordinary copper wire, No. 19	R = 2·72	„	...	13·4
Stout iron wire, No. 1	R = ·086	„	...	10·8
Ordinary iron wire, No. 18	R = 3·55	„	...	10·8
Thinnest iron wire, No. 27	R = 33·3	„	...	10·3

The copper wires seem to obstruct almost equally, and the iron wires also obstruct equally among themselves, notwithstanding their very different diameters; but the coppers obstruct more than the irons.

There is nothing absolute about these numbers; they are the record of a definite experiment, but their precise value depends on the circumstances of the experiment. It is easy to arrange so that the iron is less effective than the copper—so that, in fact, ordinary resistance becomes of consequence again. This is done by putting a long lead into the *A* circuit of the jars. But whenever the flash is made as sudden as possible—and there seems little doubt but that a lightning flash is often very sudden—then the order of the numbers is something like the figures quoted.

CHAPTER VI.

GENERAL EXPLANATION OF THESE EXPERIMENTS.

Reason of the enormous Obstruction offered by a good Conductor.

Now, what is the cause of all this astonishing obstruction offered by good conductors to a sudden rush of electricity? One may express it in a popular way, thus: It is due to electrical inertia, or what is also called "self-induction." A current cannot start in a conductor instantaneously, any more than water in a pipe can start moving at full speed in an instant. Give the water a violent blow to make it move, and it resists like a solid. The blow, if very quick and violent, may burst the pipe, but it will not appreciably propel the water. So, in a manner, is it with electricity. The flash occurs, and the conductor must either carry it off at once or not at all. There is no time to think about it, and the E.M.F. needed to overcome this inertia-like obstruction is so great, that a considerable thickness of air may be burst by it, and the discharge may flash off sideways to anything handy.

Another way of putting the matter is this: A lightning discharge is essentially a varying current; it manifestly rises from zero to a maximum, and then dies away again, all in some extremely small fraction of a second, say 100,000th or thereabouts. But that is not all; there is a certain amount of energy to be got rid of, to be dissipated, and it may easily be that a single rush of electricity in one direction does not suffice to dissipate all the stored up energy of the charged cloud. If the conductor is highly resisting, a single rush is sufficient, but if it be well conducting it is quite insufficient. What happens then? The same as would happen with compressed air or other fluid rushing out of an orifice. If it is a narrow jet, there is a one-directioned blast; but if a wide, free mouth be suddenly opened, the escaping air overshoots itself by reason of inertia, and springs back again, oscillating to and fro till the stored up energy is dissipated. Just so is it with an electric discharge through good conductors; it is not a mere one-directioned rush, it is an oscillation, a surging of electricity to and fro, until all the energy is turned into heat.

This fact is often forgotten by lightning-rod men; they speak as if there were a certain quantity of electricity to be conveyed to earth and there was an end of it; but they forget the energy of the electric charge, which must be got rid of somehow. If a great weight, or a large reservoir of water, were propped up above one's house, one would not say that, the safe thing being to get it down as quickly as possible, it was advisable to knock the props away, or to blow the bottom out of the reservoir; no, one would prefer to let it slide slowly and gradually down a well-resisting channel, so as to disperse the energy gradually.

We will remember, then, that a Leyden-jar discharge through good conductors is oscillatory, and that the oscillation continues until all the stored energy is rubbed away. The oscillations have an enormous frequency; they may be millions a second, for the whole lot of them have to cease during the excessively minute duration of the visible flash. It is well known that a flash is of far less real duration than the persistence of optical impression on the retina would lead one to believe; as is easily illustrated by illuminating a spinning wheel by an electric spark. However fast (ordinarily speaking) the wheel is spinning, it appears to be stationary, and the spokes are seen singly and clearly by the light of each spark.

It is for this reason that, although the apparent illuminating power of a powerful flash need not be greater than weak moonlight, its actual intensity, for the instant it really lasts, may exceed strong sunlight, and hence may exert a damaging effect on the retina and cause blindness.

There is another fact which it behoves us to be aware of. It is one to the importance of which the attention of scientific men has only recently been called. Experimentally it has been discovered by Professor Hughes, theoretically by Mr. Oliver Heaviside, Lord Rayleigh, and Professor Poynting; for though the necessary theory is really contained in Clerk-Maxwell, it required digging out and displaying. This has now been abundantly done, but the knowledge has scarcely yet penetrated to practical men; indeed, it has not yet been thoroughly assimilated by most physicists. The fact is this. When a current starts in a conductor, it does not start equally all through its section, it begins on the outside, and then gradually though rapidly penetrates to the interior. A steady current flows uniformly through the whole section of a conductor, a variable current does not. It is started first at the surface, and it is stopped first at the surface.

Remembering the rapidly oscillating character of an electric discharge, remembering also the fact that a rising current begins on the outside surface of a conductor, we perceive that, with a certain rate of alternation, no current will be able to penetrate below the most superficial layer or outer skin of the conductor at all. In the outer skin, of microscopic thickness, electricity will be oscillating to and fro, but the interior of the conductor will remain stolidly inert and take no part in the action.

Thus we arrive at a curious kind of resistance, caused by inertia in a roundabout fashion, and yet a real resistance, a reduction in the conducting substance of a rod, so that no portion except that close to the surface can take any part in the conduction of these rapidly alternating currents or discharges. It must naturally be better, therefore, not to make a lightning conductor of solid rod, but to flatten it out into a thin sheet, or cut it into detached wires; any plan for increasing surface and spreading it out laterally will be an improvement.

Perhaps it may be as well to guard against one favourite misconception. It has long been known that static charges exist only on the surface of conductors; it has also long been known that ordinary currents flow through the whole section and substance of their conductors. It is now beginning to be known that alternating currents may be sufficiently rapid to traverse only the outer layers of conductors, and this last piece of knowledge is felt to be rather disturbing by those who have been accustomed to dwell upon the behaviour of steady currents, and seems like a return to electrostatic notions, and an attempt to lord it over currents by their help. But the first and third facts mentioned above—the behaviour of static charges, and the behaviour of alternating currents—are two distinct facts, independent of each other; not rigorously independent perhaps, but best considered so for ordinary purposes of explanation.

We have thus mentioned two causes of obstruction

met with by rapidly oscillating currents trying to traverse a metal rod. First there is the direct inertia-like effect of self-induction to be added to the resistance proper; the resulting quantity being called by Mr. Heaviside "impedance," to distinguish it from resistance proper. For there is a very clear distinction between them; resistance proper dissipates the energy of a current into heat, according to Joule's law; impedance obstructs the current, but does not dissipate energy. Impedance causes tendency to side-flash, resistance causes a conductor to heat and perhaps to melt. The greater the resistance of a conductor, the more quickly will the energy of a discharge be dissipated, its oscillations being rapidly damped; the greater the impedance of a conductor, the less able is it to carry off a flash, and neighbouring semi-conductors are accordingly exposed to the more danger. Resistance is analogous to friction in machinery; impedance is analogous to freely suspended massive obstruction, in addition to whatever friction there may be. To slowly changing forces friction is practically the sole obstruction; to rapidly alternating forces inertia may constitute by far the greater part of the total obstruction: so much the greater part that friction need hardly matter.

This is a fairly accurate popular statement of the direct way in which self-induction aids resistance proper in obstructing an alternating current. But, in addition to these considerations, there is that other indirect way which we have also mentioned, viz., the fact that conduction of an alternating current may be confined to the surface of a rod or wire if the alternations are rapid enough. This cause must plainly increase total impedance; for the total channel open to such a current is virtually throttled, as a water-pipe would be throttled by a central solid core.

But which part of the total impedance does it affect? Does it increase the resistance part or the inertia part? In other words, does this throttling of a conductor act by dissipation of energy or by mere massive sluggishness? Plainly, it must act like any other reduction of section, it must increase the resistance, the dissipating power of a conductor, the heating power of a current. Hence the resistance of which we have spoken as entering into the total impedance has by no means the same value as it has for steady currents, and as measured by a Wheatstone bridge. It is a quantity greater—possibly much greater—than this; and in order to calculate its value, we must know not only the sectional area and specific conductivity of the conductor but also the shape of its section, the magnetic quality of its material, and the rate of alternation of the current to be conveyed.

CHAPTER VII.

APPLICATION OF THE ABOVE MODE OF EXPERIMENTING TO DETERMINE FURTHER DETAILS.

Tape v. *Rod.*

WE may here note a vigorous controversy, or difference of opinion, between Faraday on the one hand, and Sir W. Snow Harris on the other. Faraday was often consulted about lightning conductors for lighthouses, and consistently maintained that sectional area was the one thing necessary—weight per linear foot—and that shape was wholly indifferent. Harris, on the contrary, maintained that tube conductors were just as good as solid rods, and that flattened ribbon was better still. Each is reported to have said that the other knew nothing at all about the matter. Of course we know that Faraday was thinking of nothing but conduction, and conduction for steady currents. Harris had probably no theoretical reason to give, but was guided either by instinct or by the result of experience. We shall have to admit that, in this particular, Faraday was wrong and Harris was right. The following experiment may serve to illustrate the point further:

I take two copper conductors of the same length and approximately of the same weight, but one of them in the form of wire, the other in the form of ribbon, and I use them successively as an alternative path. The knobs at *A* being fixed at two centimetres apart, the *B* knobs were adjusted until the spark sometimes chose the air-gap, and sometimes the alternative conductor. (See Fig. 1 or 2.) The critical *B* spark length was then:

	Millimetres.
With the wire as alternative path	8·36
With the ribbon " "	6·26
" wire " "	8·45
" ribbon " "	6·05
" wire " "	8·21
" ribbon " "	6·06

Very distinctly showing the advantage of a flattened form of conductor over a mere round section.

The dimensions of the two conductors here compared are as follows:—Wire: No. 12 B.W.G., 218 centimetres long, weight 91·6 grammes. Ribbon: 218 centimetres long and 6·4 centimetres broad, weight 88·7 grammes.

But, it may be said, have not experiments often been made as to the advantage of tape over rod forms of lightning conductor, with negative results? Yes, but the point usually attended to is the deflagration of the conductor. But we are not examining which form of conductor is least liable to be destroyed by a flash—probably there is not much to choose between one form of section and another, for there is no time for surface cooling—we are examining which form will carry off a charge most easily, and with least liability to side-flash; and here thin ribbon shows distinct advantage over round rod.

Another form in which the experiment has been tried, by Mr. Preece, for instance, with Dr. De la Rue's battery, is to pass a discharge through rod or through ribbon and see if it is equally able to deflagrate a thin wire inserted in another portion of its path. It is found just as energetic in one case as in the other. But so it must be unless it could be held that rod *dissipates* energy easier than tape. Putting conductors in series in this way, so that the whole discharge current is bound to go through them anyhow, affords no indication at all as to its preference for one path over another if it had a choice.

Iron v. *Copper*.

You remember we have found that a rod of iron carries off a discharge more satisfactorily than a rod of copper.

But everyone will say—and I should have said before trying—surely iron has far more self-induction than copper. A current going through iron has to magnetize it in concentric cylinders, and this takes time. But experiment declares against this view for the case of Leyden-jar discharges. Iron is experimentally better than copper. It would seem, then, that the flash is too quick to magnetize the iron; or else the current confines itself so entirely to the outer skin that there is nothing to magnetize. A tubular current would magnetize nothing inside it. Somehow or other, the peculiar properties of iron, due to its great magnetic permeability, disappear.

I do not believe anyone could have expected this result. Possibly Lord Rayleigh might have predicted it, and perhaps Mr. Oliver Heaviside. It would scarcely become me to express admiration for the work of so great a master of science as Lord Rayleigh (though, parenthetically, I may mention that I feel such admiration in the highest degree), but I must take the opportunity to remark what a singular insight into the intricacies of the subject, and what a masterly grasp of a most difficult theory, are to be found among the writings of Mr. Oliver Heaviside. I cannot pretend to have done more than skim these writings, however, for I find Lord Rayleigh's papers, in so far as they cover the same ground, so much pleasanter and easier to read; though, indeed, they are none of the easiest.

Can this suggestion with regard to iron be examined and verified or disproved, in some other more direct way?

It is easy to try another experiment. I have here two conductors made of tinfoil; each is made of a long slip of tinfoil, about 3 inches broad, and 21 feet long. One is zigzagged backwards and forwards, with the interposition of three thicknesses of paraffin paper between each zigzag to secure insulation, so as to abolish self-induction as far as possible. This I call the tinfoil zigzag. The other is coiled spirally on a glass tube—again with the interposition of paraffin paper—so as to give as much self-induction as possible. This I call the tinfoil spiral. A bundle of fine iron wires—the core of an induction coil, in fact—can be introduced into this glass tube, or withdrawn, at pleasure. The resistances of these conductors, measured in the ordinary way, are: ·614 ohm for the spiral, and ·708 ohm for the zigzag. They were intended to be alike.

The connections being made as in Fig. 1, one or other of these tinfoil conductors is used as the alternative path *L*, with this result:

The length of the *A* spark being 7·3, the critical spark length at *B*, when sparks sometimes passed and sometimes failed, was 11·1 when no alternative path was provided. When the tinfoil zigzag connected the outsides of the jars instead of the wire, *L*, it was not possible to get a *B* spark till the distance was shortened to 0·6; when the zigzag was replaced by the spiral, the critical *B* spark length rose to 6·4. The iron bundle was now inserted in the spiral, and the experiment tried again. The *B* spark length remained 6·4. The iron made no perceptible difference to the impedance of the coil.

Here is a magnetic time-lag raised to an extreme. Professor Ayrton tells me he has noticed that the permeability of iron begins to diminish with very quick alternations. Here it is becoming virtually no greater than that of air. It may be said that the iron fails to get magnetized because of the opposing action of the inverse "Foucault" currents induced in it, just half a period behind the inducing currents. I thought this would be so, of course, with thick rods of iron, but with a bundle of thin wires I felt doubtful. Lord Rayleigh, however, thinks these induced peripheral currents competent to explain magnetic time-lag in every case; and I can have no doubt that he is right. Whatever the explanation, the fact of time-lag in iron is patent. Yet there is something strange about it, for that a steel knitting-needle can be magnetized by discharging a Leyden jar round it is mentioned in every text-book, and (what does not necessarily follow) it is certainly true. There are points here requiring further examination. [See further on, where it is shown that an iron core has some effect on the noise of the discharge, making it much quieter; apparently by helping to dissipate energy.]

However, if it turns out to be true that an iron rod does not get magnetized by the passage of a rapidly alternating current, it may be held a natural consequence of the fact that such currents flow mainly in its outer surface; and that such tubular currents have no magnetizing power on anything inside them.

The magnetizability of iron is, therefore, no objection to its employment in lightning conductors; but rather the reverse. Its inferior conductivity is an advantage, in rendering the flash slower and therefore less explosive. Its high melting point and cheapness are obvious advantages. It is almost as permanent as copper, at least when galvanized; and it is not likely to be stolen. I regard the use of copper for lightning conductors as doomed.

CHAPTER VIII.

FURTHER EXPERIMENTS.

Experiment of the Bye-Path.

We have seen that a conductor is more efficient in carrying off a discharge and preventing side-flash, in proportion as its self-induction is lessened; say by spreading it out into a thin sheet, or cutting it up into a number of wires, or otherwise. But no conductor is able to prevent side-flash altogether, unless it is zig-zagged to and fro so as to have practically no self-induction; in that case the side-spark is nearly stopped. But so long as a conductor is straight (and a lightning

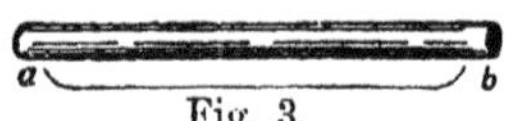

Fig. 3.

conductor must of course be straight), so long will there be some tendency to side-flash, however thick it be made. It may be a foot or a yard thick, and yet not stop side-flash.

One may easily try the following experiment. Take a yard of stout brass or copper rod an inch thick, arrange it in the path of a Leyden-jar discharge, and then arrange, as a sort of bye-path or tapping circuit, some very fine wire, such as Wollaston platinum wire (Fig. 3). It may seem absurd for any portion of the discharge to leave the massive rod and take the hair-like wire by preference, especially if an air-gap exists at *A* or at *B*, or at both. Nevertheless a portion *does* choose the fine wire path, and you get a little spark at *A* or at *B* about a sixteenth of an inch long. One may vary the experiment by trying to get a shock by holding two different parts of a thick copper rod through which a discharge is passing. The mere difference of potential between conductor and earth must of course be avoided. It is not easy to get an appreciable shock from a few yards of very stout conductor; still it can be done. Holding two points of a stout open spiral, consisting of about five yards of No. 1 copper wire, connected to earth, a faint shock can be felt with wetted hands whenever a Leyden jar is discharged through the copper. No doubt with a very large condenser the shock would be quite noticeable; and a man touching a lightning conductor, however well earthed, might perhaps receive a shock sufficient to kill him. At all events, I should not care to try the experiment with a real lightning flash.

Experiment of the Side-Flash.

Let me illustrate the tendency to side-flash by an experiment which looks more directly applicable. Fig. 4 shows a couple of tin-plates or tea-trays fixed a foot or so apart, one earthed, the other insulated. They obviously represent, one a charged cloud, the other a layer of thoroughly good conducting earth. A lightning conductor, *B L*, is provided; *L* consisting of a considerable length of stout copper wire or rod. At *C* a possible side path is provided, so that, if the flash chooses, some of it can spit off through an inch or so of air, and through an interposed resistance, *R*, to earth. A side-flash *C*, is found to occur unless the sparking distance is made too great; and the effect of increasing *R* is not to stop the side-flash, but to weaken it. Thus, even with the liquid resistance of nearly a megohm at *R*, an inch spark still passes at *C* for every flash at *B*, though the *C* spark is now very weak.

How can this tendency to side-flash be further diminished? At the end of Chapter IV., I hinted at a partial remedy—elasticity. To stop a pipe full of water from being burst by a blow given to the water, you will make the pipe elastic. An elastic cushion will ease off the violence of the shock of a water-ram.

Electric inertia was known by the other name of self-induction; electric "elasticity" is known by the other

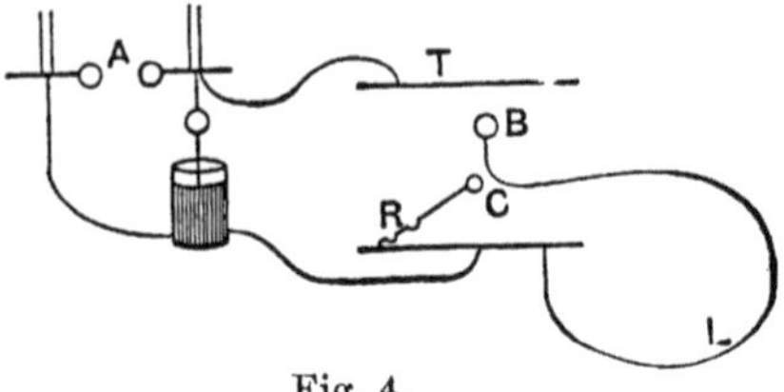

Fig. 4.

name of capacity. Increase the capacity—not the thickness or conducting power, but the electrostatic capacity of your conductor—and it will be able to carry off more.

This phrase, "the capacity of a conductor," when used by the old electricians, commonly signified merely its conducting power, this being the sole thing most of them thought of; but, using it in its modern signification, let us see what advantage is to be gained by increasing the capacity of a conductor—say, by connecting to its two ends the coatings of a Leyden jar.

Take the No. 18 wire round the room and use it as an alternative path, as in Fig. 1, first without a jar connected to its ends, then with a jar. Length of spark at *A*, 5·35 tenths of an inch; corresponding critical *B* spark-length:

Iron wire without jar, 6·5;
Iron wire with jar, 5·0.

Not a very great difference. Not so great a difference as would be gained by diminishing the self-induction by flattening the wire out into foil. Still it is in the right direction, and we see what we have to do: diminish self-induction as far as possible, and increase capacity whenever convenient. One may also try the experiment by attaching a jar to the conductor *B*, in Fig. 4: the side flash, *C*, is somewhat shortened.

But of an actual lightning conductor how is it possible to increase the capacity? There is no sense in surrounding it with an earth tube, because that, after all, would only act as an additional conductor, and might as well be so considered from the first. Neither does a great series of polarizable voltameter plates seem a feasible suggestion. No; the only practicable plan is to expand it over as much surface as possible. A lead roof, for instance, affords an expansion of fair capacity which may be easily utilized; and there should be as little mere rod projection as possible before some extent of surface begins. Flat sheet for chimneys is better than round rod—it has at least *some* more capacity and much less self-induction.

For tall isolated chimneys I would suggest a collar of sheet metal round the top, and at intervals all the way down; or a warp of several thin wires instead of a single rod, joined together round the chimney by an occasional woof; or any other plan for increasing capacity and area of surface as much as reasonably possible.

CHAPTER IX.

LIABILITY OF OBJECTS TO BE STRUCK.

Now try experiments on the liability of things to be struck. Is a small knob at a low elevation as liable to be struck as a large surface at a higher elevation? Is a badly conducting body as liable to be struck as a well conducting one? In other popular words, does a good conductor "attract lightning"?

In answering this question experimentally, we must draw a careful distinction between the case of a flash occurring from an already charged surface, which has strained the air close to bursting point before any flash occurs, and the case of a flash produced by a rush of electricity into a previously uncharged conductor too hastily for it to prepare any carefully chosen path by induction. The two cases are—1st. Steady strain; 2nd. Impulsive rush. Take them separately.

First with Steady Strain.

First an experiment on the liability of things to be struck when the air above them is in a state of steady strain gradually increased. I take the two tin plates arranged one over the other, and stand between them three conductors, one ending in a large knob, a second ending in a small knob, and the third ending in a point (Fig. 5).

The experiment consists in working up the charge of a jar attached to the top plate until discharge occurs, and in adjusting the three conductors so that they may be indiscriminately struck. One finds that the point, even when very low, prevents discharge altogether. It may indeed be too low to be effective; or again it may be insufficient to cope with the supply of electricity; but we see here the well-known function of a point—prevention of discharge. Remove or cover up the point now, and attend only to the large and small knobs. If the knobs are negative and the plate above them positive, brush discharge goes on from the knobs, and it is not easy to get a long flash; but by reversing the connections the tendency to brush is greatly lessened, and we now get flashes some three or four inches long. But always to the small knob. The small knob has to be pulled down about three times as far from the charged surface as the big knob is, before it ceases to protect the big knob; and the latter is then for the first time struck. This occurred, for instance, when the distance of the big knob from the top plate was 9, and the distance of the small one $29\frac{1}{2}$ (tenths of an inch). They were then either of them struck indiscriminately. If the little knob was any higher than $29\frac{1}{2}$, it alone was struck.

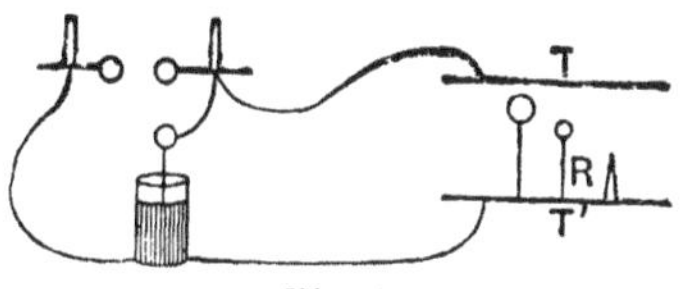

Fig. 5.

Now what is the effect of resistance upon the protecting power of the point or small knob? Scarcely any. Instead of connecting the small knob direct to the bottom tray, connect it through a capillary liquid megohm, *R*, and it gets struck just as easily as before.

Here are distance readings when they get struck with

equal ease: Large knob distance, 7·5; small knob distance, with high resistance, 22·0. The point protects both, up to distance 60·0, until covered up by a thimble.

Thus the flash actually prefers to jump three times as much air, and encounter a megohm resistance, rather than take the short direct path offered by the bigger knob. The sizes of the knobs in this experiment are: 1·27 inch diameter for big knob, and 56 inch for small knob.

Of course, the cause of this is well known. It is merely that the air breaks down at the weakest point, viz., on the surface of one of the knobs, and the tension on the small knob is much greater than that on the big one, for a given difference of potential. The fact that the discharge begins in the air above the conductor explains why it is that adding resistance—even enormous resistance—to a conductor makes no difference to the length of the spark which strikes it. The path is prepared inductively in the air, and the resistance of the path which the discharge must ultimately take makes no difference, provided it is not so nearly infinite as to prevent the free adjustment of the static charges and inductions set up as the machine is worked. But though high resistance makes no difference to the length of the spark, it does affect its noise and violence. The discharge striking the knob when much resistance is interposed has only a soft velvety noise. Its energy, or heating effect, is much the same, but its suddenness, and therefore its noise and violence, are enormously lessened.

This water resistance is equivalent to a shocking bad earth; and its effect is, we see, to make the spark gentle. But it is an evident advantage to have a discharge take this quiet and manageable form. The worse a conductor is, the quieter will be the flash, and, up to a certain limit, it will protect just as well apparently. Hence, surely, a bad earth is an advantage. But wait a bit; we have not yet considered the other case—the impulsive rush.

Second, with Impulsive Rush.

Let us now modify the experiment to try the second form of the experiment—the impulsive rush.

Fig. 6 shows the connections. There is no difference of potential between the trays up to the very instant of discharge. The jars gradually charge up (they stand on the same wooden table), and ultimately discharge at A; a violent rush then takes place into the two plates, and the conductors between are struck. Adjust them till they are all struck with equal ease, as before; we find the conditions utterly different. No longer does the small knob protect the taller big knob; no longer does the point exert any special protective influence. All three bodies—large knob, small knob, point—are equally liable to be struck if at the same height, and no one is more liable than another; simply the highest is struck if they are at all equally conducting. It is easy to get all three struck at once. *Now* make one badly conducting, and its protective virtue is gone. Put the liquid megohm into any one of the three conductors, and that one is no longer struck. It ceases to protect the other two even if taller than they; nay, even if it be raised so as to *touch* the top tray, thus establishing direct conducting communication of a poor kind between cloud and earth, still next to none of the flash chooses that path, and the other two conductors get struck with apparently just the same ease as

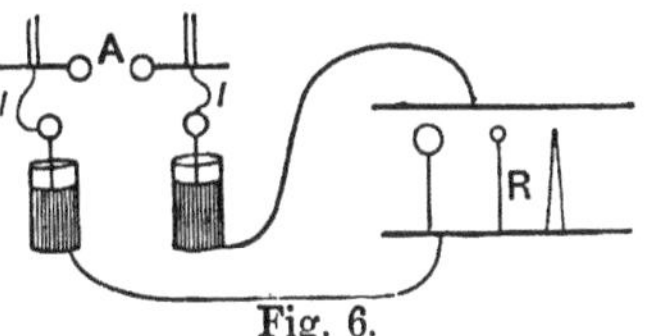

Fig. 6.

before. This is the real objection to a bad earth; it cannot protect well against these sudden rushes.

Sudden rushes are liable to occur; the clouds spark first into one another, and then, as a sort of secondary effort or back kick, into the earth. For instance, two clouds one above the other; the top cloud sparks into the lower, and this at once overflows to the earth. In these cases the best conducting and highest objects are struck, quite irrespective of any question of points and knobs. Points are no safeguard against these flashes, as you see. The point gets struck by a vivid flash, exactly of the same character as that which strikes one of the knobs; it has no time to give brushes or glows; its special efficacy in preventing discharge exists only in the case of steady action, where the path is pre-arranged by induction. In the case of these sudden rushes, the conditions determining the path of discharge are entirely different. No doubt they have to do with what is called the "impedance" of the various conductors, but they have nothing to do with the shape or size of their terminals.

CHAPTER X.

EXPERIMENTS BEARING ON THE "RETURN STROKE" AND OTHER UNEXPECTED VAGARIES OF LIGHTNING.

Experiment of the Recoil Kick.

It will have been noticed that in the experiment of Fig. 1, the spark obtained at B is longer than the spark at A. And the question arises why this should be so. Plainly what is happening is this: the discharge at A sets up electrical oscillations, and the charge of the jars is rapidly reversed. The difference of potentials of the inner coats changes from, say, $+V$ to nearly $-V$, and back again; the difference of potential of the outer coats changes therefore from 0 to nearly $2V$, and hence the B spark may be expected to be nearly twice as long as the A spark; and so it is.

These electrical oscillations are of considerable interest, and have sundry practical bearings; let us proceed to make them more conspicuous. Fig. 7 shows a couple of long leads, L and L', reaching round the room (No. 18 wire in two 95-feet lengths was actually employed), insulated from one another and from the earth, but attached to the two poles of a machine; the machine having also a couple of Leyden jars attached to it in the given arrangement of main discharge circuit, A. The customary manner when supplied by the maker. A discharger, *B*, can be arranged to bridge the gap between these leads, either near the machine, as B_1, or about the middle, as B_2, or at the far end, as B_3. Now, of course, sparks can be obtained either at *A* or at any of the *B* knobs, and all about the same length; but supposing the *A* knobs to be brought nearer than the *B* knobs, the spark would be expected to occur at *A* only. Nevertheless, on trying the experiment, one finds that every time a spark occurs at *A*, a longer spark occurs at *B*; it is, as it were, precipitated with a rush; and the longest spark of all is obtainable at the far end, viz., at B_3.

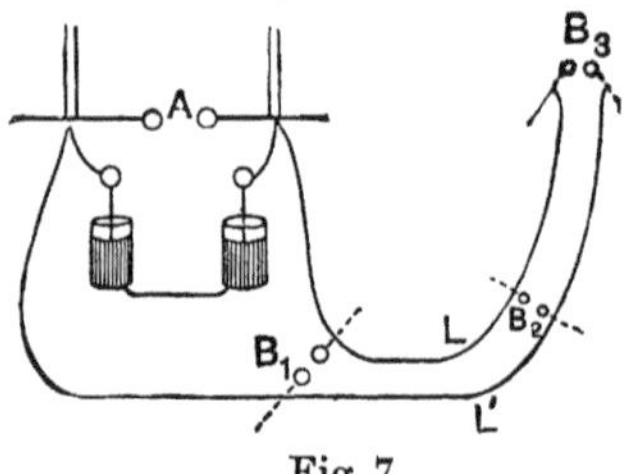

Fig. 7.

Here are some figures, in the obtaining of which, however, for convenience of manipulation, the *B* length remained constant in each position, and the least length of the determining spark, *A*, was the thing observed:

Nearest position.	Spark length, *A*	3·20	4·15	5·12
	Corresponding spark length, B_1	3·22	4·80	6·18
Middle position.	Spark length, *A*	1·92	2·37	2·70
	Corresponding spark length, B_2	3·22	4·80	6·18
Furthest position.	Spark length, *A*	1·60	2·2	2·45
	Corresponding spark length, B_3	3·22	4·80	6·18

The electricity in the long wires is surging to and fro, like water in a bath when it has been tilted; and the long spark at the far end of the wires is due to the recoil impulse or kick at the reflection of the wave. Evidently there are some quantitative relations to be specified here, and there will be some best capacity of the jars corresponding to a given length of conductor, and to a nearer the length of the conductors corresponds to a half-wave length, or some multiple of a half-wave length, of the oscillations produced by the discharging jars, the more perfect will be the synchronism between the pulses, and a longer recoil kick may be expected. The arrangement may in fact be compared to a resonant tube excited by a tuning fork or reed, to Melde's experiment with vibrating strings, or to any other case of forced vibration.

The following numbers will just serve to show some difference of effect caused by different sizes of jars. Using two pint jars in series, and fixing the A spark at 4·5, the B_3 spark at the far end just ceases when the knobs are pulled out to 7·8. But, replacing the pint jars by gallon jars, the B_3 spark does not cease till the knobs are separated to 10·9; the corresponding spark in the position B_1 being only a trifle longer than the A spark, viz., 4·7.

It is not to be supposed, however, that by increasing the capacity of the jars still farther, a still better effect will be obtained; for on replacing the two gallon jars by the very large condenser of alternate glass and tinfoil sheets, we find the spark at B_3 fails when the knobs are only 5·9 tenths of an inch apart: the length of the A spark remaining 4·5, as before.

There is, as I say, a quantitative relation; and it is a relation which the modern theory of electricity makes known. I cannot go into it here, but I may just say that, very approximately, the wave length of the electric oscillation of a discharging jar is 2π times the geometric mean of the static capacity of the jar and the electromagnetic inertia of the discharger. (See Chapter XIII. and XIV.)

The capacity of the two gallon jars in series (this being the capacity which gave the best result with 95 feet of leads) was about ·002 microfarad, or, say, 1,800 centimetres; hence, supposing the wave length of the discharge through the A knobs to be something like 190 or 200 feet (twice the length of the leads), we should calculate the self-induction of the circuit formed by the jars, short connecting wires, and A wires, as something like five metres, which is a reasonable enough value.[1]

Repeating the experiment (Fig. 7) with the two gallon jars in series, but insulated this time from the earth, a still longer spark at B_3, the far ends of the long wires, can be obtained; viz., $A = 4{\cdot}5$, $B_3 = 14$; and even when the knobs of the discharger are separated beyond this distance, a brush still passes between them for every spark at A.

Removing the discharger altogether, and making the experiment in the dark, a very interesting effect is seen; the further end of each wire glows with a vivid brush light; showing the exceedingly high potential to which they are raised by the recoil. I do not see the effect with thick No. 1 wires, but with No. 18 wires it is very marked. The glow on each of the two wires is independent of the position of the other; thus, if the connections are made so that the wires run opposite ways right round the room from the machine, the distant end being insulated, it is still the end of each furthest from the machine that glows, although they are separated from one another, for most of their length, by the whole width of the room. With the two gallon jars the wires glow over fully three-fifths of their entire length. With jars

[1] Since the delivery of the lecture, a great number of quantitative observations on these lines have been made. Evidence of electromagnetic waves three yards long has been obtained. I expect to get them still shorter. [The shortest waves of which I have now obtained evidence are 3 inches. 1890.]

of much larger or much smaller capacity the length of glow is conspicuously less.

Connect a small pint jar to the far ends of the wires, and all these effects cease. The increase of static capacity reduces their potential below the brushing point. Arranging the jar so as to leave an air space between it and one of the wires, a spark passes into it at each *A* spark; but the jar is not the least charged afterwards—proving that the spark is a double one, first in and then out of the jar, a real recoil of a reflected pulse; hence it is also that the appearance of the brush is the same on the two wires, one is not able to say which is the positive and which the negative wire, for each is both.

Experiment of the Surging Circuit.

What seemed to me, when I first made it, a curious illustration of these electrical surgings or oscillations going on in a conductor which is being suddenly discharged at one end, is afforded by the following extremely simple experiment.

Attach one end of a long insulated wire to the machine, and connect the other pole of the machine to earth. Jars connected up also to the machine do no harm, but they are unnecessary. The wire now practically constitutes one coating of a condenser, of which the walls of the room are the outer coat. The wire is made to form a nearly closed circuit, and its further end brought within an inch or so of the near end, as at *B* (Fig. 8).

Under these circumstances one would at first sight say that a spark at *B* was absurd, for the two knobs are metallically connected through a stout conductor, which may be No. 1 wire, and not necessarily many yards long. Nevertheless, it will be found that sparks at *B* can be obtained quite as long, though not quite as strong, as those at *A*. Every *A* spark is accompanied by a *B* spark, unless, of course, the knobs are too far separated.

One may surmise that it is the static charge on the distant portion of the conductor, which having to rush toward *A*, prefers the short path, *B*, instead of the longer path *viâ* the wire. But this is not the whole account of the matter, as can be shown by interposing high resistance in the conductor at various points; say at the places 1, 2, 3, or 4 (Fig. 8). Introducing a quarter of a megohm at the point 1 or at 3 weakens the *B* spark very much, and apparently about the same whether it be at 1 or at 3; the strength (*i.e.*, the noise and appearance) of the *A* spark remaining much the same. Introducing a high resistance at the point 4 weakens both the *A* and the *B* sparks. Introducing a high resistance at the point 2 leaves both sparks pretty much as strong as they were without it.

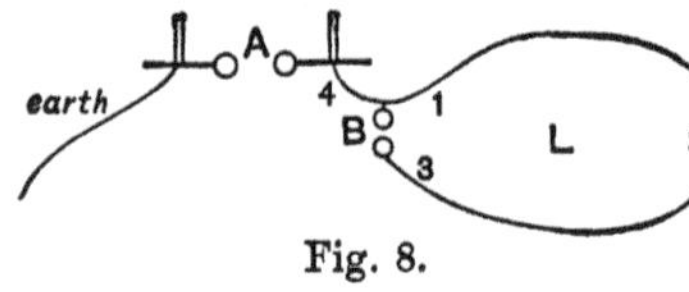

Fig. 8.

The spark at *B* is caused by electrical oscillation—a surging of the charge of the wire to and fro like water in a pipe. One might liken it to an elastic pipe pumped very full of water and its end closed with spring valves. If, then, one end is suddenly opened and shut, a pulse is transmitted through the pipe, which may force open the valve at the far end and let some water escape.

It is these electrical oscillations, I doubt not, which account for the long spark sometimes obtained by the use of a "Winter's ring."

This last experiment, Fig. 8, however it may be explained, has an obvious application to the question of connecting roof-gutters to lightning conductors. It is most desirable to connect them, but both ends should be connected. If only one is connected, the far end is very likely to spit off a flash.

Again, we see how, when a flash strikes a system, the electricity goes rushing and swinging about everywhere for no apparent reason, just as water might surge about in a bath or system of canals into which a mass of rock had just dropped, splashing and overflowing its banks. Just so with electricity. Bell-wires, gas-pipes, roof-gutters, conduct side-flashes in a way most puzzling to the older electricians; and thus gas may get ignited in the most unexpected places, and passengers in a train may feel a shock because a charge has struck the rails. In powder magazines it is apparent how dangerous this lawless sparking tendency may be; for even the hinge of a door may furnish opportunity for some trivial spark sufficient to ignite powder. By no means, it seems to me, should high rods be stuck up to invite a flash to such places. Build them, or line them, with connected iron, barb them all over the roof, connect them to the deep ground in many places, and I do not see what more can be done.

Experiments on Overflow of Jar.

Another way of making these electrical surgings more conspicuous is by their effect in causing a jar to overflow, *i.e.*, to spark round its edge. The overflow of a common jar is in fact very like a lightning flash, for it is a discharge direct between two coatings. That is just what a lightning flash is. There is ordinarily no conductor present except the two coatings—the clouds and the earth. I found a curious remark in Franklin[1] about the overflow or fracture of Leyden jars. He found that when

[1] "Franklin's Works," by Sparks, 1840. Vol. v. p. 462.

F

one broke from overcharging, a great number went together; *and also that the spark in the ordinary circuit did not fail.* He was led by this observation to doubt if the breakage of the jar was really due to a discharge of the electricity through it, and he surmises that it may be due to a sudden expansion of air bubbles in the interior, suddenly relieved of the strain. (See page 151.)

No doubt he is wrong there, but the observation of the facts is good and noteworthy. We will repeat the experiment. It is not necessary to burst the jars; overflow round the edge is just as good, and cheaper. The overflow experiment can be put into a variety of forms; perhaps it will be sufficient if I show the simplest.

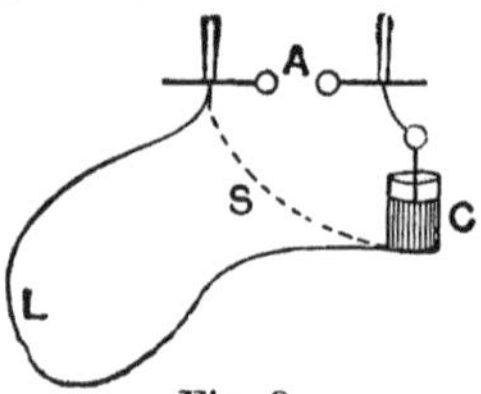

Fig. 9.

Fig. 9 shows the arrangement. It consists in nothing but establishing the connection between the machine and one or both of the coatings of a jar, through a long wire instead of through a short one as usual. If one connects the jar C by the dotted line, it does not overflow until the spark length A is very great; but with a long lead, L, to make the connection, a very short spark at A will cause the jar to overflow or discharge round its edge.

Here are a few numbers. The jar is one of the gallon jars, with the glass fully three inches above the tinfoil, so that, when it overflows, the spark has to strike along fully six inches of glass. When L is the thick No. 1 copper circuit round the room the jar overflows every time an A spark occurs, even though the length of this A spark is only ·64 of an inch. Short circuit out the long lead, as shown by the dotted line, and the jar refuses to overflow until the A spark length has been increased to 1·7; and when it does overflow now the violence is very considerable. Remove short circuit again, and the jar overflows in ever so many places at once with great violence, a perfect cascade of flashes leaping round the edge. Bring the *A* knobs nearer together, and the overflow does not wholly cease till their distance apart is again ·62.

On another occasion one got, for the *A* spark length, sufficient to cause overflow of jar—

·56 with the long lead.
1·7 with an ordinary short wire.

With a small pint jar, and a less height of glass above its coatings, I took the following readings: Length of *A* spark sufficient to cause overflow—

·67 with iron wire round room.
·52 with copper „
1·40 with short circuit.

Fig. 10 shows very well, in a diagrammatic fashion, the effect of long leads in causing a jar to overflow, or of course to burst if the glass edges are too tall for the thickness and homogeneity of the glass to stand.

Fig. 10.

A are the machine terminals, *B* those of a discharger, and J_A and J_B are two jars connected up by fairly long leads. Now of course sparks may be obtained either at *A* or at *B*, one as easily as the other, and one of the jars is liable to overflow, but not both. It is the jar J_B which overflows when a spark occurs at *A*. It is the jar J_A which is made to overflow by a spark at *B*. It might be thought that if the *B* knobs were fairly near together, a spark at *A* might precipitate a spark at *B* instead of making J_B overflow; but it is not so. The event is not indeed impossible by any means, if the *B* knobs are pretty near together; but it is easier for the jar to overflow by direct discharge between its coatings over a space of some six inches, than it is for it to discharge through the wire and rods of the discharger and an air-space of an inch or even less. It is not easy to help a jar to overflow by discharging tongs; even a foot of conducting wire is a great obstruction to the passage of the flash; it greatly prefers direct discharge through air unobstructed by the self-induction confinement of a conductor.

This is also illustrated by the extraordinary length of insulating surface which is found necessary in the Leyden jars supplied to Holtz and other such machines, and by the fact that such jars often flash, even over a foot, instead of through a few inches of air space led up to by the proper discharging knobs. Though indeed it must be noticed, as it was by Franklin, that the overflow of a jar by no means necessarily robs the proper circuit of its flash. The two things occur together. It is usually the spark which causes the overflow. Perhaps one may say that the ease of discharge direct between coatings without any conductor, accounts in some measure for the extraordinary length and unexpected paths of some lightning flashes.

These electrical oscillations and overflows, which it is thus easy to set up in a charged conductor, manifestly explain what is known as the "return stroke." I pointed out in Chapter I. that the ordinary explanation of the return stroke, the recovery of electrical equilibrium disturbed by static induction, was by no means able to account for effects of the least violence; but this fact, that a discharge from any one point of a conductor may cause such a disturbance and surging as to precipitate a much longer flash from a distant part of it, at once accounts for any "return stroke" that has ever been observed.

It is for this reason that I think it possible that a tall chimney or other protuberance in one's neighbourhood may be a source of mild danger; inasmuch as if it is struck it may be the means of splashing out some more discharges to other smaller prominences, which otherwise were beyond striking distance. It is in this way, also, that I imagine multiple flashes, such as those referred to in Chapter II. are produced. I liken them to the cascade of flashes rushing over the sides of a jar, when connected up with a long lead, and when the *A* spark is pretty long.

CHAPTER XI.

CONCLUSION OF THE SOCIETY OF ARTS LECTURE.

Experiment of the Gauze-House.

FINALLY, we have to ask is it possible for the interior of a thoroughly inclosed metal room to be struck; or, rather, can a small fraction of a lightning flash find its way into a perfectly inclosed metal cavity; for instance, a spark strong enough to ignite some gun-cotton in a metal-covered magazine which might happen to be struck?

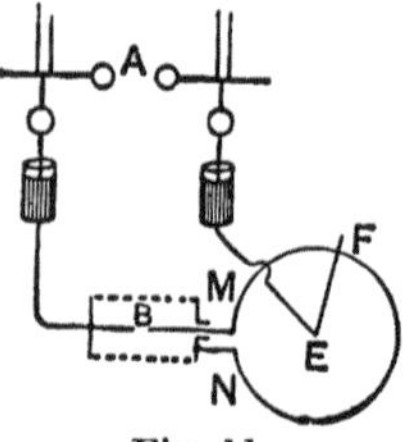

Fig. 11.

It is not easy, but it can be done; at least under such conditions as are likely to obtain in practice it can be done. My friend Mr. Chattock, to whom I am indebted for much kindly assistance and suggestion, has made the experiment in the form shown in Fig. 11.

A metal gauze cylinder, with tinplate ends, has a couple of conductors, one soldered to each end, protruding into the interior, so as to approach very near each other, and these are the conductors put into the path of the discharge. If both conductors are entirely inclosed in the metal chamber, we have not yet succeeded in getting a spark between them; but if either of them protrude any portion of their length outside the chamber, then sparks in the chamber can be obtained.

In Fig. 11 the conductor *N M B* is shown thus protruding, penetrating the chamber through a small glass tube near *M*, but soldered to it near *N*. *E F* is a movable arm or radius, making contact with this conductor. If contact is made near *N*, it takes a very strong *A* spark to give any spark at *B* inside the gauze cylinder; but as the contact is shifted towards *F* or *M*, it becomes very easy to get a small spark at *B*.

The application of this to powder magazines is that if any conductor (like a gas-pipe) pass out of a building before being thoroughly connected with its walls, it is possible for a spark to pass from something in the interior of the building to this conductor whenever a flash strikes the building; even though it be perfectly connected to that same conductor outside. We thus find that the complete and certain protection of buildings from lightning is by no means so easy a matter as the older electricians thought it.

Suggested Practical Recipes.

In many cases we may be content to fail of absolute security and be satisfied with the probable safeguard of a common galvanized iron rod or rope. But for tall and important buildings, for isolated chimneys and steeples, and for powder magazines, where the very best arrangement is desirable, what is one to recommend? I prefer to call attention to principles rather than to advocate any particular nostrum; and I return to the matter at greater length later, but there is no harm in my saying that I see nothing better than a number of lengths of common telegraph wire. I think a number of thin wires far preferable to a single thick one; and their capacity must be increased when possible by connecting up large metallic masses, such as lead roofs and the like. But the connection should be thorough, and made at many points, or sparks may result. Balconies, and other prominent and accessible places, should not be connected.

The earth should be deep enough to avoid damage to surface-soil, foundations, and gas and water-mains. As to the roof, I would run wire all round its eaves and ridges, possibly barbed wire, so as to expose innumerable points; and the highest parts of the building must be specially protected; but I would run no rods much above the highest point of the building, so as to precipitate flashes which else might not occur, in search for a delusive area of protection which has no existence.

The conductors must not be so thin as to be melted or deflagrated by the flash; but really, melting is not a very likely occurrence, and even if it does occur, the house is still protected; the discharge is over by the time the wire has deflagrated. The objection to melting is two-fold. First, the red-hot globules of molten metal, which after all are not usually very dangerous out-of-doors; and secondly, the trouble of replacing the wire. I should be content to put up a great number of telegraph-wire conductors, and wait till one is melted, before thinking too much of the likelihood of such an occurrence. The few instances ordinarily quoted of damage to lightning conductors by a flash do not turn out very impressive or alarming when analyzed. The only place really liable to be melted is the place where the flash strikes the conductor.

In conclusion, I trust that men of experience in

these matters will consider the facts and suggestions I have brought forward as objects worthy of attention and further inquiry, and will study them in the light of their experience.[1]

[1] At my second lecture to the Society of Arts I learnt that some experiments on lightning protectors, something like those of mine on "the alternative path," had been made previously by Professor Hughes and M. Guillemin (see Professor Hughes' presidential address to the Society of Telegraph Engineers). Also I find that Dr. Hertz has made experiments on electric oscillations very like those of mine on "the surging circuit." (Wied. "Ann.," 1887.) In a subsequent communication I hope to notice some of the observations of Professor Hughes.

Instances of side-flash referred to in footnote to page 15.

"Lightning," says Prof. Arago, "having struck a rather thick rod erected on a house in Carolina, U.S., afterwards ran along a wire carried down the outside of the house to connect the rod on the roof with an iron bar stuck in the ground. The lightning in its descent melted all the part of the wire from the roof to the ground-storey without in the least injuring the wall down which the fire was carried. But at a point intermediate between the ceiling and the floor of the lower storey things were changed; from thence to the ground the wire was not melted, and at the spot where the fusion ceased the lightning altered its course altogether, and, striking off at right angles, made a rather large hole in the wall and entered the kitchen. The cause of this singular divergence was readily perceived, when it was remarked that the hole in the wall was precisely on a level with the upper part of the barrel of a gun which had been left standing on the floor leaning against the wall. The gun-barrel was uninjured but the trigger was broken, and a little further on some damage was done in the fireplace." Prof. Arago concludes that indoor lightning conductors would be dangerous. Another case is that of a banker at Lyons whose house had a conductor which had proved itself perfect by carrying off a previous flash. But he happened to place in the wall near it a large iron safe, and next time his house was struck the lightning shattered the wall, got at this safe, and (so it stated) partially melted its contents.

CHAPTER XII.

PREVIOUS EXPERIMENTS OF MESSRS. HUGHES AND GUILLEMIN, AND OF ROOD.

THE footnote at end of last chapter, with which I concluded the lectures in 1888 to the Society of Arts, demands expansion; and first I append extracts from letters which I received from Prof. D. E. Hughes shortly after the delivery of the lectures, wherein he calls attention to some previous experiments on telegraphic lightning protectors made by himself and M. Guillemin somewhat on the lines of what I call the "alternative path" method, and which were not at the time received with the attention they deserved.

The experiments he describes are conducted on the "steady strain" principle, rather than on the "impulsive rush" method; for any practicable motion of discharging tongs must be regarded as infinitely slow. For the rest, the experiments are like mine on the "alternative path," except that instead of measuring directly the E.M.F. needed to drive a current through a conductor by observing the length of spark corresponding to it, these experimenters make use of a fine wire as their shunt circuit. [I also have experimented in this way, but found it less rapid and convenient in practice.] The slight divergences of results are interesting. The letters, I think, speak for themselves, and require no further comment. I would certainly have quoted these experiments in my lectures had I known about them, as I ought to have done.

Just one remark about one of the concluding sentences in the third letter, "however much we may doubt the Wheatstone-bridge method I employed." Any observations of mine derogatory to Wheatstone-bridge methods as applied to lightning-conductor-testing relate solely to the time-honoured practice as learnt by schoolboys from what seems time immemorial. The methods with alternating currents devised by Prof. Hughes stand on a different footing altogether, and if only the frequency of alternation were made great enough (I fear, however, that the telephone would be then giving a note immensely too high to be audible), they might very well be used to give the information about any conductor which is needed. But a Leyden jar system of testing would be easier and better. Though, indeed, it by no means follows that the path followed by a weak spark will also be chosen by a strong one. I can get sparks to jump in altogether different places according to their strength, nothing at all being varied except energy of spark.

The following are the letters referred to above:

"108, Great Portland Street, W., March 17, 1888.

"DEAR PROF. LODGE,—I was very much interested in the beautiful experiments you made in your lecture at the Society of Arts this afternoon, and I regret that I was unable to be present at your first lecture, which no doubt would have enabled me more completely to understand your present view of 'lightning conductors.'

"The great portion of the experiments showed by you yesterday entirely agree with those obtained by Prof. Guillemin and myself, whilst those relating to the use of iron and copper seem to disagree. Your experiment certainly demonstrated your view, whilst mine seems to me to demonstrate just the opposite.

"In 1864 the Administration des Lignes Télégraphiques of France, through the Commission de Perfectionnement, of which Prof. Guillemin and myself were members, charged us with the mission of verifying experimentally the merits of different lightning protectors in use for the protection of the telegraph instruments from lightning. These experiments were carried out at

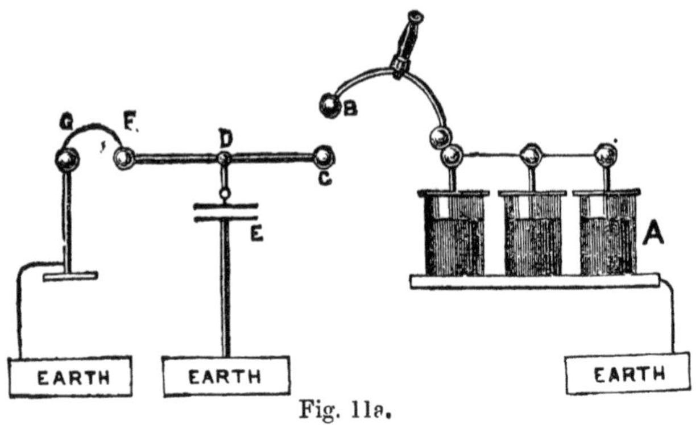

Fig. 11a.

the laboratory of the Ecole de St. Cyr, of which Prof. Guillemin was the Professeur de Physique.

"We tried to imitate as nearly as possible the conditions which occur in practice.

"The general idea of our experimental arrangement was the following:

"In order to have a powerful source of electricity at high tension we made use of a large battery, *A*, of 12 large jars, and as we found it took too long to charge these with the ordinary electrical machines, we used a powerful Ruhmkorff coil, which would give a spark of

15 centimetres. The jars were charged to a fixed degree, and could be discharged through our apparatus by means of the universal discharger *B*. An insulated rod of brass, *C D F*, served to conduct the charge to the experimental protector *E*, and by a derivation, *F G*, to earth. At *G F* we could place a telegraphic electro-magnet, a fine iron wire, or an electrometer, the object being to try and save the fine wire at *G F* from destruction by the protective action of the protector *E*.

"We found that with a high charge no protector would entirely protect *G*, even if a small air space intervened between *F* and *G*; and we also found that if *D* was directly connected to earth by a copper rod of one centimetre diameter, there would still be sufficient electricity pass *D* to *G* to burn a fine iron wire.

"This result proved that we could not in all cases expect absolute protection from any known protectors.

"By gradually diminishing the charge we arrived at a point where we could make comparative experiments.

"The protectors have all a small air space, in order not to weaken the ordinary working current.

"The received idea that two plates with innumerable points near each other would protect in a far greater ratio than simple plates separated by a thin sheet of paraffin paper or gutta-percha was disproved; for the simple flat plates known as the 'American protector' were invariably better than those with points. The plates formed a small condenser, the surfaces of which could be brought far nearer to each other than is possible in practice with points.

"In a report which I made as reporter, I pointed out all these experimental facts, and it was published in the 'Annales Télégraphiques,' 1865, and I again gave these facts in my inaugural address to the Society of Telegraph

Engineers, January 28th, 1886, and published in their 'Journal,' No. 60, p. 17.

"We tried numerous experiments as to different kinds of wire at *E*, and if we placed a very large condenser at *E* in place of a protector it seemed to have a far greater effect.

"Having been called to Russia I was unable to continue these experiments at that time, but Prof. Guillemin having taken a great interest in them, he continued them, and published his results in the 'Comptes Rendus' (I forget what date). He found that if he united *D* to earth by a thin flat plate, even tinfoil, a very large measure of protection had been obtained, and that flat conductors were evidently superior for a sudden charge of high-tension electricity. He supposed that this effect was due to its electrostatic capacity, thus forming a species of condenser. Later experiments by myself have pointed out the advantages of the flat conductors, but my results lead me to suppose that, in addition to the reasons of Prof. Guillemin, it is because a flat conductor has far less self-induction.

"The experiments which you showed yesterday also seemed to point out the advantage of flat conductors, and your views seemed to agree with those of Prof. Guillemin, and you have experimentally proved this in the beautiful experiment where you show the different results obtained from a tinfoil sheet conductor where it is arranged so as to have more or less self-induction.

"The greatest and perhaps the only difference in your results was the superiority of iron over copper. The methods used by you fully demonstrated it, and it also resembled the method used by Prof. Guillemin and myself, except that we had one large battery discharged to earth, whilst you seemed to have two batteries, one at

A and the other at *G*. If I am correct, there may be some peculiar phenomena due to our different arrangements.

"That self-induction is injurious to rapid transmission of electrical currents of low potential is shown in practice by their different behaviour. Practice has shown that copper alone is suitable for long-distance telephony and telegraphy, and I still believe that the same marked difference holds good with high-tension currents for the same reason.

"However, there is nothing like experiment to prove a fact, and I shall read your paper and future experiments on this subject with the greatest interest, knowing that you are a most careful experimenter. The subject is one of vast importance, and it is astonishing that with the innumerable experiments that have been made, from Franklin to Faraday, in all countries, and with apparently irreproachable methods, there should be still such a wide division of opinion as to the merits of round or flat conductors and of iron or copper."

Second Letter.

"You suggest that the difference between our results might perhaps be due to the use of a discharger, and this may explain the difference.

"The object of Prof. Guillemin and myself was to send a powerful charge through a wire with the utmost possible suddenness.

"By the use of the discharger we could charge our battery to the desired point, then disconnect the Ruhmkorff, and then bring our discharger as suddenly as possible in connection with the experimental apparatus.

"It is evident that we could never do this with the rapidity that we desired, as the spark would jump across the air before full contact was made. We believed, however, that we came nearer than if we had a fixed distance, allowing the charge to spark across the interval, as then our experimental apparatus or any change made in it seemed to have an influence on the *time* discharge of the battery. It is so long, however, since these experiments were made that I cannot discuss this point, and probably your views or method may be more in accordance with the actual conditions of a sudden discharge than ours were.

"We certainly agree in our results and views on the most important points. For I quite agree with you that Wheatstone's bridge, or any apparatus used for steady currents, docs not indicate what passes at the first moment of contact.

"Judging from our experiment, where we could not protect the fine iron wire by a direct thick copper, iron, or any short metallic contact, and from later experience on self-induction with low-tension currents, I believe that at the very first instant all bodies offer an infinite resistance, copper being no better than air. I believe you also take this view, as you well illustrated it by the flow of water, the hammer blow, inertia, etc.

"The effect is so marked with discharge from a good Leyden jar battery that it becomes a serious question if any lightning conductor yet made could convey the whole of a lightning discharge quietly to earth without its seeking other channels through substances having an infinitely higher ohmic resistance; and the whole question would be discouraging were it not for the fact that we have undoubted proof that lightning conductors do protect in spite of the experiments cited.

"I have tried to gain some exact information from the Report of the Lightning Rod Conference, 1882, in the numerous cases there cited, and I find that in most cases the rods have protected, but many have failed, mostly from bad earths; but numerous cases are there cited in which rods were melted, both of copper and iron—in fact, the cases cited do not show any marked preference for either metal, except that there seems a stronger testimony in favour of Sir W. Snow Harris's copper tape conductor than any other.

"It is quite possible that all the researches that have been made have never yet approached the true condition of things. Evidently experiments made with simple sparks from machines never approach the true conditions. A very large Leyden jar battery or powerful condenser seems nearer; but even this must be weak compared with a true lightning discharge. But still it ought to teach us, as it has done in your case and others, myself among the number, that an enormous inertia or apparent resistance is offered by the best of metallic conductors at the first instant of discharge.

"I shall read your experiments and views on this interesting subject as soon as they appear in print, and I hope that your new series of experiments will clear up such mystery as still remains."

Third Letter.

"I find that I have not fairly represented Prof. Guillemin's views—due to only reading an abstract. I have found his Paper, entitled 'Sur la Décharge de la Batterie Electrique et sur l'Influence de la Configuration des Conducteurs,' 'Comtes Rendus,' 1866, pp. 1,083 to 1,085.

"In this Paper he believed the advantage of thin flat sheets to be due to the mutual reactions of each portion of the current on each other, and although he does not use the term he evidently means self-induction.

"I find also that he demonstrated this, noting the difference of several copper wires when twisted together or separated as in a sheet; in fact, an identical result which I obtained later with low-tension currents and Wheatstone's bridge (No. 60, p. 15, 'Journal' Society of Telegraph Engineers, 1886). Thus, he did not attribute the advantage of sheet conductors to electrostatic or mere surface conduction. The experiments made by Prof. Guillemin and myself, which I have already mentioned, were made in 1864, but not published until 1865 ('Annales Télégraphiques,' 1865, pp. 290 to 302, Tome VIII.).

"I believe that Sir W. Snow Harris also attributed the advantage that his experiments showed in sheet conductors to the same reason and not to mere surface conduction, as I had previously read, for I find in an abstract of his Paper, 1843, published in 'Report of the Lightning Rod Conference, 1882,' p. 86, these words:—'He says the beneficial effect of superficial conductors appears to depend on the removal of the electrical particles further out of the sphere of each other's influences.'

"This evidently means the same explanation as given later by Prof. Guillemin, and still later by myself in my Paper on 'Self-Induction,' so in justice to them both I feel it my duty to point these out to you; and I find now that however much we may doubt the Wheatstone bridge method I employed, it certainly gave identical results for a similar experiment with those obtained by sudden discharges from Leyden jar batteries. I have suggested to the editor of 'The Electrician' that he should translate and reprint Prof. Guillemin's remarkable Paper, as it would be only justice to him, as I know he suffered severely from the sneers of Messieurs *les Savants* on the publication of his Paper.

"Do not trouble yourself to reply to this, as my only object is to correct my remarks as to Prof. Guillemin's views made in my first letter.

"With best wishes, sincerely yours,

"D. E. Hughes."

The following is a translation of the Paper referred to by Prof. Hughes in his last letter:

Leyden Jar Discharges and the Influence of the Shape of Conductors.[1]

"Some lightning-guard experiments, conducted last year by MM. Hughes, Bertsch, and myself at the instance of the "Commission de Perfectionnement des Lignes Télégraphiques," afforded us an opportunity of observing a phenomenon which it seemed to me could not be reconciled with the laws of electric conductivity. A continuous copper wire did not conduct perceptibly better than a similar wire into the circuit of which a lightning guard of the point pattern had been introduced. I now present to the Académie des Sciences the result of researches undertaken by me with a view of ascertaining to what extent this phenomenon really does deviate from the ordinary law.

"According to Ohm's law, the intensity of a stable current is independent of the surface of the conductors. My experiments show that in the case of Leyden jar discharges (where the current is variable and never sensibly

[1] Translation of a Paper read by M. C. M. Guillemin before the Académie des Sciences, 14th May, 1866. "Comptes Rendus," pp. 1,083-85.

steady) increasing the surface of the conductors facilitates the passage of the discharge.

"To show this, I arranged two conductors in parallel so that they should be simultaneously traversed by the discharge from a large 6-jar battery, having a total condensing surface of about 1 sq. m. One conductor was an iron wire, ·1 mm. in diameter, and of variable length; the other conductor was a thin sheet of tin, 2 metres long and 6 cm. broad, insulated on a glass table. This conductor was altered in shape without alteration of cross-section. The effect of these alterations could be gauged by the length which could be given to the iron wire without its being melted.

"The first thing I did was to arrange matters so that the sheet of tin took a sufficient proportion of the discharge to prevent 15 cm. of the iron wire attaining a red heat. Then the sheet was bent back upon itself lengthways, diminishing the surface without alteration of length or cross-section; the iron wire then became dull red. If the surface of the sheet was still further reduced the iron wire fused along its entire length. The two conductors were then put in series, and the tin sheet still had the best of it.

"This phenomenon arises apparently from the inductive action of conductors upon one another. Increasing their surface facilitates the discharge by increasing the distance between the mutually opposing forces. This view is confirmed by the following experiment:

"Sixty wires, 2 m. long, and 25 mm. in diameter, were now placed in parallel. When the wires were 1 cm. apart the iron wire was well protected, and did not heat to 400° C.; the closer, however, the sixty wires were brought together, the hotter the iron became; first it got warm, then red hot, and finally, when the wires were in close proximity, melted. When the wires were twisted into a cable the effect was a maximum. By exaggerating the conditions of the experiment, it is easy to see that a conductor of large surface may have a much higher resistance than a round wire, so far as voltaic currents are concerned, and yet be a better conductor of static discharges.

"The original phenomenon was this. In the case of two short, thick conductors, the interposition of a thin layer of air in the circuit of one did not greatly affect the relative quantities of electricity sent through them by an instantaneous discharge.

"These results naturally lead me to think that it would be advantageous to substitute copper bands, to 3 cm. broad and at least 1 mm. thick, for the copper wires at present used on our land lines to connect lightning guards to earth. The presumption is that the protection afforded would be far more thorough.

"In these experiments the Leyden jars were charged from a Ruhmkorff coil of large size, and five or six seconds sufficed to highly charge the six jars. The great power of this apparatus permitted me to carry out with ease experiments which it would be very difficult to perform with ordinary apparatus. The Reiss thermometer was used in conjunction with these researches.

"C. M. GUILLEMIN."

In connection with a subsequent lecture of mine at the Royal Institution on the discharge of a Leyden jar,[1] Lord Rayleigh in 1889 directed my attention to a series of researches by the American experimenter Ogden N. Rood, on the nature and duration of Leyden jar and

[1] Reprinted at the end of "Modern Views of Electricity" (Macmillan).

lightning discharges, and on rapidly-revolving mirrors, an account of which will be found in the "American Journal of Science" for 1869, 1871, 1872, and 1873.

These observations are of considerable interest, but they are too long to quote here.

The following brief analysis of Rood's papers may be useful:

First paper ("Silliman's Journal," vol. 48, p. 153), concerns an application of the revolving mirror to determine the duration and character of the Leyden jar spark. It is practically Feddersen's method modified.

Second paper ("American Journal of Science," vol. 1, p. 15), concerns the duration of lightning. He drew black spokes on a card, twirled it rapidly on a shawl pin, and looked at it during a thunderstorm. Each flash brought out the pattern sharply, though in some cases there was a broadening of the pattern. In these cases he estimated the average duration of a flash, as about $\frac{1}{480}$th second. Often it was much less. He suggests phosphorescence as possibly accounting for some of this duration.

Third paper ("Am. J. Sci.," vol. 2, p. 150). He applies the revolving mirror to a small jar, and gets a practically instantaneous spark.

Fourth and fifth papers ("Am. J. Sci.," vol. 4, pp. 250 and 371). He studies the multiple discharges excited by a coil.

Sixth paper ("Am. J. Sci.," vol. 5, p. 163). He observes multiple flashes of lightning.

CHAPTER XIII.

ON THE THEORY OF LIGHTNING CONDUCTORS.

THE more mathematical development of the "Mann" lectures was published in the "Philosophical Magazine" for August, 1888, and from that article I quote here the following portions for convenience:

That a condenser discharge is oscillatory has been known ever since 1853, when Sir William Thomson's great paper "On Transient Currents" appeared. Quite recently it has been recognized, first quite explicitly perhaps, by Mr. Heaviside in the "Electrician" for January, 1885,[1] that rapidly alternating currents confine themselves to the exterior of a conductor;[2] and Lord

[1] See also "Phil. Mag.," August, 1886, *et seq.*

[2] It is not possible, I think, to give Mr. Heaviside the credit of the original discovery of this theorem (though doubtless he discovered it for himself), for it had been virtually anticipated by so many persons. Not counting a wide general mechanical theorem of Sir William Thomson, which may be held to include this as a special case, a great part of it is clearly indicated by Clerk-Maxwell in his paper in the "Phil. Trans." for 1865. It then reappears in a more or less developed form in several papers of Lord Rayleigh, specially perhaps that in the "Phil. Mag." of May, 1882; and it is clearly stated for electrical oscillations in a spherical or cylindrical conductor by Prof. Horace Lamb ("Phil. Trans.," 1883). There are also several papers by Oberbeck tending in the same direction. It is well known that some ingenious experiments of Prof. Hughes first excited public interest in the matter and quickened the mathematical abstraction into life.

Rayleigh ("Phil. Mag.," May, 1886) has developed an expression of Maxwell's so as to give the real resistance and inductance of a conductor for any frequency of alternation.

I propose to apply these considerations to the case of a lightning flash.

A lightning flash is the discharge of a condenser through its own dielectric, and is more analogous to the bursting or the overflow of a Leyden jar than to any other laboratory phenomenon. The condenser-plates may be two clouds, or they may be a cloud and the earth. The discharge occurs mainly through broken-down air, but a lightning rod may form part of its path.

The particular in which lightning transcends ordinary laboratory experiments is difference of potential or length of spark. The quantity of electricity is very moderate, the capacity of the condenser is quite small, but the potential to which it is charged is enormous. Flashes are often seen a mile long, and there is said to be a record of one seven miles long. Allowing 3,000 volts to the millimetre, a mile-long spark means a potential of 16 million electrostatic units.

The capacity of a condenser with plates a square mile in area and a mile apart is roughly about $\frac{2}{3}$ of a furlong, or say 10^4 centimetres.

The energy of such a condenser charged to such a potential is enormous, being over 10^{20} ergs, and there is no need to assume that so much as a tenth of this is ever dissipated in any one flash.

We may not be far wrong if we guess the capacity emptied by a considerable flash as about 10 metres, or one thousandth of a microfarad. The total *charged area* is commonly much greater, but it is not all well connected, and it does not discharge all at once.

To make the problem a definite one consider the following case :

An air-condenser with plates of any size separated by a distance h (height of cloud) and charged up to bursting strain ($\frac{1}{2}$ gramme weight per square centimetre; the less strength of rare air is hardly worth troubling about). Let a small portion of this condenser, of area πb^2, now discharge itself; being separated from the rest after the trap-door and guard-ring manner. A volume of dielectric $\pi b^2 h$ is relieved of strain, and the energy of the spark is $E = \frac{981}{2} \pi b^2 h$ ergs.

The capacity discharged is $S = \frac{Kb^2}{4h}$, and the maximum potential can be reckoned by putting

$$V = \sqrt{\left(\frac{2E}{S}\right)} = \sqrt{\left(\frac{4\pi \times 981 h^2}{K}\right)}$$

$$= 110h \text{ electrostatic units.}$$

Let the discharge pass straight down the axis of the cylindrical region of length h and radius b, and let the channel occupied by it have a sectional radius a. If the path is a metal rod, then a is the sectional radius of that rod.

We have now to calculate the self-induction of such a discharge. A discharge of this kind differs from ordinary cases in having no obvious return circuit. What is happening is a conduction or disruption current down the axis of the cylindrical region considered, and an inverse displacement current in concentric cylinders all round it. I shall assume that this inverse displacement current is uniformly distributed over the whole area. A conduction rush is not uniformly distributed over the section of its conductor, but is concentrated by mutual induction of the parts towards the periphery; similarly, but inversely, there will be a tendency for the displacement currents to be stronger near the central axis than far away; but there is this difference, that whereas in a conductor currents are able to distribute themselves how they please, they will not be so free in an insulator. It is not quite correct to take the distribution as uniform, but it will not make very much difference probably. (That it is not correct may be seen by considering the initial and final stages of the dielectric. Either it is the whole of a condenser that is being discharged or it is a trap-door portion. If only a portion, the initial state is one of equal strain, but lines from surrounding charged areas spread in laterally to all the outer regions, and so finally there is an unequal distribution of strain in it. If the whole is being discharged, then the initial state of strain is not uniform, while the final is.)

Calling the whole current down the axis C_0, we have an equal inverse displacement all over the area $\pi(b^2-a^2)$, so that the density of its distribution is σ, where $C_0 = \pi(b^2-a^2)\sigma$.

The intensity of magnetic force at any distance r from the axis is

$$f = \frac{2(C_0 - C)}{r},$$

where C is that portion of the displacement recovery which lies nearer to the axis than r. This is accurate, for the distribution of the current matters nothing so long as it is in coaxial cylinders; the portions external to r have no effect.

On the hypothesis of uniform distribution,

$$C = \pi(r^2 - a^2)\sigma = \frac{r^2 - a^2}{b^2 - a^2} C_0.$$

Hence
$$f = \frac{2C_0}{r} \cdot \frac{b^2 - r^2}{b^2 - a^2},$$

where the a may in practice be neglected as usually too small to matter. This is the number of lines of force through unit area at the place considered; and the whole magnetic induction in the cylindrical space considered outside the conductor is

$$h\int_a^b \mu f dr = \mu h C_0 \left(2 \log\frac{b}{a} - 1\right).$$

For the part inside the conductor there is an extra term to be added, which, on the hypothesis of uniform distribution in the conductor, comes out

$$\mu_0 h \int_0^a \frac{2\frac{r^2}{a^2} C_0 dr}{r} = \mu_0 h C_0,$$

and which may really have any value between this and zero, according to the rapidity of the alternations, and the consequent deviation from uniform distribution.

The entire magnetic induction may be written LC_0; hence we get the value of L, the coefficient of self-induction, or the inductance, of the circuit,

$$L = h\left(2\mu \log\frac{b}{a} - \mu + \mu_0\right) \quad . \quad . \quad . \quad (1)$$

This I shall write for convenience $h(\mu u^2 + \mu_0)$, so that u^2 is an abbreviation for $\log\frac{b^2}{a^2} - 1$.

In practice u may be a number not very different from 4 or 5.

Of the three terms in equation (1), the first and most important depends on no hypothesis as to distribution at all; the second depends on the assumption of uniform distribution of displacement-recoil in the dielectric, and may therefore really be greater, but not less; the third term depends on the magnetization of the conductor itself by a uniformly distributed current, and if the current keeps itself to the exterior surface, as a very rapidly alternating one will, this term vanishes.

Now that we know S and L, we can easily find the criterion for the discharge to be oscillatory, and can determine the rate of alternation.

The discharge will be oscillatory unless the resistance it meets with exceeds a certain critical value, viz.:

$$R_0 = \sqrt{\frac{4L}{S}} = \sqrt{\frac{\frac{4h\mu u^2}{Kb^2}}{4h}} = \frac{4h\mu u}{b\sqrt{\mu K}} = \frac{4hu\mu v}{b} \quad . \quad (2)$$

where $v = \dfrac{1}{\sqrt{(\mu K)}} =$ the velocity of light $= \dfrac{30}{\mu}$ ohms;

so the critical resistance is

$$R_0 = 120\frac{h}{b}\sqrt{\left(2 \log \frac{b}{a} - 1\right)} \, ohms . \quad . \quad (2')$$

And inasmuch as in practice h is likely to be much greater than b, and b much greater than a, this is a big resistance, which is not likely to be exceeded by the discharger. For if the line of discharge is a metallic conductor, a is moderate, but then so is R; whereas if the flash occurs through air, and it is not easy to say what the equivalent R is, then a must be considered extremely minute.

Suppose h to be a mile, b 50 metres, and a a millimetre; the critical resistance R_0 comes out about 15,000 ohms.

I think we shall be right in saying that this far exceeds any reasonable value that can be attributed to the resistance met with by a disruptive discharge. It is generally supposed, indeed, that a conductor and earth must have a resistance of only a few ohms, unless they are to form a considerable portion of the whole resistance a flash meets with.

In so far as the path consists of different conductors in series, it is a mere matter of summation to take them all into account.

If the actual resistance falls greatly below the critical value R_0 the discharge is thoroughly oscillatory, and the strength of the current at any instant is

$$C = \frac{V_0}{nL} e^{-mt} \sin nt, \quad . \quad . \quad . \quad . \quad (3)$$

where $m = \frac{R}{2L}$, and $n^2 = \frac{1}{LS} - m^2$. The impedance is, therefore, nL.

When the discharge is thoroughly oscillatory n is greatly bigger than m, so that the above is practically

$$C = \frac{V_0}{\sqrt{\left(\frac{L}{S}\right)}} e^{-\frac{Rt}{2L}} \sin \frac{t}{\sqrt{(LS)}} \quad . \quad . \quad (3')$$

The time-constant of the dying-away amplitude is $\frac{2L}{R}$; the period of the alternation is $2\pi\sqrt{(LS)}$.

The frequency constant,

$$n = \frac{1}{\sqrt{(LS)}} = \frac{2v}{bu}, \quad . \quad . \quad . \quad . \quad (4)$$

is very great, being usually something like a million a second, more or less.

Now Lord Rayleigh has shown ("Phil. Mag.," May, 1866) that with excessive frequencies of alternation the resistance of a conductor acquires the following greatly modified value, R being its ordinary amount,

$$R' = \sqrt{\{\tfrac{1}{2} nh\mu_0 R\}}, \quad . \quad . \quad . \quad . \quad . \quad (5)$$

$$= \sqrt{\left(\frac{vh\mu_0}{bu} . R\right)}.$$

Or, taking the permeability of the conductor the same as that of the space outside,

$$R' = \frac{1}{2u} \sqrt{(RR_0)} \quad . \quad . \quad . \quad . \quad . \quad . \quad (6)$$

The actual resistance is, therefore, some fraction, something like, say, an eighth, of the geometric mean of the ordinary resistance of the conductor and the critical resistance (2).

Under the same circumstances the value given by Lord Rayleigh for the inductance is

$$L' = (L \text{ for space outside conductor}) + \sqrt{\left(\frac{\mu_0 Rh}{2n}\right)},$$

or, as we shall now write it,

$$L' = \mu h u^2 + \frac{R'}{n} \quad . \quad . \quad . \quad . \quad . \quad . \quad (7)$$

The second term has to do with the magnetization of

the conductor, and is, for high frequencies, very small. It is interesting as showing that of the two terms in the quantity "*impedance*,"

$$\sqrt{(R'^2 + n^2 L'^2)},$$

or, as it becomes for condenser discharges,

$$\sqrt{\left\{\tfrac{3}{4} R'^2 + \frac{L'}{S}\right\}},$$

the second is always the larger; because, by (7),

$$nL' = R' + n\mu h u^2.$$

Practically the second term is so much the larger that it is the only one that matters, and so

$$\text{impedance} = nL' = nL = n\mu h u^2 = \frac{2v\mu h u}{b} = \tfrac{1}{2} R_0 = \sqrt{\frac{L}{S}} \quad . \quad . \quad . \quad . \quad . \quad . \quad . \quad . \quad . \quad . \quad . \quad . \quad . \quad . \quad (8)$$

or

$$\text{impedance} = 60\frac{h}{b}\sqrt{\left(2 \log\frac{b}{a} - 1\right)} \textit{ohms.} \quad . \quad . \quad . \quad (8')$$

The total impedance, therefore, to a condenser discharge is half the critical resistance which determines whether the discharge shall be oscillatory or not; it has no important connection with the ordinary resistance of the conductor; neither does it depend appreciably on the magnetic permeability of its substance.

Hence, so long as the specific resistance of the conductor does not rise above a certain limit, its impedance depends almost entirely upon the amount of space magnetized round it, and upon the capacity of the discharging condenser; and is barely at all affected either by the magnetic permeability, or the specific resistance, or even the thickness, of the conductor. The one thing that does matter is its length. True the diameter of the conductor does appear in the expression for impedance, but only under a logarithm; hence the effect of varying the thickness is only slightly felt.

The fact that impedance to a condenser-discharge is equal to half the critical resistance, or $\sqrt{(L/S)}$, and depends not at all upon the ordinary resistance of the discharging circuit (provided this keeps well below the critical resistance for which the discharge ceases to be oscillatory), is manifest also from equation (3′).

Thus, then, we find that a lightning conductor offers an obstruction to a discharge as great as what a resistance of several thousand ohms would offer to a steady current of corresponding strength; the actual obstruction being given by equation (8′).

Another way of putting the matter, is to say that for the first few oscillations the damping term, e^{-mt} in equation (3), has no appreciable effect; and that, accordingly, the E.M.F. applied to the conductor alternates rapidly from V_0 to $-V_0$ and back again.

But V_0 is the initial potential of the condenser, diminished (so far as the conductor is concerned) by the E.M.F. needed to jump whatever thickness of air it has jumped before reaching the conductor. Hence this V_0 may be something quite comparable to the potential needed to jump through air all the rest of the way, and it may depend on a mere nicety whether it prefers the conductor or not.

Thus arises the difficulty experienced in helping a jar

to overflow by means of discharging-tongs brought near the two coatings. Sometimes the flash will make use of the tongs, sometimes it will prefer to go all the way through air; the fact being that the obstruction offered by a metal requires a large portion of the potential needed to break through a corresponding length of air. Undoubtedly the metal rod offers some advantage; but it is much less than has been usually supposed.

During the instant of discharge, therefore, the upper part of a lightning rod experiences enormously high potentials in alternately opposite directions. Any conductors in the neighbourhood may easily receive side flashes, and even the bricks into which its supports are driven may be loosened and disturbed; and all this quite irrespective of any question as to the goodness or the badness of the "earth." It becomes, therefore, quite a question whether it is not, after all, advisable to try and confine the discharge to the conductor by means of insulators, or whether it is better to reduce the excessive potential by lateral extensions of considerable static capacity. The advantage of sharing the discharge among a number of well-separated conductors, instead of concentrating it all in one, is obvious.

Theory of Experiments on "Alternative Path."

In the lectures to the Society of Arts, reprinted above, I describe some experiments I have made on the E.M.F. needed to force a discharge through various conductors, by seeing what length of air-space it will prefer to jump. The original potential of the condenser being able to jump, say, two inches without any alternative path, it remains able to jump, say, $1\frac{1}{2}$ inches when offered as an alternative a copper rod a quarter of an inch thick and six or seven yards long. This gives a rough notion of the kind of results obtained, and it shows that the extremities of the rod experience almost the whole of the original E.M.F. of the condenser.

Some experiments on much the same lines had been previously made by Prof. Hughes and M. Guillemin (see "Comptes Rendus," 1886, "Annales Télégraphiques," 1865, Address to Society of Telegraph Engineers, 1886), also Chapter XII. above; but they used a fine wire instead of an air-space, and tried what conductor would protect the fine wire from being deflagrated.

Under these conditions the experiment is practically a comparison between the impedances of two conductors—one of which has its inertia term the bigger, while the other has its resistance term the bigger.

The general theory of divided circuits has been given by Lord Rayleigh ("Phil. Mag.," 1886, pp. 377 *et seq.*), and inasmuch as in the present case there is practically no mutual induction between the two conductors, and the frequency of alternation is very great, the resultant resistance and induction take the following forms :—

$$R = \frac{R_1 R_2}{R_1 + R_2} + \frac{(L_1 R_2 - L_2 R_1)^2}{(R_1 + R_2)(L_1 + L_2)^2}; \quad . \quad . \quad . \quad (9)$$

$$L = \frac{L_1 L_2}{L_1 + L_2}. \quad . \quad . \quad . \quad . \quad . \quad . \quad . \quad . \quad . \quad . \quad . \quad (10)$$

The resultant impedance is, as usual,

$$P = \sqrt{\left\{\tfrac{3}{4} R^2 + \frac{L}{S}\right\}} \quad . \quad . \quad . \quad . \quad (11)$$

Perhaps, however, it is hardly fair to assume that the discharge will remain oscillatory when one of the branches of the divided circuit is permitted to have a high resistance. Certainly one cannot apply to such degenerate formulæ for criterion conditions.

The general expressions are :—

$$R=\frac{R_1R_2}{R_1+R_2}+\frac{n^2}{R_1+R_2}\cdot\frac{(L_1R_2-L_2R_1)^2}{(R_1+R_2)^2+n^2(L_1+L_2)^2}; \quad (12)$$

$$L=\frac{L_1R_2^2+L_2R_1^2}{(R_1+R_2)^2}-\frac{(L_1R_2-L_2R_1)^2}{(R_1+R_2)^2(L_1+L_2)}$$

$$+\frac{(L_1R_2-L_2R_1)^2}{(L_1+L_2)\{(R_1+R_2)^2+n^2(L_1+L_2)^2\}}; \quad (13)$$

$$n^2=\frac{1}{LS}-\frac{R^2}{4L^2}. \quad . \quad . \quad . \quad . \quad . \quad . \quad . \quad . \quad . \quad . \quad . \quad (14)$$

To get the frequency these three equations must be treated simultaneously; and even so the solution is not complete, for n appears also in the true expression for R_1 and R_2, so that the complete solution for a case of divided circuit condenser-discharge is by no means simple.

The experiment with an air-gap as the alternative path is better; because one may be sure then that *none* of the discharge chooses that path, when it is properly adjusted for its sparks just to fail.

Liability of Objects to be Struck.

There are also described in Chapter IX. above, some experiments on the liability of objects to be struck. A distinction is drawn between two possible cases :—

(1) Where the air above the object is subjected to a steadily increasing strain till breakdown occurs.

(2) Where the strain is thrown instantaneously upon air and conductors with a sudden rush.

In the first case the path is prepared inductively in the air, and the breakdown occurs at the place where the tension first reaches its limiting value; this is generally on a small knob or surface, and so this is struck and carries off all the discharge independently of its resistance. If its resistance is great the flash may be feeble; if its resistance is small the flash may be noisy; but the place of occurrence of the flash is not determined by these considerations. Glow and brush discharges from points and small surfaces may readily prevent any noisy flash from occurring.

The second case is different. When a sudden rush occurs, the discharge shares itself among several conductors in something like inverse proportion to their impedances, quite independently of any considerations of maximum tension or pre-arrangement of path by induction; so that no distinction is observable between points and large knobs, in this case. Points cease to have any protective virtue; they can be struck by a noisy spark as readily as can a knob. The highest object will, in general, be struck most easily, provided its impedance is not very great. If it has a very high resistance it is barely struck at all, and it does not then protect the others.

Experiment of the Recoil Kick.

Among other experiments described above are some which appear to be of considerable theoretical interest, wherein a recoil kick is observed at the ends of long wires attached by one end to a discharging condenser-circuit.

Fig. 7 or 12 or 13 shows the arrangement.

The jar discharges at *A* in the ordinary way, and

simultaneously a longer spark is observed to pass at B at the far end of two long leads. Or if the B ends of the wire are too far apart to allow of a spark, the wires glow and spit off brushes every time a discharge occurs at A.

The theory of the effect seems to be that oscillations occur in the A circuit according to equation (3′) with a period

$$T = 2\pi\sqrt{(LS)},$$

where L is the inductance of the A circuit, and S is the capacity of the jar. These oscillations disturb the surrounding medium and send out radiations, of the precise nature of light, whose wave-length is obtainable by multiplying the above period by the velocity of propagation.

This velocity is known to be

$$v = \frac{1}{(\sqrt{\mu K})};$$

so the wave-length is

$$\lambda = vT = 2\pi\sqrt{\left\{\frac{L}{\mu}\cdot\frac{S}{K}\right\}}. \quad . \quad . \quad . \quad (15)$$

Now $\frac{L}{\mu}$ is the electromagnetic measure of inductance, and $\frac{S}{K}$ the electrostatic measure of capacity. Each of these quantities is of the dimension of a length, and the wave-length of the radiation is 2π times their geometric mean.

The propagation of these oscillatory disturbances along the wires towards B goes on according to the following laws;—

Let l_1 be the inductance per unit length of the wires; let s_1 be their capacity, or permittance as Mr. Heaviside calls it, per unit length; and let r_1 be their resistance per unit length.

Then, for the slope of potential along them we have

$$-\frac{dV}{dx} = l_1 \frac{dC}{dt} + r_1 C, \quad . \quad . \quad . \quad . \qquad (16)$$

and for the accumulation of charge, or rise of potential with time,

$$-\frac{dV}{dt} = \frac{1}{s_1} \cdot \frac{dC}{dx}. \quad . \quad . \quad . \quad . \quad . \quad . \quad . \qquad (17)$$

These are equations to wave-propagation, and will give stationary waves in finite wires of suitable length, supplied with an alternating impressed E.M.F.

The solution for a long wire, for the case when r_1 is small and the frequency big,[1] is

$$V = V_0 \, e^{-\frac{m}{n}x} \cos n \left(t - \frac{x}{n_1}\right) \quad . \quad . \quad . \quad . \qquad (18)$$

where

$$m_1 = \frac{r_1}{2l_1}, \text{ and } n_1 = \frac{1}{\sqrt{(l_1 s_1)}}.$$

The velocity of propagation is therefore n_1, and the wave-length is $\frac{2\pi n_1}{n}$.

Now, for two parallel wires as in Fig. 12 (page 112),

[1] Mr. Heaviside has treated the problem in a much more general manner, see "Phil. Mag.," 1888, especially February 1888, p. 146.

$$l_1 = 4\mu \; log\frac{b}{a} + \frac{r_1}{n},$$

and

$$s_1 = \frac{K}{4 \; log\frac{b}{a}},$$

while r_1 = the geometric mean between its ordinary value and $\frac{1}{2} n\mu_0$; where the μ and K refer to the space outside the substance of the wires, μ_0 refers to their substance, a is their sectional radius, and b their distance apart.

The second term of l_1 is, we have seen, practically zero for these high frequencies. Hence (n_1) the velocity of propagation of condenser discharges along two parallel wires is simply the velocity of light, the same as in general space; because $l_1 s_1 = \mu K$.

The pulses rush along the surface of the wires, with a certain amount of dissipation, and are reflected at the distant ends; producing the observed recoil kick at *B*. They continue to oscillate to and fro until damped out of existence by the exponential term in (18). The best effect should be observed when each wire is half a wave-length, or some multiple of half a wave-length, long. The natural period of oscillation in the wires will then agree with the oscillation-period of the discharging circuit, and the two will vibrate in unison, like a string or column of air resounding to a reed.

Hence we have here a means of determining experimentally the wave-length of a given discharging circuit. Either vary the size of the *A* circuit, or adjust the length of the *B* wires, until the recoil spark *B* is as long as possible. Then measure, and see whether the length of each wire is not equal to

$$\pi\sqrt{\left(\frac{L}{\mu}\cdot\frac{S}{K}\right)}.$$

Further on I record some numerical results of observations made in this way.

It is interesting to see how short it is practically possible to make waves of this kind. A coated pane can be constructed of, say, two centimetres electrostatic capacity, and, by letting it overflow its edge, a discharge circuit may be provided of only a few centimetres electromagnetic inductance. Under these circumstances the radiated waves will be only some 20 or 30 centimetres long, corresponding to a thousand million alternations per second. Some beautiful diffraction experiments have been described by Lord Rayleigh in a Friday evening discourse to the Royal Institution (reprinted in "Nature," June 1888), and some of these might be used to concentrate the electromagnetic radiation upon some sensitive detector—possibly one of Mr. Boys's radiomicrometers, more likely some chemical detector—some precipitate or other that can be shaken out of solution by the impact of long waves, or some of Captain Abney's photographic agents.

Certainly the damping-coefficient $R/2L$ is high, and the radiation has a very infinitesimal duration; but a rapid succession of discharges can be kept up by connection with a machine.

No doubt much shorter waves still may be obtained by discarding the use of any so-called condenser, and by causing the charge in a sphere or cylinder to oscillate to and fro between its ends, as might be done by giving it a succession of sparks. These oscillations, it is to be feared, however, would have too small energy to be detected by ordinary means. If they could be made quick enough to affect the retina, no doubt we could detect them with the greatest ease; but it is manifest that this can only be done by reducing the circuit to a size less than the wave-length of light. The wave-length of the electrical radiation is six times the mean of the inductance and capacity, and each of these quantities is very comparable with the linear dimensions of the conductor concerned. By setting up electric oscillations in a body as small as a molecule, no doubt they would be rapid enough to give ordinary light waves; but the probability is that this is precisely what light waves are.

Either the atoms are made to vibrate relatively to the æther, by the effect of heat, and so to produce radiation; or else electrical oscillations are set up in comparatively quiescent atoms, not by heat, but by the impact of radiation from other sources, or by some organic process set in play by living protoplasm.

It is thus I would seek to explain phosphorescence and other direct production of light from cold sources.

This direct production of light we have not yet learned artificially to accomplish; we can only heat bodies and trust to their emitting light in some unknown manner as a secondary result; but the direct process has been learnt by glowworms and Noctilucæ, and it is for us, I believe, one of the problems of the immediate future.

University College, Liverpool,
July 7, 1888.

Postscript.—Since writing the above I have seen in the current July number of Wiedemann's "Annalen" an article by Dr. Hertz, wherein he establishes the existence and measures the length of æther waves excited by coil discharges; converting them into stationary waves, not by reflection of pulses transmitted along a wire and reflected at its free end, as I have done, but by reflection of waves in free space at the surface of a conducting wall. My friend Mr. Chattock has also written to me about a recent experiment exhibited to the Physical Society by Dr. E. Cook, which (when interpreted) shows that the same discharge as can excite æther waves a kilometre long can excite air waves of one millimetre.[1] The whole subject of electrical radiation seems working itself out splendidly.

Cortina, Tyrol, July 24, 1888.

[1] The experiment consisted in showing that, in the neighbourhood of a Leyden jar spark, fine powder, such as silica, throws itself into a "ripple-mark." Mr. Chattock repeated the experiment with powder in a tube, and by discharging a known capacity through a circuit of known self-induction verified approximately that the æther wave-length and the air wave-length bore to one another approximately the ratio a million to one, as they should.

CHAPTER XIV.

PROCEEDINGS AT THE BRITISH ASSOCIATION MEETING IN BATH.

Referring again to the footnote at the end of Chapter XI., some early experiments of Dr. Hertz are mentioned as analogous to mine on the surging circuit. In these experiments an open rectangle and other shapes of wire were connected as lateral offshoots to the terminals of a sparking coil, and supplementary or surging sparks were seen to occur in the gap of the rectangle and were taken as evidence of electric oscillation thus excited.

This observation was the beginning of the celebrated series of discoveries by Hertz, and it closely resembles what I had been observing too. In fact we were both on the same tack, but Hertz was a little ahead. In July, 1888, as mentioned at the end of the last chapter, written a week after the publication appeared, Dr. Hertz crowned his researches by the brilliant memoir in Wiedemann's "Annalen," wherein he described the detection of the waves by an electric resonator, and the measurement of their length by an interference method.

I also had measured wave-length by an interference method, but had only worked with the waves along wires, whereas Hertz had worked in free air. Hertz had also worked on wires to some extent either then or later, and found an apparent discrepancy between the velocity along wire and the velocity in air.

I, however, found no such discrepancy. The wave-length as calculated with the ordinary velocity of light in air, and as measured on the wires, agreed within the limits of accuracy of that kind of observation; and subsequent experience has led Dr. Hertz to accept this conclusion, and to suppose that reflection from walls, or some other disturbing cause, vitiated the quantitative accuracy of those particular experiments of his.

When the British Association met in Bath, which it did in September of this same year, Prof. Fitzgerald, who was President of Section A, directed world-wide attention to the discoveries of Hertz, and hailed them as conclusive proof of the truth of Clerk-Maxwell's theory of Light.

The next day it was my business to communicate to the section a brief summary of the results of my experiments in the same direction; and the following gives the gist of my two communications.

Measurement of Electro-Magnetic Wave-Length.[1]

When returning thanks for the reception of his address yesterday morning, our President[2] thought fit to disclaim too much share in the suggestion of an oscillatory Leyden jar circuit as a practicable source of radiation of moderate wave-length. But such a disclaimer he must not be allowed to make. He mentioned as coming before himself the names of Feddersen and other German experimenters on these alternations. But to them belongs not this credit, but another. What they accomplished was the experimental verification of the existence, and the

[1] Paper read before the British Association Meeting (Section A) at Bath, September, 1888.

[2] Professor G. F. Fitzgerald.

counting of the frequency, of these already predicted oscillations. Prior credit must therefore be put further back. As is very well known, Helmholtz in 1847 suggested that a Leyden jar discharge must be oscillatory in a circuit of small dissipation, and Sir William Thomson in 1853 calculated out the whole details of the process and gave its equation; an equation which, though we now write it down glibly enough, was at that date unintelligible or difficult of comprehension to all but a very few—perhaps to all without exception. Not the mere equation, of course, but the coefficient A which it contained, "the electro-dynamic capacity of the discharger," or, as we now call it, the self-induction or inductance of the circuit. Every important detail of this prediction has now been verified by the labours of the German experimenters, Feddersen, Schiller, and others. But any idea that radiation was propagated through the æther by such oscillations I am sure had never occurred to them. I remember the history of the thing pretty well, because I had been thinking much myself on the direction of how directly to manufacture light, *i.e.*, how to construct an electrical oscillator of a given and sufficient frequency. In fact, a year or so before the Southport meeting I suggested to this section a plan whereby I fancied it might be done. That particular plan I saw soon afterwards to be incapable of working, and the suggestion has therefore no merit. But it made me able to clearly appreciate the suggestion, or rather the certainty, which was expressed by Prof. Fitzgerald at Southport, that the oscillations in a discharging Leyden jar circuit were just the things that would be practicable, and which would generate light of some measurable, though no doubt still considerable, wave-length.

More recently Lord Rayleigh showed at a Friday

evening discourse in the Royal Institution some remarkably beautiful interference experiments in sound, whereby inaudible sound waves could be detected, measured, converged, and, in fact, treated in just the same manner as one is accustomed to treat light waves.

Putting the two things together, I was hopeful of being able to attack the problem experimentally and to prove the existence of radiation with measurable wavelengths from a Leyden jar circuit by interference experiments conducted after the model of Lord Rayleigh's. Meanwhile I happened to be experimenting on lightning conductors, and somewhat to my surprise, as an outcome from those experiments, I hit on an arrangement which, without any thought or scheming at all, gave me evidence of the very waves I had been thinking so much about, and enabled me to measure their lengths, though not in a previously planned-out way. I described these experiments quite hastily and briefly, in the midst of other matter, in a lecture to the Society of Arts last March on lightning conductors, as "the experiment of the recoil kick."

I continued the experiments after the lectures, and proved the existence of æther waves of various lengths, the shortest wave that I obtained distinct evidence of being three yards. I intended to describe those experiments at this meeting with a reference to the Southport suggestion of Prof. Fitzgerald, quite unknowing at the time that we should have the pleasure of seeing him in the chair. However, when going away from Liverpool on a holiday this summer, I read in the train the July number of Wiedemann's "Annalen," and there found that Dr. Hertz had obtained much better and more striking evidence of these electromagnetic waves, and had measured their length by an interference experiment exactly like one of those used by Lord Rayleigh for sound.

I hasten to acknowledge the superiority of Hertz's method of demonstration to my own; and so far as evidence of the waves is concerned a description of my experiments is now superfluous. Nevertheless, the mode of propagation of the pulses and their mode of reflection is different in my experiments from what they are in those of Hertz, and although mine is not so good a method, yet it may have some interest as confirming the view taught us by Poynting and others concerning the mode of propagation of energy through the æther, and the theory of Kirchhoff and Heaviside concerning the rate of transmission of signals by a telegraph wire.

The theory of the experiment is given in my Paper in the August number of the "Philosophical Magazine."

The velocity of a pulse along a wire is $\frac{1}{\sqrt{l_1 s_1}}$,

where, for very high frequencies, $l_1 = 4 \mu \log \frac{b}{a} + \frac{r_1}{p}$;

$r_1 = \sqrt{\frac{1}{2} p \mu_0 R_1}$; and $s_1 = \frac{K}{4 \log b/a}$.

So velocity of pulse along a wire $= \frac{1}{\sqrt{(\mu K)}}$, the same as in free space.

Now take an oscillatory Leyden jar circuit and apply Thomson's theory to it,

$$p = \frac{1}{\sqrt{LS}}.$$

If this circuit emits waves of velocity V, their length in free space is

$$\lambda = \frac{2\pi V}{p} = 2\pi \sqrt{\frac{LS}{\mu K}}.$$

Along two parallel wires they go at the same pace, and therefore have the same length. Apply a discharging Leyden jar circuit with a pair of long insulated wires (Fig. 12), and we have waves of known length propagated along the wires.

Now take the wires of finite length. At the free ends we shall have reflections, a sort of recoil or kick, evidenced

Fig. 12.

by a brush discharge or spitting off from the ends of the wire; and in the dark the far ends of the wires glow. Supply them with knobs, and we get surprisingly long sparks. For instance, referring to Fig. 13, the *A* spark being 4·4 the *B* spark was 15·0 in one case.

When may we expect to get the longest sparks? When the period of oscillation of the circuit and of the appendages to it are the same. This will be when the length of each wire is half a wave-length, or some multiple of half a wave-length. The two will then vibrate together like a resonant column.

My experiment consisted, then, in so varying the current or the resonant wire that the recoil kick is a maximum.

The results of varying L are shown by some such curve as this (Fig. 14), where horizontal distance represents distance of A knobs from jars, and where vertical ordinates represent length of B spark; and are such as entirely to confirm the theory so far as the accuracy of the experiments go.

It is much easier to get good measurements with fairly

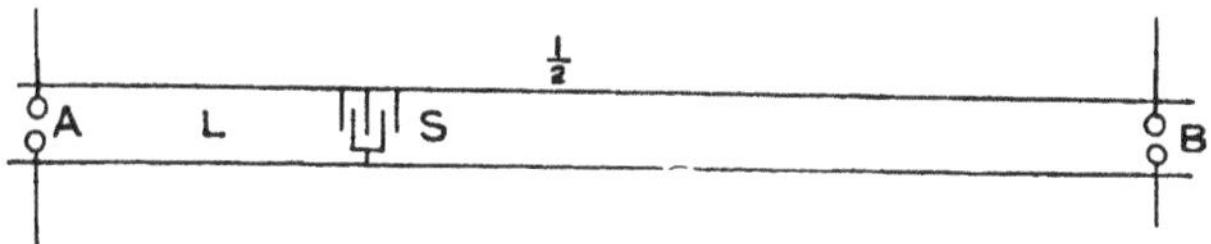

Fig. 13.

big condensers, and therefore long waves, else the energy is not sufficient to give very marked results. With very long wires arranged out of doors on a dark night I hope to get evidence of nodes by the periodic glow. [This has now been done.]

The experiments with shorter waves were made with

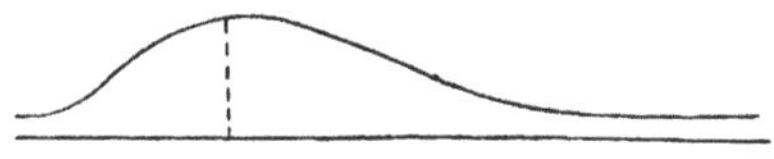

Fig. 14.

an adjustable tube condenser, various lengths of which could be used; and on the shortest waves with a mere coated disc condenser the size of a penny.

To get frequencies of the rapidity of visible light the linear dimensions of the circuit must be something comparable to 10^{-6} centim.; and so electric oscillations in atoms will give rise to ultra violet rays.

On the Impedance of Conductors to Leyden Jar Discharges.[1]

Among other experiments which I made in connection with lightning conductors were some on what I call "the alternative path." These gave some surprising results, showing that iron and copper acted equally well as conductors, and that sometimes the iron was better than copper; also that thickness of conductor and ordinary conductivity mattered very little.

Since then I have more completely applied the theories of Clerk-Maxwell and Lord Rayleigh concerning the impedance of conductors to very high frequencies of alternation, and have been able to arrive at a complete understanding of all, or nearly all, the alternative path results. I bring forward these experiments, therefore, as a verification of Lord Rayleigh's modification of Maxwell for extremely high frequencies.

Impedance being defined as

$$I = \sqrt{R'^2 + p^2 L'^2}$$

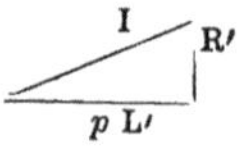

I R′ p L′

Rayleigh develops an expression of Maxwell's for E.M.F. into expressions for R' and L', which take the form, when p is infinitely great,

$$R' = \sqrt{\;p\, l\, \mu_0\, R}$$
$$L' = L + \frac{R'}{p},$$

these being the limits of a Bessel function series; where L refers only to the medium outside the substance of the

[1] Read before the British Association Meeting (Section A) at Bath September, 1888.

wire, and μ_0 is the magnetic permeability of the material of the wire, l being its length.

Wherefore, $$I=\sqrt{(pL+R')^2+R'^2}.$$

$$=pL\sqrt{1+\frac{2m}{\sqrt{p}}+\frac{2m^2}{p}},$$

where $m=\frac{\sqrt{\frac{1}{2}\,l\,\mu_0 R}}{L}.$

L depends on the size and shape of the conductor but not at all on its material, and it only depends slightly on its thickness; m depends on the conducting substance,

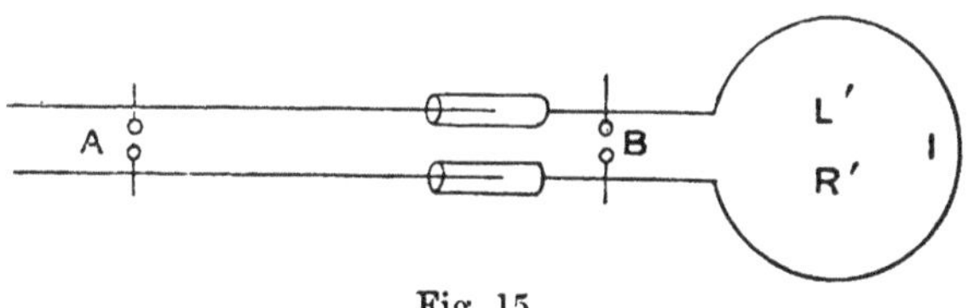

Fig. 15.

and increases with both μ and R. R is the ordinary resistance of the wire.

At very high frequencies the first term of those under the root in the expression for I alone matters, and therefore all substances which conduct at all behave in the same way. At less frequencies the terms involving R and μ_0 begin to be of importance.

All this is borne out by the experiments, and the theoretical impedance is found to be very nearly proportional to the observed difference of potential needed to drive a given charge through any given conductor.

The experiment is conducted thus:

Two jars are arranged to be charged and discharged at A. The circuit is completed either by an air-gap at B, or by the conductor to be experimented on. The conductor is arranged in the form of a circle and connects the *B* knobs so as to afford an alternative path. The experiment consists in adjusting the *B* knobs until the spark jumps the air-gap just about as often as it misses and chooses the conductor; the size of the jars, the length of the *A* spark, and the position of *A* on the two wires connected to the inner coatings being the things varied in successive series experiments. A whole lot of circular conductors, *I*, are tried one after the other with each adjustment. (Fig. 15.)

The length of spark *B* measures the E.M.F. needed to effect discharge through the conductor.

This is found proportional to the theoretical impedance for

No. 2 Iron,	No. 5 Copper,	No. 2 Brass,
No. 18 Iron,	No. 18 Copper,	No. 24 Brass,
No. 25 Iron,	No. 23 Copper,	
	No. 40 Copper,	

with frequencies varying in different experiments from twelve million vibrations per second to a quarter million vibrations per second.

The current amplitude during the discharge was 3,000 ampères, sometimes more, sometimes less.

The impedance of a conductor $2\frac{1}{2}$ metres long bent into a circle was

180 ohms for thick wire, No. 2 } at 12 millions
300 ohms „ thin „ „ 40 } per second,

the ordinary resistance being ·004 ohm and 2·6 ohms respectively. Taking a lower frequency, the impedance was

43 ohms for thick .... } at 3 millions.
78 ohms „ thin .... }

At a quarter million vibrations the material just begins to matter, and iron has a trace more impedance than copper; but even now they are practically equal, being about four ohms for thick and six ohms for extremely thin wire.

These are the results of the latest experiments. In some of the older experiments iron came out distinctly better than copper—even thin iron better than thick copper. Theory does not, so far as I see, throw any light upon this, and I am going to repeat those experiments in something like their original form, so as either to confirm or to modify these few anomalous results. [For complete expansion of this, see Chaps. XX. and XXVI.]

The main point established is that for frequencies comparable to a million per second, such frequencies as occur in jar discharges and probably in lightning, the impedance of all reasonably conducting materials is the same, both by theory and by experiment, independently of their magnetic properties as well as of their conductivity; and that even the sectional area of a conductor only affects its impedance in a minor degree, so that a change to $\frac{1}{1000}$ the area scarcely doubles the impedance.

DISCUSSION ON LIGHTNING CONDUCTORS.

In the course of the meeting a joint sitting of the Physical and the Engineering section was held, and a long discussion on the subject of lightning conductors was opened by Mr. Preece.

It would be tedious to refer to this discussion at length; it was reported almost verbatim in the "Electrician"[1] for

[1] An uncorrected and very erroneous report appears in the British Association volume for the same year, but the "Electrician" and "Electrical Review" reports are far more correct and authentic

September 21 and 28, 1888, volume xxi., also in the "Electrical Review" for the same dates; and the main points at issue were summarized by the present writer as follows: selecting those statements of Mr. Preece and his supporters which seem most generally accepted, or likely to be accepted, on that side; and numbering the opposing statements so as to correspond with them. No statement is here quoted or suggested which to the writer seems entirely absurd; because absurd statements may easily be made in debate without sufficient thought, and because such statements are not likely to be generally or weightily accepted, even if pressed by their propounder.

Statements made by Upholders of the Older or Orthodox Views.

1. Properly constructed lightning rods never fail. When existing rods fail it is because there is something the matter with them—usually an insufficient earth.

2. Leyden jar discharges have nothing oscillatory or alternating about them, or at least the existence of such alternations is an unproved assumption. [This position was soon abandoned as untenable.]

3. Even if Leyden jar discharges should turn out to be oscillatory, there is no reason why lightning flashes should be of the same character. Lightning flashes have an apparent duration, and transmit telegraph signals, deflect compass needles, and do other things which alternating currents could not do.

4. The one thing needful for an efficient lightning protector is conductivity, sufficient conducting power to convey the whole charge quickly and harmlessly down to the earth, with which the conductor must make elaborate contact.

5. No danger is to be feared from a lightning conductor

if only it be well earthed and be sufficiently massive not to be melted by a discharge. All masses of metal should be connected to it, that they may be electrically drained to earth.

6. The shape of the sectional area of a conductor is quite immaterial; its carrying power has nothing to do with extent of surface; nothing matters in the rod itself but sectional area, or weight per foot run, and conductivity.

7. Points, if sharp, should constitute so great a protection that violent flashes to them ought never to occur.

8. Lightning conductors, if frequently tested for continuity and low resistance by ordinary galvanic currents, are bound to carry off any charge likely to strike them, and are absolutely to be depended upon. The *easiest* path protects all other possible paths.

9. A certain space contiguous to a lightning rod is completely protected by it, so that if the rod be raised high enough a building in this protected region is perfectly safe.

Contrary Statements made by the present writer.
(Numbered so as to correspond with the preceding.)

1. Rods as at present constructed, though frequently successful, may and do sometimes fail, even though their earth is thoroughly good; the reason being that they offer to a flash a much greater obstruction—a much worse path—than is usually supposed: an obstruction to be reckoned in hundreds or thousands of ohms, even for a very thick copper rod.

2. When a Leyden jar is charged it corresponds to a bent spring, and its discharge corresponds to the release of the spring. Its discharge current alternates, there-

fore, in the same way and for much the same reason as a twitched reed or tuning fork vibrates. The vibrations decay in either case because of frictional heat production, and because of the emission of waves into the surrounding medium. A single spark of a Leyden jar, examined in an exceedingly fast revolving mirror, is visibly drawn out into a close succession of oppositely-directed discharges, although its whole duration is so excessively minute.

3. A lightning flash is a spark between cloud and earth, which are two oppositely electrified flat surfaces, and the flash corresponds therefore to the internal sparking between the two plates of a great air condenser. All the conditions which apply to a Leyden jar under these circumstances are liable to be true for lightning. Sometimes the resistance met with, either in the cloud itself or in the discharger, may be so great that the spark ceases to be oscillatory, and degenerates into a fizz or rapid leak; but there can be no guarantee that it shall always take this easily manageable form; and it is necessary in erecting protectors to be prepared for the worst and most dangerous form of sudden discharge. The apparent duration of a lightning flash is due to its frequently multiple character, and indicates successive discharges, not one long-drawn-out one. Nothing that lightning has been found to do disproves its oscillatory character; because Leyden jar discharges, which are certainly oscillatory, can do precisely the same.

4. Although some conductivity is necessary for a lightning conductor, its amount is of far less consequence than might be expected. The obstruction met with by an alternating or rapidly varying discharge depends much more on electro-magnetic inertia or self-induction than upon common resistance. So much obstruction is

due to this inertia that a trifle more or less of frictional resistance in addition matters practically not at all.

It is desirable to have a good and in dry weather a deep earth, in order to protect foundations and gas and water mains from damage, and in order to keep total impedance as low as possible.

5. The obstruction offered by a lightning rod to a discharge being so great, and the current passing through it at the instant of a flash being enormous, a very high difference of potential exists between every point of the conductor and the earth, however well the two are connected; hence the neighbourhood of a lightning conductor is always dangerous during a storm, and great circumspection must be exercised as to what metallic conductors are wittingly or unwittingly brought near or into contact with it. When a building is struck the oscillations and surgings all through its neighbourhood are so violent that every piece of metal is liable to give off sparks, and gas may be lighted even in neighbouring houses. If one end of a rain-water gutter is attached to a struck lightning conductor the other end is almost certain to spit off a long spark, unless it also is metallically connected. Electric charges splash about in a struck mass of metal, as does the sea during an earthquake, or when a mountain-top drops into it.[1] Even a small spark near combustible substances is to be dreaded.

6. The electrical disturbance is conveyed to a conductor through the æther or space surrounding it, and so the more surface it exposes the better. Better than a single rod or tape is a number of separate lengths of wire, each thick enough not to be easily melted, and well separated

[1] See "The Eruption of Krakatoa," a pamphlet by E. Douglas Archibald, Tunbridge Wells; or the Report of the Krakatoa Committee.

so as not to interfere with each other by mutual induction.

The liability of rods to be melted by a flash can be easily over-estimated. A rod usually fails by reason of its inertia-like obstruction, and consequent inability to carry off the charge without spittings and side-flashes; it very seldom fails by reason of being melted. In cases where a thin wire has got melted, the energy has been largely dissipated in the effort, and it has acted as an efficient protector; though, of course, for that time only.[1]

[1] The following is a typical case of the way in which a very thin wire may exert a protective influence with success although it itself gets wholly destroyed. (See also p. 195.)

Case of the Action of a Small Conductor.

Franklin, in a letter to Collinson read before the Royal Society, Dec. 18, 1755, describing the partial destruction by lightning of a church-tower at Newbury, Mass., wrote, "Near the bell was fixed an iron hammer to strike the hours; and from the tail of the hammer a wire went down through a small gimlet-hole in the floor that the bell stood upon, and through a second floor in like manner; then horizontally under and near the plastered ceiling of that second floor, till it came near a plastered wall; then down by the side of that wall to a clock, which stood about twenty feet below the bell. The wire was not bigger than a common knitting needle. The spire was split all to pieces by the lightning, and the parts flung in all directions over the square in which the church stood, so that nothing remained above the bell. The lightning passed between the hammer and the clock in the above-mentioned wire, without hurting either of the floors, or having any effect upon them (except making the gimlet-holes, through which the wire passed, a little bigger), and without hurting the plastered wall, or any part of the building, so far as the aforesaid wire and the pendulum-wire of the clock extended; which latter wire was about the thickness of a goose-quill. From the end of the pendulum, down quite to the ground, the building was exceedingly rent and damaged. . . . No part of the afore-mentioned long, small wire, between the clock and the hammer, could be found, except about two inches that hung to the tail of the hammer, and about as much that was fastened to the

Large sectional area offers very little advantage over moderately small sectional area, such as No. 5 B.W.G.

7. Points, if numerous enough, serve a very useful purpose in neutralizing the charge of a thunder-cloud hovering over them, and thus often preventing a flash; but there are occasions, easily imitated in the laboratory, when they are of no avail; for instance, when an upper cloud sparks into a lower one, which then suddenly overflows to earth. In the case of these sudden rushes, there is no time for a path to be prepared by induction, no time for points to exert any protective influence, and points then get struck by a violent flash just as if they were knobs. Discharges of this kind are the only ones likely to occur during a violent shower; because all leisurely effects would be neutralized by the rain-drops better than by an infinitude of points.

8. The path chosen by a galvanic current is no secure indication of the course which will be taken by a lightning flash. The course of a trickle down a hill-side does not determine the path of an avalanche. Lightning will not select the easiest path alone; it can distribute itself among any number of possible paths, and can make paths for itself. Ordinary testing of conductors is therefore no guarantee of safety, and may be misleading. At the same time it is quite right to have some system of testing and of inspection, else rust and building alterations may render any protector useless.

9. There is no space near a rod which can be definitely styled an area of protection, for it is possible to receive

clock; the rest being exploded, and its particles dissipated in smoke and air, as gunpowder is by common fire, and had only left a black smutty track on the plastering, three or four inches broad, darkest in the middle, and fainter towards the edges, all along the ceiling, under which it passed, and down the wall."

violent sparks or shocks from the conductor itself. Not to speak of the innumerable secondary discharges which, by reason of electro-kinetic momentum and of induction, and of the curious recently-discovered effect of the ultra-violet light of a spark, are liable to occur as secondary effects in the wake of the main flash.

Just one word on the subject of iron *versus* copper. The writer last year thought and stated that, in so far as the substance of the conductor was magnetized by a discharge, iron would obstruct a lightning flash or any other rapidly-varying current enormously more than copper does. But the fact is, that the substance of a conductor is, by sufficiently rapidly alternating currents, not magnetized at all. The current is tubular, keeps wholly to the outer surface, and magnetizes nothing inside. Hence the magnetizability of the substance of the conductor is of no moment at all; and iron, therefore, will do every bit as well as copper. Mr. Preece's experience with half a million iron wire telegraph-post protectors leads him to uphold iron as entirely satisfactory. So, on this one point, as well as on the necessity existing for a good earth, the upholders of the older and of the newer views have been able to agree.

Immediately after the discussion Dr. Janssen exhibited some preliminary attempts at photographs of lightning taken with a double-nozzled camera, having two sensitive plates, one fixed, the other revolving thirty times a second.[1] The same flash was depicted on both plates, but on the moving plate it was separated into two or three distinct streaks, showing its multiple character. Each constituent, however, was as clear and distinct on the

[1] The plan was suggested by the present writer in a letter to "Nature," July 12, 1888.

rotating as on the fixed plate, and had, in fact, exactly the same shape and appearance, so that one could be superposed on the other exactly; thus proving its instantaneous character. The rate of spin was naturally nothing like sufficient to exhibit the alternating character of each constituent.

It is to be hoped that many more photographs of lightning will be taken on this plan, because there are evidently many kinds of composite flashes, and it is an excellent way of analyzing them and correcting the impressions, often erroneous, formed by the eye.

The result of the discussion was to impress everyone with the desirability of making observations on actual lightning, and ascertaining how far its behaviour accords with small-scale laboratory experiments.

Appendix to the Bath Discussion.

Among many letters from correspondents in the scientific journals which followed this discussion, on the subject of lightning photographs, the dark flash, etc., I take the annexed direct observations from the "Electrician" for October, 1888.

Lightning.

TO THE EDITOR OF THE "ELECTRICIAN."

"I feel inclined to believe with Prof. Lodge that some flashes are oscillatory and some are not. Those that meander through the air, striking at nothing in particular and vanishing into space, should, I think, be classed as non-oscillatory, while those which strike direct from the clouds to the earth—thick heavy flashes—probably oscillate until the potential between the points is equalized.

While on this subject I should like to learn if the following phenomena can be accounted for:—One evening last year, after sunset, I was watching a dark black cloud, with clearly defined outline, high above the horizon, when from the top flashes of lightning shot upwards into quite clear atmosphere. There was nothing visible for it to strike at. I understand from a gentleman who has spent some years in India that it is not uncommon for flashes to strike across the clear blue sky in which there is no sign of a cloud.

In conclusion, I should recommend all those who are interested in the study of lightning flashes to refer to negatives or glass positives for their information, as there is much interesting detail lost in silver prints.—Yours, etc., W. P. ADAMS, A.K.C., A.S.T.E."

Springwell, Barnes, October 8, 1888.

"SIR,—The phenomenon to which Mr. W. P. Adams refers in his letter to the "Electrician," p. 775, is one that I have observed on one occasion only, and though I have described it to several people, I have not hitherto found that it has been noticed by others.

"It is now so long ago that I cannot remember all the circumstances, but the impression produced at the time was so vivid that I am certain of the following particulars:—

"A thunderstorm travelling from S. to N. passed immediately over the village of Wing, in Rutland, at about 9 or 10 p.m. When it had travelled so far away that the thunder-cloud was low on the horizon, and the altitude of the top of the cloud was about half or a third, or possibly less, that of the lower bright stars in the Great Bear, at that time in their lowest position (and this fixes roughly the time of year), I noticed that while the true flashes of lightning between the cloud and the earth were often visible themselves, they were mostly accompanied by long thin flashes which extended up well into the space between the bright stars of the Great Bear, which were shining brightly in a perfectly clear sky. Sometimes but one of these lines of light would be formed, generally two or three, and I remember distinctly that one flash of real lightning, in or below the cloud, was accompanied by seven of these mysterious lines, which simply went up from the cloud into the clear sky and left off.

"These flashes had all the usual character of ordinary lightning, except that they seemed much thinner or less luminous. Their light, though seen direct, was much less than that in or below the cloud. I was sure that they were not the only flashes, but that they accompanied what I have called above the real lightning.

"Though I have often recalled the circumstances I can offer no explanation. Perhaps they will be of interest to Dr. Lodge.—Yours, etc., C. V. BOYS."

Physical Laboratory, South Kensington.

SIR,—The interesting phenomenon observed by Mr. Boys and by Mr. W. P. Adams, and recorded by them in your two last issues, suggests a case of discharge by overflow. I mean that the upper layers of the atmosphere—the region of virtual high conductivity (really weak dielectric strength)—may have discharged down to the cloud until that overflowed to the earth. Or it might be that the flash from cloud to earth so lowered the potential of the cloud as to permit a discharge or set of discharges from the high-potential regions above. I should thus regard it merely as a case of a couple of sparking intervals in series.

Many things seem to suggest that it is the upper

layers of atmosphere that are primarily charged, and that clouds are either conducting protuberances or else stepping-stones; probably sometimes one, sometimes the other.

Mr. Boys would, perhaps, say whether there was anything in his observation irreconcilable with this simple idea.

I may take this opportunity of remarking how much work can be done at meteorological stations and observatories in the matter of accurately observing and recording lightning; photographic records, obtained by proper appliances for distinguishing multiple from successive flashes, being, of course, superior to all others.

An experimental lightning conductor on a flagstaff near every meteorological observatory would also be a most desirable addition. It need not be associated with danger; a system of fuses or cut-outs, or an east or west steel bar, might be used to record the passage of a flash, and the rod need not be examined until after the cessation of violent disturbances. By having the conductor of different thickness at different parts one could learn what size is really likely to be melted. One could also arrange so as to gain information about side-flashes.—Yours, etc., OLIVER J. LODGE.

I also received the annexed important letter from Prof. Ewing, wherein he shows that what is often referred to as anomalous magnetization, *i.e.*, the alternations of magnetism set up by Leyden jar discharges in steel needles near their path, can be explained by a gradual lengthening of period in the oscillations, combined with the facts concerning magnetization discovered by himself.

"DEAR LODGE,—I see in the report of the lightning

rod discussion (which I was very sorry not to be able to attend) that you referred to the difficulty there is in seeing how an oscillatory current can leave permanent magnetic effects in an iron conductor or in neighbouring iron, and Lord Rayleigh spoke of experiments showing that when a Leyden jar is discharged through a helix with a steel bar inside it, the bar may be magnetized *in cylindrical layers* with opposite directions of magnetization at different distances from the surface. Does not this suggest oscillations in which the period lengthens while the amplitude decays?

"Current *A* is so transient that its effects are extremely superficial. The inner layers are protected by induced currents in the iron (as Lord Rayleigh remarked), but it magnetizes the surface layer strongly. Current *B* is not strong enough to reverse the effect of *A* on the outermost layer, but it is slow enough to magnetize a layer that lies a little nearer the axis. *C*, again, is too weak to reverse the effect of *B*, but its effects reach deeper still, and so on.—Yours, etc., J. A. EWING."

University College, Dundee, September 29, 1888.

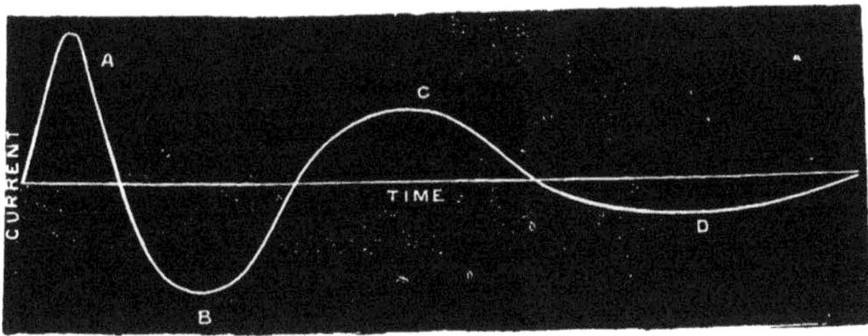

Fig. 15a.

CHAPTER XV.

EXPERIMENTAL LIGHTNING CONDUCTORS AND OTHER OBSERVATIONAL MATTERS.

The editor of the "Electrician" (Mr. Snell) made a valuable suggestion ("Electrician," November 2, 1888, p. 810), that at many telegraph stations abroad, where thunder-storms are frequent and violent, it might be possible to set up lightning conductors for experimental purposes, and thus accumulate experience concerning their behaviour more rapidly than by waiting for storms and visiting damaged buildings in this country. He asked me to make some suggestions as to how they should be rigged up, and what should be tried; and I am glad to make such suggestions if they are likely to be any help. Skilled observers, who are frequently favoured with storms, will soon accumulate experience enough to render unnecessary any suggestions from me. But to begin with it may be useful to draw up a few notes on the subject, and I accordingly print the following memoranda.

Memoranda with regard to the Erection of Experimental Lightning Conductors at Telegraph Stations Abroad.

1. It would not be wise to erect anything experimental on an inhabited building, still less on a building utilized

for telegraph purposes. A detached shed or a flagstaff is a more suitable support for an experimental rod. A shed offers rather more scope for experimenting than does a flagstaff, though perhaps it is less likely to be struck. There are, however, a few experiments which can properly be made *in the neighbourhood* of an actual lightning conductor, always provided that it be not tampered with in the least. (See § 10 *et seq.*)

2. To gain experience as to the advantage or disadvantage of very thick conductors, a couple of conductors might be erected on the same shed or on a pair of adjacent poles:—one a very stout copper rod, the other a quite thin iron wire, even so thin as No. 27.

3. A lightning conductor might be erected on a pole, consisting of various thicknesses of wire in series. Suppose, for instance, iron wire is used. It should be No. 1, or very thick, at the top where the lightning strikes, because that is the place where local melting is almost certain; then it might change to No. 5, No. 8, No. 12, No. 16, No. 18, No. 20, and then once more to No. 5, to take it properly to earth. It would be instructive to find some of these wires melted and not others.

4. An obvious modification of the above experiment is to join copper and iron in series alternately, to use strip in some places instead of wire, and otherwise to vary the circumstances, but always making thoroughly good joints.

5. A distinct conductor might be erected, say of No. 5 iron, with a purposely bad joint, say $\frac{1}{4}$ inch gap, to see what happens there. At another part of the same conductor there might be an imperfect joint, the parts in contact but not joined properly.

6. Since it is the object to get these conductors struck, rather than to dissipate atmospheric electricity silently,

it may be well to terminate them with knobs instead of points.

7. As detectors, to record whether a given rod has been struck or whether a flash has passed in a given direction, one may suggest—

(*a*) A small steel bar or needle placed pointing east and west at right angles to the conductor; to be tested after a storm for magnetism with a compass needle.

(*b*) Some form of lead or tinfoil fuse or cut-out inserted in the length of the rod itself.

(*c*) A branch cut-out adjusted in a tapping circuit, say a bit of wire joined to two points of the conductor a few yards apart, and having included in it an exceedingly fine piece of platinum or other wire to be destroyed by the branch current.

(*d*) A similar branch or tapping circuit, say of No. 16 wire, coiled several times round a quill glass tube containing a sewing needle; to be tested for magnetization.

(*e*) Some form of registering Cardew or other voltmeter might be similarly connected to two points on the conductor.

(*f*) An Abel fuse or other convenient explosive might be used.

(*g*) A small bulb containing a mixture of hydrogen and oxygen, with a pair of Pt wires close together, would furnish evidence of a very minute spark.

8. To observe side-flashes, a short supplementary conductor, with a detector in its course, could be led from an independent earth to near a point on the conductor a few yards above ground, leaving, say, half an inch gap. A side-spark might very well occur across it.

9. Side-flashes may also occur to quite insulated conductors—for instance, a rod hung inside the shed on silk or glass or paraffin, with one end near the conducter;

only the amount passing in this case, being small, is not so easily detected. A gas leak, or an Abel fuse, is the readiest way of obtaining a record of the passage of a small spark.

10. Near a main conductor—a real one or an experimental one—supplementary short rods might be erected, some earthed, some insulated, and at varying distances from the main conductor. Each of these conductors should have a detector in it, and induced surgings in totally disconnected conductors may then be observed, without any side-flash to them.

11. A horizontal incomplete circuit of wire may have one end connected with the lightning rod and the other end arranged anywhere. Spittings off are likely to occur from this free end, even though it be curved upon itself and brought back close to its beginning.

12. Along the walls of a shed a series of detached conductors, tinfoil strips, or thin wire, may be arranged, with moderate gaps between. They are very likely to spark into one another when a flash occurs, whether the final one of the series be connected to earth or not.

13. A pair of insulated wires arranged parallel and close together anywhere near the conductor, but quite disconnected from it—one end of one being connected to the ground, and the other end of the other being connected with anything fairly insulated—are very likely to have a spark pass between them when a flash occurs. One way of proving the passage of such a spark is to have a powerful battery permanently joined to the two wires through a cut-out, for the spark will then start an arc.

14. Entirely uninsulated things, such as gas and water-pipes, are liable to spark into each other when a flash occurs; and this also may be detected by a sufficiently

sensitive recording arrangement, such as a suitably arranged gas leak, or the detonating bulb mentioned in 7 (*g*).

15. In all cases the best possible earth should be made for the experimental conductor, and by frequent testing no loop-hole should be left for any extraordinary and interesting occurrence to be explained away by the stupid and commonplace exclamation, "bad earth."

16. It seems to me at present of most interest to prove the existence under real lightning circumstances, and with a *perfect* conductor, of electric surgings in all manner of unlikely places—between insulated bodies, between earthed bodies, between bodies at a considerable distance from the disturbance. It will then be desirable to ascertain whether or not such surgings are not more violent when the lightning conductor is a stout rod than when it is a very thin wire, which is deflagrated while acting as a protector.

Manifestly the best protector is one which conducts the flash to earth and which yet sets up least side-flashing and surging. The object of the experiments, therefore, is largely to determine what it is that best satisfies those conditions, and at the same time to note and record all concomitant circumstances of interest.

OLIVER LODGE.

Modes of Damage by Lightning.

Some recent occurrences suggest that what experiment indicates to be a possibility does sometimes happen, viz., that electricity surges up from an earth contact by an underground route and causes damage without having any obvious entrance or exit to the sky. When discharge occurs to the ground, conductors, like gas-pipes buried in it are so disturbed that they easily give off sparks; not sparks which will charge a Leyden jar—the electricity splashes up and subsides again in an instant,—but sparks which very readily light gas and give shocks.

This is a straightforward experimental fact. One would not, therefore, be surprised to hear of lightning producing a similar effect on a larger scale; so that, when the ground is struck, earth connections of all kinds to neighbouring houses and telegraph stations become sources of danger rather than safety, and bring up from the main flash echoes or splashes which may in some cases do damage.

Three instances, or apparent instances, of this have been reported to me recently. The first is the case of a house at Wavertree, wherein a gas-pipe was melted in an underground cellar and the gas ignited, while there was no trace of any kind indicating how the lightning had got to the cellar, except that the owner saw a fireball or glow travelling up the drive, along what turned out to be the course of the gas-main. (A fuller account is subjoined.)

The second instance is that of a lighthouse in India, reported to me by Colonel Fraser, R.E., in which while flashes were striking the lightning conductor, a man was killed inside the building, though no apparent branch flash reached him from anywhere. Various explanations of this case may be suggested, and I do not press it, but quote it below.

The third case is that kindly communicated by the Electrician of the Commercial Cable Company, where instruments not connected with any line-wire or cable, but only connected to earth through the lightning protector, were completely destroyed. A record of this may be found in another part of the book.

Returning to the first, or Wavertree case, the following

is a statement by the occupier of the house, E. Lennox Peel, Esq.:

"One stroke, *A* (Fig. 16), melted and twisted up 8ft.

Fig. 16.

of roof-gutter round the corner, and also knocked off 2ft. of ornamental finishing, *C*. At first we thought this was the same stroke as the one that landed (*B*) in the cellar, *under* drawing-room window, *D*, fusing 3ft. of gas-pipe, firing the gas, and scorching the cellar ceiling; but as we can find no trace of *A* coming *inside* the house we fancy now that we were hit twice, and that we never saw *A* at all from the dining-room window, *E*.

"There were two tremendous crashes, one of which we put down to a stroke on the church exactly opposite, the steeple, with conductor, being just opposite dining-room window, *E*, across road from the gate shown. At the first crash (but we now believe *before* we heard it) a ball of fire seemed to come away from the church, cross the road, rush through the gates, up to the drive, and pass *E* as if to *D*. I was standing inside the window *E*, and saw it coming, but had to shut my eyes. There is a gas-lamp opposite right-hand gate-post, and our gas-pipes are laid under the drive up to *B*. The next crash came thirty seconds later, and was probably *A*. Both were distinctly heard in church, as if the house had been struck twice. The steeple was surrounded by black cloud, high up, swirling round, and white vapour low down. The lightning came from every quarter, about nine flashes in less than two minutes, with very large hailstones."

I myself examined the premises some weeks after the event, and it is clear to me that the church was not struck, or, at least, not struck violently. The points of its conductor are sharp, and no part of the conductor or earth shows the least trace of damage. The church steeple is lofty and its centre is only fifty-two yards from the house. The conductor is a copper tape by Newall of Gateshead, and seems in excellent condition. The house is an ordinary building, not so tall as the body of the church itself; it is, therefore, entirely dwarfed by the spire, and that it should have been struck in preference to the spire is remarkable. Service was going on in the church at the time. There were seven flashes all pretty close together, and two of them were specially loud and alarming. One of them Mr. Peel believes to be connected with the damage to the roof at *A*; the other to the damage in the cellar at *B*.

The telephone wires from church to a neighbouring house are quite thin and uninjured, and evidently have nothing to do with it. The path of the flash which struck *A* is not at all clear. It may have gone down a rain side-gutter and water-butt not far off, but if so it has left no trace of its passage. No connection of any kind is apparent between the two damages. And it seems to me much more likely that the second flash got into the gas-pipes outside the house altogether. Several feet of pipe have been very effectually fused, but this may have been mostly done by the ignited gas, which burned some time before discovery. It was $\frac{3}{4}$ in. pipe—not compo—and it must have been a pretty strong disturbance to melt it.

The ball of fire seen travelling along the ground which often makes its appearance in these stories is too frequently set down to imagination, because it is not a thing to be expected on ordinary theories; but when we know so little of the phenomena, it is surely wiser to accept provisionally even incredible statements, if often repeated, and try and amend our theories by them, rather than to ignore what may turn out to be perfectly true.

It may be remembered that during some of the experiments I showed at the Institute of Electrical Engineers, a wire conveying the discharge glowed with a luminous brush all along its length (it is an effect I have frequently obtained in the laboratory), and a correspondent ("Electrician," May 3, p. 749) wrote to ask about what he likened to a "luminous mouse" which he saw travelling along the wire, and taking an appreciable time in transit.

The *progressive* appearance is doubtless an optical effect, but the luminosity is a reality, and, since it appeared to travel then, so it may easily appear to travel when caused by lightning.

I should suppose that gas-pipes conveying the discharge are at so high a potential that the brush from them may be sometimes visible above the ground. If this be a fact, it is not one that can be regarded as at all established as yet; but the suggestion may direct observers' attention to the possibility both of this brush effect and also of damage being done by a charge which surges up and subsides again whence it came, without having any separate egress, or, as it has hitherto been usually expressed by observers, without any perceptible ingress to the place where the damage is done.

A fourth instance I have heard of quite recently as occurring at Harvey Road, Cambridge, where damage was done to roof and to basement, but no apparent connection or path from one to the other. I should suppose that the roof damage was done by a side-flash spitting off from the main-flash and returning to it again without finding any separate earth. It is very necessary to realize that this lateral expansion or temporary overflow may occur, so that things are momentarily charged and discharged again without their necessarily affording any thoroughfare whatever.

The following are the communications from Colonel Fraser:

Malabar District, May 29, 1889.

"PROF. O. J. LODGE, F.R.S.

"Dear Sir,—Having several lightning conductors to

do with, and rather inclined to your expressed distrust of them, it may be of interest to mention the particulars of an occurrence at the Mangalore Lighthouse on the 14th inst. This lighthouse (Fig. 17) consists of a masonry shaft about 10ft. diameter, standing on a cubical plinth of 12ft. sides or so, with two small store-rooms having terraced roofs on the flanks. There is a passage in the plinth having an entrance door and doors at the side leading into the store-rooms, and an inner door giving admission to the base of the column, which has flights of wooden ladders up to the capital, on which there is a gallery and the lantern of glass and bronze.

"The Public Works have lately been arranging the lightning conductor, the upper part of which down to the ground was supplied by the English makers of the lantern, and consisted of a 2in. copper wire rope fastened to the exterior of the column and its plinth.

"The earth connection only remained, and as the lighthouse is on a dry rocky hill plateau, I had a cable of 49 telegraph wires made at Madras to reach down to a well to be sunk to moist earth in the valley below. When at the place last month this cable had been soldered to the copper rope and laid in an open trench more than 200ft. long to close to the well. As moist ground was not reached at the depth expected the cable was too short, and a piece of extra copper rope had to be waited for as well as the further deepening of the well. From the ground to the top of the lantern is, I suppose, about 150ft., or less.

"At 7.30 p.m. a thunderstorm came on it is reported, and as heavy rain fell, the trench and cable were wet for a length of more than 200ft., so that the earth connection must have been complete, notwithstanding the cable not ending in the well.

"The following extract from the Post Officer's report states what happened:—'A group of natives had been

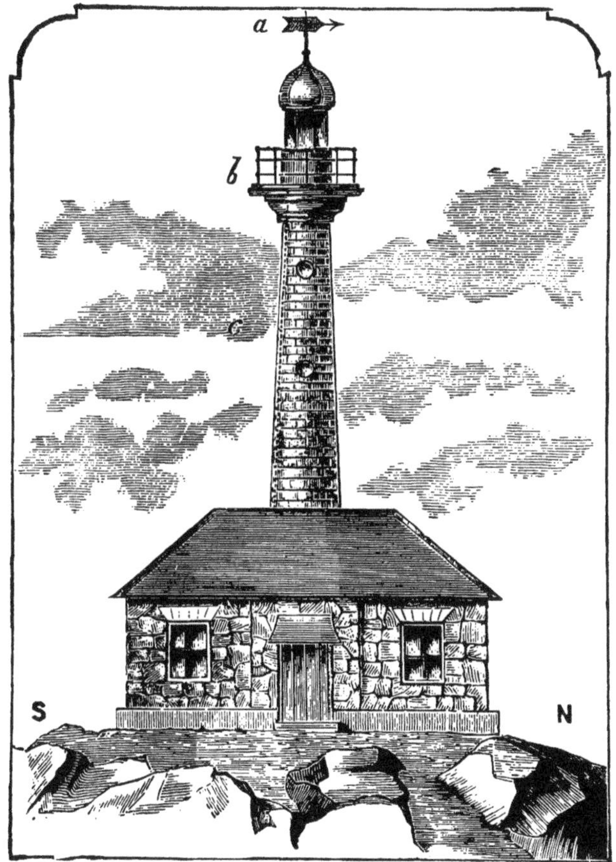

Fig. 17. MANGALORE LIGHTHOUSE.
a, copper vane; *b*, wooden railing; *c*, copper wire rope at the back of the tower.

sitting near the lighthouse, and all but two ran for their homes when they felt rain. These two clamoured at the closed door for admission into the lighthouse, in which two keepers had been on duty. On entering the inner room one remained standing, and the other, a youth of twenty-one years, who had lately come up from the Madras College, sat on the sill of the inner door. Before the keeper had time to close the outer door the flash came; the sitting youth fell backwards, and feebly calling for water, died almost instantaneously; his friend was struck up the legs, and was partially paralyzed the next day, but afterwards got over it; and the second keeper was hurled across the room. The left arm of the keeper certainly showed marks of exterior burns. No external marks appeared on the dead man.' Another account I had from my P. W. subordinate alluded to a small piece of zinc sheeting over the outer doorway as the point the lightning struck.

"However, in writing to the head of the Marine Department in reply to inquiries about the lightning conductor, I quoted from your article in 'Nature' of March 14, on 'Leyden Jar Discharges,' which I had been much interested in reading, and its footnote.

"I said, in my opinion, as the force of discharge was not in the wire, but in the open air and space (water included) round it, the inside of a lighthouse was a most dangerous place to be in if a flash went through the conductor, as some of the disturbance would go through you.

"It is also a question with me if a person, being of conducting material—water, etc.—themselves, they would not spark from induction, much as a piece of metal; and to what extent small masses of metal, such as door-locks, watches in the pocket, etc., get charged by a lightning conductor; and there is a third element, the wetted exterior plaster in patches not necessarily continuous.

"In my official letter I alluded to the radiations from an electric discharge, and that some of these may have an effect similar to sunstroke upon the human organism—so that radiation and not electricity may have caused the fatal accident.

"This was evidently a very bad thunderstorm, as a tree was struck as well as the lighthouse, but the building was unaffected.

"The addition of a conductor may have made it proof against being shattered, and at the same time not to be occupied without danger when there is electricity about.

"I am inclined to think that the noise of thunder is the radiations made audible much in the mode of your Royal Institution experiment on the C note.—I remain, yours very faithfully, A. T. FRASER."

Calicut, June 13, 1889.

"PROF. O. J. LODGE, F.R.S.

"My dear Sir,—After writing, I have received detailed accounts for which I called, and as no conclusions can be drawn from an inaccurate description, enclose nearly all my official letters, putting the Marine Department at Madras right in their facts.

"The statements of the two educated natives, one of whom was struck, are interesting, but contain so little more than I have extracted that I do not trouble you with them.

"The lightning conductor evidently was acting, and is said to have been struck twice, as well as a tree three or four hundred yards away.

"I find the lighthouse rooms are tiled, not terraced. It is possible they were terraced within my recollection, as the present tiles are certainly new. An amended sketch is enclosed.

"There is a sheet zinc hood $4\frac{1}{2}$ feet by 3 feet over the front door, and there were twenty-three of the patent clay tiles, each $6\frac{1}{4}$ lb. weight, blown away at the eaves just above the hood.—I am, yours very faithfully,

"A. T. FRASER."

Extract from official letter, June 12, 1889, *embodying the narrative of two native eye-witnesses in Mangalore Lighthouse on May* 14 *when struck by lightning, to correct the first account sent to the Marine Department, Madras.*

"1. The detailed account of Inspector Ram Rao, one

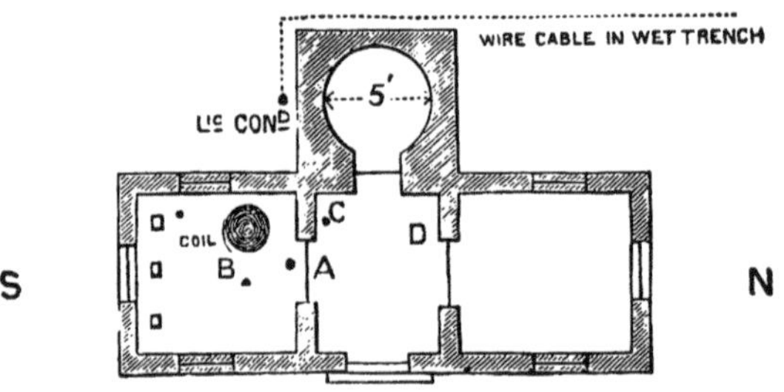

Fig. 18. GROUND PLAN OF LIGHTHOUSE.
A, assistant master's position; *B*, deceased, standing; *C*, inspector; *D*, two other natives.

of those in the Mangalore Lighthouse during the thunderstorm of 14th May, differs in some particulars from the information the Post Officer was able to obtain and communicate.

"2. Including the keeper, there were ten natives inside the basement,—seven in the passage, two in the south room, and one upon its doorway or sill (Fig. 18).

"3. The inspector was on a seat in the passage, near the door leading to the column, which was closed. The front (outer door) was also shut and bolted.

"4. The first thing he observed was a 'spark'—no flash struck anywhere—on the ground before him about a foot and a half from his seat. No lightning was seen by him to come through the front door. The electricity was on the floor of the passage, not far from the door of the column.

"5. Assistant-master Narani Rao, Government College, also gives an account of what he saw happen, he being on the door-sill.

"6. He says two people in the passage suddenly saw a light creeping on the floor near their feet, and about the same moment a long spark in the south room; two saw the luminous body on the floor, and two the spark in the room, but only one saw both.

"7. Simultaneously with the latter, the narrator and the native who died both dropped on the ground. His impression was that a current of electricity was flowing from his feet to his head, and he soon recovered. He is satisfied that the current did not come through the front door. Singularly enough, the deceased was talking to him about lightning conductors, and asked if his keys were likely to attract it, just before he was struck, without noticing he was standing close to a large coil of wire.

"8. Those who saw the spark in the room are positive it was not parallel to the floor. One thinks it was perpendicular, and the other slightly inclined to the lightning conductor outside.

"9. It seems there was a coil of 94 feet of galvanized iron fencing wire lying in the corner of the south room, some 5 feet in direct distance from the conductor outside, with the wall between.

"10. I would be inclined to attribute one of the individuals being fatally struck to his standing near this loose metal, and there were also some kerosene tins in the room, all either inductive or conductive surfaces.

"11. But I have to notice that, strictly speaking, none of them had a right to be admitted into the lighthouse on the occasion, and it was, unluckily, one of the worst places they could have chosen in which to take refuge, and there is no system of lightning conductors, which, in my opinion, can make it safe.—A. T. F."

A Singular Case of Damage by Lightning.[1]

BY A. P. CHATTOCK.

"In the afternoon of the 23rd of May 1889 the wall of a garden in Windmill Hill, Hampstead Heath, the property of Mrs. Inman, was damaged during a thunderstorm in the manner shown in the accompanying drawing (Fig. 19), which is a copy of a sketch I made on the spot an hour or two after the storm. The wall was completely perforated, the place chosen being at one of the buttresses; but the interesting feature of the case lies in the presence of a lamp-post just opposite, which overtops, and, owing to the narrowness of the footway, almost overhangs the wall. The actual dimensions are given in the diagram (Fig. 20).

"According to ordinary theories the wall should have enjoyed peculiar immunity from attack at this point, yet the discharge singled out this very spot, and apparently preferred to pass through bricks and mortar under the shadow of a good conductor rather than enter the projecting end of the ladder-rest, which was not $1\frac{1}{2}$ ft. distant.

"There seems to me only one moral to be drawn from

[1] "Electrician," vol. 23, p. 621.

these facts, viz., that the flash was what Dr. Lodge calls an "impulsive rush" discharge; either it struck the top of the lamp-post and rebounded sideways along the ladder-rest, as such a discharge probably would, or else, which is quite possible, it entered the wall direct, without noticing the post at all. The presence of the lamp just

Fig. 19.

opposite the hole inclines one rather strongly to the former view, but in opposition thereto is the fact that, so far as I could see, there was no marked indication of melting or burning at the end of the ladder-rest. Either way the case is an interesting one, taken in connection with the late discussions on the subject of discharges, and one only wishes the flash had been actually seen. That this was not the case appears from the following answers to questions, which, through the kindness of Mrs. Inman, I obtained from her gardener, who was in the garden when the accident occurred:

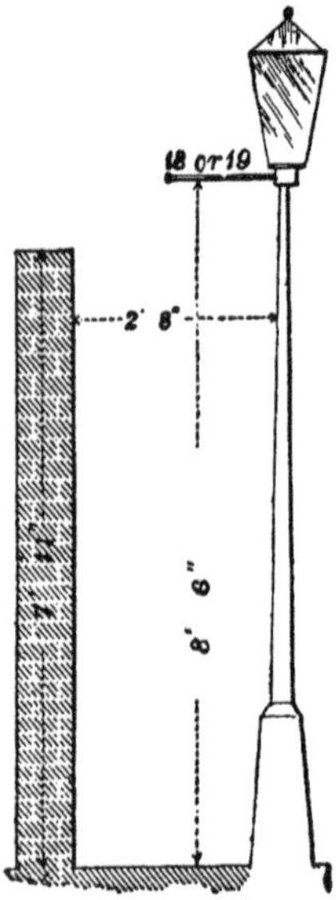

"1. So far as he knew, no one saw the damage done.

"2. The bricks fell chiefly into the garden, *i.e.*, away from the lamp-post.

"3. The damage could not have been brought about by mechanical means. The wall was in just as efficient repair there as elsewhere.

"4. No metal object touched the wall, but some galvanized wire holding up the trees was very near—about 4 ft. to 5 ft. from the wall, attached to a tree.

"5. The time of the occurrence was about a quarter to five p.m."

Fig. 20.

Note on the Bursting of Leyden Jars.

The circumstances attending the electrical fracture of Leyden jars are of some interest. So far as my experience goes, they do not often burst by being merely overcharged to too high a potential—whether they ever do so I am not sure—but they burst at the instant of discharge by the recoil and oscillations set up in the discharging circuit. When this discharging circuit is a long and thick one I should have expected their fracture to be more probable, were it not that they then readily overflow. A discharge spark-length of an inch or less will easily make a jar spark a distance of six inches over its edge if it be joined up to a suitable circuit, say a No. 0 or even a No. 12 wire round a large room, and this ease of overflow may perhaps save them from breaking.

With a short discharge-circuit it may be possible to take sparks several inches long without their overflowing, and the question is whether they are then more liable to burst or not. In other words, are bursting and overflowing the same thing, or does one act as a safety-valve to the other? Fortunately my jars, though they frequently overflow, scarcely ever burst, and accordingly my experience is too limited to be any use.

When I take very long sparks from jars I use two in series; their liability to overflow is then greatly diminished. This is no doubt the empirical reason of the pair of jars used in all forms of inductive machine.

I expect jars usually burst when used singly or joined in parallel, seldom when joined in series. The double thickness of glass in the latter case is an obvious consideration, but there is more in it than that. The main reason of the greater safety of jars in series seems to me to be that the electric surgings do not operate on both coatings simultaneously to the same extent as they do with a single jar.

Apropos of the bursting of jars I have permission from the writers of the two following letters to communicate them, and I also quote a statement by Franklin on the same subject.

The suggestion at the end of Mr. Boys' letter, about the period of longitudinal vibration of the glass possibly synchronizing with the electric oscillation when fracture occurs, is a good one, and future observations should be so recorded that the truth of the hypothesis can be examined. As a convenience I have, therefore, drawn up the following list of things which it is well to record whenever a jar breaks electrically. They are all such as can be easily obtained after the accident.

Data desirable to know in future observations of Broken Jars.

(1) Mode in which jars are connected if more than one.

(2) Total area of effectively coated surface.

(3) Average thickness of effectively coated glass (say by weight).

(4) Quality of the glass (viz., its specific inductive capacity roughly, or at least its specific gravity).

[Or, instead of 2, 3, 4, total capacity of jars, as arranged, before fracture.]

(5) Thickness of glass near fracture.

(6) Total length of discharge-circuit.

(7) General arrangement of discharge-circuit (*e.g.*, rough scale plan of same).

(8) Thicknesses of wire or rod used in discharge-circuit.

[Or, instead of 6, 7, 8, total self-induction of discharge-circuit].

(9) Length of air gap in discharge-circuit, with size of knobs used.

Answer by Dr. Franklin to a question put by Dr. Ingenhousz about 1777.

By the circumstances that have appeared to me, in all the jars that I have seen perforated at the time of their explosion, I have imagined that the charge did not pass by those perforations. Several single jars that have broken while I was charging them have shown, besides the perforation in the body, a trace on both sides of the neck where the polish of the glass was taken off the breadth of a straw, which proved that great part at least of the charge, probably all, had passed over that trace. I was once present at the discharge of a battery containing thirty jars, of which eight were perforated and spoilt at the time of the discharge; yet the effect of the charge on the bodies upon which it was intended to operate did not appear to be diminished. Another time I was present when twelve out of twenty jars were broken at the time of the discharge, yet the effect of the charge which passed in the regular circuit was the same as it would have been if they had remained whole.

Were those perforations an effect of the charge within the jar forcing itself through the glass to get at the outside, other difficulties would arise and demand explanation. 1. How it happens that in eight bottles, and in twelve, the strength to bear a strong charge should be so equal that no one of them would break before the rest, and thereby save his fellows, but all should burst at the same instant. 2. How it happens that they bear the force of the great charge till the instant that an easier means of discharge is offered them, which they make use of, and yet the fluid breaks through at the same time.—*Franklin's Works,* edited by J. Sparks, vol. v. p. 462.

The following two letters from Mr. Boys and from Mr. Bottomley tell their own tale:

Science and Art Department, March 5, 1889.

"It may interest you to hear how some of the large Polytechnic jars were broken here when we tried the 7ft. Wimshurst machine on them.

"On connecting up 1, 2, 3, or 4 "for quantity," or in pairs "for intensity," with the machine, and charging until the brush discharge from the conductors was such as to make further increase of potential unattainable, and then causing the knobs to approach one another, there used to be a deafening explosion when they were from 12 to 16 inches apart, and almost every time one (or more ?) jars were found shattered in one or more places. At each place of rupture the centre portions of the glass were white, being completely pulverized. This part would be a $\frac{1}{4}$ inch or so in diameter. Outside this radiating but irregular cracks spread in all directions over a space of two inches or so, and these cracks were joined by other cracks, more especially near the central powder.

"This never happened during the charge, but only when the discharge was made at the knobs, and no doubt was largely due to electric oscillations breaking down the glass, which, while it could stand the steady pull one way, could not resist the see-saw. This may, for anything I know, have synchronized with the natural period of vibration of a bar of glass as long as the jar was thick where it broke.

"Finding that the proceedings seemed destructive of jars we left off.—Yours, etc., C. V. Boys."

13, University Gardens, Glasgow, March 22, 1889.

"The experiments, of which I wrote to you very hurriedly on February 6th, were made with a very large influence machine which my friend Sir Archibald Campbell, of Blythswood, has been building. I will tell you something more about the machine a little later; but meantime, in a very imperfect state—indeed almost at first trial—we got $\frac{1}{3}$ of a milliampere from it, as measured by decomposition of water. I have no doubt we can get a great deal more when we know the conditions necessary for keeping the potential of the parts which act as inductors from falling away as we draw off current.

"While the machine was giving a spark of $5\frac{1}{2}$ or 6 inches we sparked through two very large jars, 12 in. high by 7 in. in diameter, the glass about $\frac{1}{8}$ to $\frac{3}{16}$ thick, tinfoil 9 in. up. On examining them it turned out in each case that the jar had been perforated in many places. In one jar there are six or seven holes, in the other I can count thirteen. The tinfoil was blown outward *on both sides* in each jar in six or seven places. It stood up from the glass in little protuberances bigger than a split pea, and in many cases, at any rate, the summit of the protuberance is pierced. On removing the tinfoil the glass showed unmistakable signs of many holes arranged in a line (in the case of the jar with the largest number of holes in two lines, one branching off from the other). The holes are about half an inch apart; the distances tolerably regular. There is in one of the jars one very heavily pulverized place and two in the other;—round spots about a quarter of an inch in diameter, where the glass is a mass of white powder, in each case at one end of the line of holes and seemingly a starting place; and in the very centre of the most thoroughly pulverized spot there is a small clean round hole, through which I can put a wire one-hundredth of an inch in diameter right through, quite loosely, and without rubbing on the sides of the hole. (I have not tried the largest wire possible, as I do not want to disturb the glass powder just yet.) The remaining holes are small perforations, a crack joining them all, and round about each there is a small spot covered with metal driven off the tinfoil on to the glass.

"One of the sets of perforations is just round the upper

edge of the tinfoil coatings. The other is round the bottom of the jar.

"I shall be greatly interested to see what you say about the matter, and when we break any more—which I doubt not will be soon—you shall hear of it.—Yours, etc.,

J. T. Bottomley."

Note added by Mr. Boys.

"The first of Mr. Bottomley's letters reminds me that the position of the holes in the glass was indicated by large blisters in the tinfoil, and that in the centre of the white powder there was, as he describes, a clean hole. The glass in these jars was not, as far as I can remember, as much as $\frac{1}{8}$ inch in thickness at the places where they broke.

C. V. B."

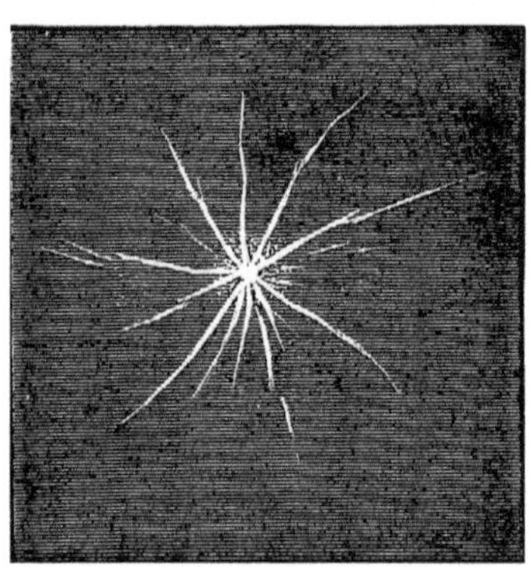

Fig. 20a.

CHAPTER XVI.

SUMMARY AND REPETITION OF IMPORTANT POINTS.[1]

So far I have abstained, as far as possible, from practical recipes and from anything like authoritative advice, contenting myself with calling attention to certain aspects of the subject which had been overlooked. I have ventured to imply that none of the older electricians had any notion of the real conditions of the problem; that they all, from Franklin to Faraday and down to the present day, treated it as a much easier matter than, in fact, it is; and that there had been very little real progress in this particular department since the time of Franklin.

Recent advances in electrical theory made it easy for me to see further into the matter than the far greater men of the past had had any chance of doing; and a few very simple and easy experiments soon brought the conditions of the problem clearly before me.

It was these conditions upon which I laid emphasis in my lectures to the Society of Arts. Any practical outcome I left to a later period, and very likely to other hands. The first requisite seemed to be to grasp the conditions of the problem as illuminated by theory; the

[1] Being a communication to the Institution of Electrical Engineers, 25th April, 1889.

second, to carry out practical reform in the light of a large experience.

Before proceeding to suggestions toward this end I wish to emphasize further some of the matters already hastily touched upon, by running over some of the conclusions at which I arrived, and re-establishing and demonstrating their correctness either by theory or by fresh experiment, whichever may seem the most simple and satisfactory under the special circumstances.

Two Main Cases of Lightning Flash.

1. All discharge is virtually that of a Leyden jar. There are always two conductors separated by dielectric, and the discharge is the breaking down of the dielectric at its thinnest or weakest place.

In a thunderstorm the charged conductors are obvious,

Fig. 21. CASE *a*.

The accompanying seven figures represent respectively the same conditions, as they may occur in Nature, and as they can be arranged artificially. Case *a* in each figure is the steady-strain case. The others are varieties of the impulsive rush, where a spark at *A* precipitates a spark at *B*, the place where *B* occurs having been subjected to no preliminary strain. Clouds correspond to coatings of jars; spaces between them correspond to glass or other insulating space. The position of the charging machine *M* is indicated in Figs. 22, 24, 26, for convenience. The leak, or imperfect conductor, in Fig. 26, is needful in order that the jar may charge. It takes no part in the discharge. Its place is taken in Figs. 25 and 27 by the rain-shower. In Fig. 23 the rain-shower, or leak, is permissible, but unnecessary.

being either two clouds, or else a cloud and earth, and the dielectric is the air between.

2. It must sometimes happen, when one cloud discharges into another, that the potential of this other is suddenly raised high enough to cause it to discharge into the earth, even though no strain previously existed in the air between it and earth. The same thing may happen in various other ways when two clouds spark into each other, as indicated by the diagrams (Figs. 23, 25, 27).

3. There are, therefore, two main cases—(*a*) When

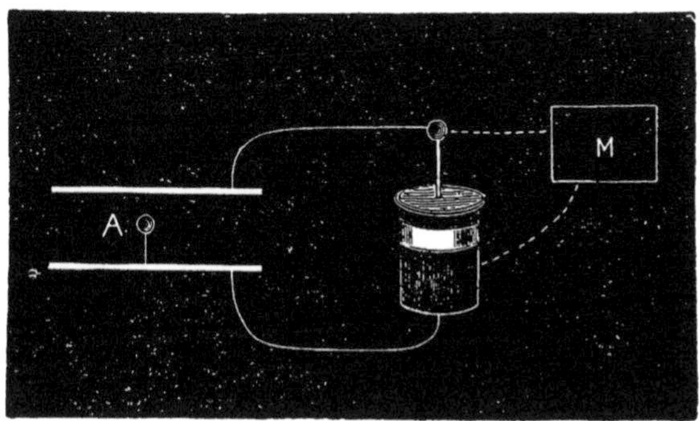

Fig. 22. CASE *a*.

the strain in the dielectric near the earth has been of gradual growth, in which case the path of discharge will be prepared inductively beforehand; (*b*) when the strain arises so suddenly that there is no time for any prearranged path. The first I call "steady strain"; the second, "impulsive rush." It is most important to recognize these two cases, and to understand the extremely different conditions attending the two. The first case only was ever contemplated by the older electricians; in fact, so far as I know, it was my experiments in 1888 which first called attention to the other case.

Conditions of Protection and of being Struck, under the Circumstances of each Case.

4. I will now illustrate experimentally [1] the conditions under which discharge occurs in each of these two main cases; and I will take first the case of steady strain, or case *a*. This insulated sheet of tin plate is supported horizontally a foot or two above another plate lying on the table, and it is electrified by a Wimshurst machine. It represents a charged cloud hovering over the land, the lower plate representing the earth. Between the two I can erect buildings, and lightning conductors terminated in various ways. The typical terminals which I will here use to illustrate the conditions are four—viz., a large knob, or, as I shall call it, a "dome"; a small knob, which I shall call "knob"; a sharp point; and a gas flame, to represent a chimney or other furnace current of rarefied air. Putting the knob and the dome between the plates, we find the knob struck by preference, even though the dome stands at a much higher elevation. Introducing the point, we find it protects both, by a silent discharge, until it is lowered very considerably; and that then several points may protect when one does not. Replacing the point by the flame, we find that it protects too, but not so efficiently as the point, and that it gets curiously beaten down and darkened in the act of protection. The point is not struck by a noisy flash until it is raised pretty close to the upper plate, when it is struck; but a bunch of points is even then not easily struck, sometimes continuing to discharge with a constant fizz right up almost into the cloud.

[1] These experiments were shown with a splendid machine most kindly brought over and erected for the purpose by Mr. Wimshurst himself.

5. Try the effect of a bad earth or other very high resistance interposed in the path of the conductor, say a capillary water tube or a bit of wet rag or a wet string.

The violence of the spark is greatly lessened, and the sound is now gentle, but that which was struck before is still struck: resistance, so long as it be something short of infinite, makes no difference to the ease with which a given object is struck under the circumstances of this case *a*. Insert the wet rag in path terminated by point, and it still protects, practically as well as before. Insert it in path terminated by knob, and it gets struck at the same elevation as before.

The fact is that the path of the disruptive discharge is all negotiated and pre-arranged in the air above, especially on the surface of any small conductor reared into this space, and the resistance which the flash may ultimately have to meet with in its passage to earth is a thing of subsequent consideration.

6. So much for the conditions attending case *a*, the steady strain. Now attend to case *b*, the impulsive rush. We shall find everything very different.

Alter the connection so that a charged Leyden jar must, when it discharges, discharge direct into the upper insulated tin plate, and thence overflow to the ground if it is able to raise the potential high enough. If the plate is too far above ground for a flash to occur, the Leyden jar does not completely discharge; it only produces a number of fizzes and spits, and the greater part of its charge remains in it. But when the upper plate is within sparking distance of the ground (and the sparking distance under these circumstances is surprisingly great by reason of the impetus with which the electricity rushes into the top plate), then the jar discharges completely, and we have a violent crack both between the knobs of its discharger and in the air gap between the two plates, which is in the path of the discharge; the arrangement being really two condensers in series, but only one charged (Figs. 23 and 24).

The only object of the Leyden jar in case *a* is to give more body to the flash. In case *b* the Leyden jar, or its equivalent capacity, is essential (Figs. 25, 26, and 27 are varieties of case *b*).

7. Putting the dome and the knob between the plates arranged as in case *b*, we find that one gets struck as

Fig. 23. CASE *b*1.

easily as the other; the knob has now no advantage: whichever is the higher, that gets struck, without reference to other considerations. Introducing the point also, we find precisely the same is true for it; its protective virtue, so much insisted on by the older electricians, is entirely non-existent. It gets struck no more easily, and no less easily, than the dome, and it gets struck by a flash of precisely the same noisy character as the others get struck with. A bunch of points acts in exactly the same way. A comb of 24 needle-points protects nothing, and gets struck just the same as anything else.

8. Now introduce the flame, and one notices a marked difference. In case *a* it protected less well than the point, but it was not struck noisily any more than the point was. In the present case it gets struck with violence, and it gets struck much more easily than anything else. Adjust dome and knob and point at about the same level, and they get struck, one or other, at random. Adjust the flame a great deal lower, and it

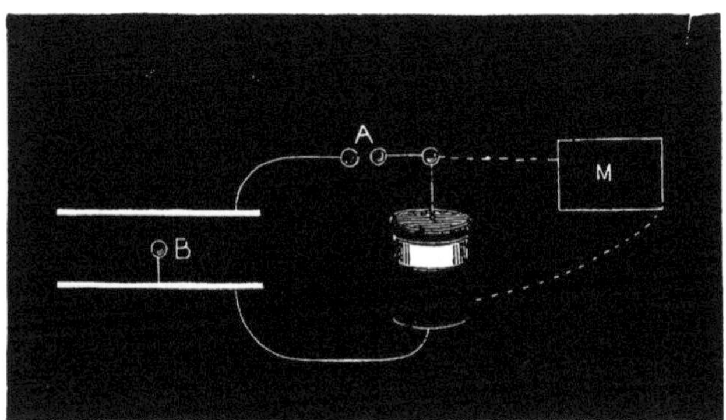

Fig. 24. Case *b*1.

protects them all, not by silent discharge, but by getting struck itself instead.

"Protection," however, is in this case not the word to use. The flame better represents a chimney requiring protection, while the point corresponds to the pointed terminal of a lightning conductor raised a good deal higher with the intention of protecting it. Protect it it does not, however; the flash strikes down the column of hot air and through the flame, while it avoids the more lofty pointed terminal altogether.

Bring a point or a knob, or anything, *into* the column of hot air above the flame: *then* it gets struck easily enough, and protects the flame, but not if it is on one side of the hot-air column. The experiment has an obvious moral in relation to the protection of chimneys: it suggests that the Continental plan of a bar or arch across the mouth of the chimney (Plate XIV.) may after all be justified.

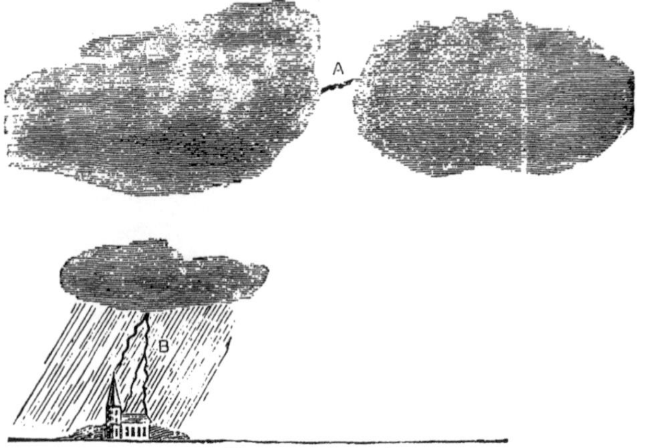

Fig. 25. CASE *b*2.

9. Try the effect of resistance in the path of the discharger *now*, and we find it is altogether different to what it was in case *a*.

In case *b* things get struck according to their height, independent of the shape of their terminals, but not independent of their resistances.

Interpose a wet rag in the path of any one, and that one fails to be struck; it is not struck, and it fails to protect the others from being struck, even though it be reared up till it touches the top plate. The top plate need not therefore be insulated at all carefully for this case *b* experiment.

Imitation of Lightning.

10. Now let us modify case *b* by making the top plate a sieve full of water, so as to get the flashes in a shower of rain. One cannot well try case 1 in a rain shower: the plate would be discharged too rapidly and continuously by the water-drops for its potential to fully rise. But in case 2 the top plate is not necessarily

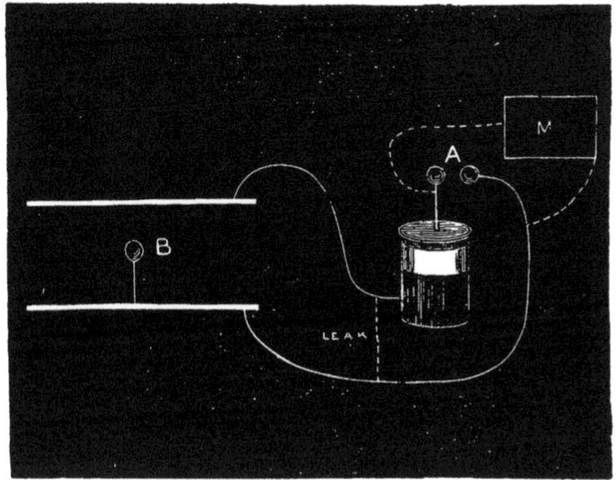

Fig. 26. CASE *b*2.

charged at all until the rush comes, and so the rain shower does no harm.

Flashes in the rain can be got of surprising length and shape, for they make use of the water-drops as stepping-stones. By adding salt to the water they become longer still, but there is no need thus to improve its conductivity for what I want to show.

Notice how the flash contorts itself, taking sometimes extraordinary paths as it jumps from drop to drop, but yet exhibiting its instantaneous character by showing the drops as stationary in its illumination.

11. Remove the things standing on earth-plate just beyond fair striking distance, and what do we see? Appearances precisely like those which are observed in many lightning photographs.

A crowd of violet discharges fill the rainy air—forks, and branch and multiple flashes, not very bright or very noisy, but extraordinarily numerous, and striking on innumerable places at once. A rot is set up in this air

Fig. 27. Case *b*2.

at every attempt of the jar to discharge, exactly as happens in one of the most striking of the photographs belonging to the Royal Meteorological Society.

Masts and spars and deck and ocean may be simultaneously struck by these interesting flashes; and, though they do not here appear very violent, yet I expect that on a larger scale of Nature they are not very safe, and may easily have a heating effect sufficient to ignite bodies. These experimental ones are able to ignite gas in the midst of the rain.

12. While we have this arrangement at work we may as well try an interesting little experiment on what happens when lightning reaches water. The rain water has here been collected in a zinc tray some three inches deep, and by bringing a knob from the top plate, some six inches or so above the water, a flash strikes it. It prefers to strike anything metallic if it can, but if there is nothing else within reach, it will strike the water. On reaching the water the flash forks out and ramifies in all directions in a crow-foot pattern, giving the same sort of appearance, only coarser, as that obtained by Mr. J. Brown by taking sparks on to a photographic dry plate and then developing.

Bring the knob nearer and nearer, the same thing happens, until the water is touched by the knob, and even after it is submerged. But as soon as the metallic conductor is submerged the ramifications get less and less vigorous, and when a sufficient surface is immersed they cease. I have never seen these ramifications spread of themselves *below* the surface of the water. They appear to me to keep entirely to the surface. But I have not yet finished investigating these appearances.

13. Immersing a half-full beaker in the water, sparks can be got to the water inside it, though they prefer to curl round and go outside. The noise the sparks make when they go inside is a curious one, and sounds as if the glass cracked each time, but it does not. If the glass is moist, a brush cascade round its edge can be seen in the dark. If it be dry, the water inside gets charged, and fizzes audibly back to the knob for a second or so after the spark has ceased, a dimple being visible in the water below the knob.

14. Live things in the struck water—worms, flies, fish, etc.—will most certainly get struck; but so they

do under far less violent disturbances than what they would here be subject to. There is nothing surprising in the fish of a pond or lake being killed by a flash of lightning; and it has often happened.

When H.M.S. "Conway" was struck many years ago, and protected by its lightning conductors, it is related that the sea-water was seen to be luminous on all sides of the ship. This is exactly the effect I now imitate.

15. There is one more experiment on discharge in water which I have just tried, and which it is interesting to show. I take a pointed rod, and, protecting it by a glass tube, immerse its point under water in a beaker containing a plate connected to the other coat of the jar, and pass a spark. With the point negative, there is a bright glow region round it every time, but the discharge is quiet. With the point positive, the flash is of a dazzling white, and is accompanied by a great deal of noise and violence, threatening to smash the Leyden jar, and throwing down the copper plate towards the bottom of the beaker with fury.

Oscillatory Character of Lightning.

16. Before leaving the outdoor department of our subject I must say a few words on one branch of it concerning which there is evidently considerable uncertainty and haziness abroad—I mean the oscillatory character of a lightning flash.

That a Leyden jar discharge is usually oscillatory must now be regarded as so extravagantly proved that any doubts that may have existed on the subject must surely by this time be cleared away, at least for the case where the discharge has to utilize a wire circuit. But perhaps it is still doubted for the case when a jar overflows its edge, or, still more, when it merely sparks through its own dielectric, straight between the coatings.

Now, as I have insisted all along, a lightning flash is a spark through the dielectric of a jar whose two coatings are either two clouds, or else cloud and earth. Hence, if any importance is attached to the fact (as I believe it) that lightning flashes are oscillatory, it is necessary to prove it for a Leyden jar sparking direct between its coatings, especially when the coatings are not very close together.

17. The reason I do attach importance to the oscillatory character of a discharge is because I have worked out the quantitative behaviour of conductors on that aspect of the matter; and though, as Professor Fitzgerald said at Bath, everything would hold just as well for a single oscillation—viz., one violent rise and decay of current (which without any doubt *must* accompany a lightning stroke or any other quick discharge whatever) —if rapid enough, yet the rapidity of such a charge as this does not seem to me probably at all sufficient to account for some of the effects. The rapidity of variation of current in that case would be directly connected with the total duration of the flash; and though we have evidence that it is very momentary, yet we have no evidence that it is so instantaneous (say a millionth of a second) as the semi-period of one of the oscillations may be, a dozen or more of which may accompany an entire flash. However, I admit, of course, that all I want is a tremendously rapid variation of current; and if I can be given this by one oscillation, the rest are unnecessary, and may be dispensed with.

18. When I speak of the oscillatory character of a flash, let it be understood once for all that I do not mean

in the least such a thing as can be analyzed by waggling the head. Flashes analyzable by waggling the head must be multiple ones, and the interval of time between their constituents (which may be, say, the fiftieth of a second or thereabouts) is a long period compared with that of an oscillation such as I mean, bearing the same ratio to it as a quarter of a century bears to an hour.

19. A direct experimental proof that lightning is oscillatory will be obtained when photographs of it are taken on a sensitive plate revolving 1,000 times a second. Something short of that speed would cause the image of the flash to blur, but that speed might be sufficient to analyze out the oscillations, when examined carefully with a magnifier, the focussing being good.

Till then the easiest proof that it is oscillatory is a theoretical one, and it can be put in a few words.

20. Consider an air condenser with two coatings, each of area A, separated by the distance h, and let it burst its dielectric. It is well known that the discharge is oscillatory when the whole resistance met with by the discharge is anything less than a critical value.

$$R_0 = 2\sqrt{\left(\frac{L}{S}\right)}.$$

Now, attending only to the straight part of the discharge, and ignoring the current rushing up in the plates to the spark path, the self-induction of a straight conductor of length h and sectional radius a is very approximately

$$L = 2\mu h \log \frac{4h}{a}.$$

The capacity of the discharged condenser is

$$S = \frac{KA}{4\pi h}.$$

Hence the critical resistance which must not be exceeded is given by

$$R_0^2 = \frac{32\pi\mu h^2}{KA} \log \frac{4h}{a};$$

or

$$R_0 = 10\mu h \text{ “}v\text{”} \sqrt{\left(\frac{1}{A} \log \frac{4h}{a}\right)}$$

$$= 300 \text{ ohms} \times \sqrt{\left(\frac{1}{A} \log \frac{4h}{a}\right)} \times h.$$

The important thing to notice in this value of $\sqrt{\frac{L}{S}}$ is that it is approximately proportional to the first power of h, the distance between the plates of the condenser.

21. Next consider the resistance of the discharge path. It too will be proportional to the length h, and it may be written

$$R_1 h,$$

where R_1 is the resistance per unit length of the discharge path.

The condition for oscillation, then, is that this $R_1 h$ shall be less than R_0; or *the resistance of unit length of the spark must be less than*

$$300 \text{ ohms} \times \sqrt{\left(\frac{1}{A} \log \frac{4h}{a}\right)}.$$

22. The term under the square root will take different values according to circumstances, and it may be greater or less than 1 per metre. Ordinarily, however, it will be greater than 1 per metre, unless the area of charged surface is considerable. The important thing is the way

in which h, the distance between the plates, enters into the expression. It does not come in very prominently at all, but so far as it does influence the result it permits the discharge to be oscillatory more easily when the plates are a good distance apart than when they are close together.

23. Take a couple of typical examples.

First, a Leyden jar bursting its glass. A fine needle may be just put through the hole usually made in these cases, so we can take the sectional radius a as something like a tenth of a millimetre. The thickness of the glass may be 2 millimetres, hence $4^h/_a = 80$ or thereabouts; and the natural logarithm of this is about 4. Suppose the area of coated surface is half a metre square, then the critical resistance which a metre of the spark must not exceed is 1,200 ohms, and so the 2 millimetres of it must not exceed 2·4 ohms.

It is difficult to say whether this is or is not a large resistance for such a short spark, and hence it is difficult to be sure that such a spark is anything more than a mere one-directional discharge. To go into it more fully the currents in the metal coatings would have to be considered.

Next take as example a cloud area at an elevation of one kilometre; and because a lightning discharge usually makes perforations of fair diameter, we may suppose a to be about a millimetre. (A widely erroneous estimate in this quantity makes but little difference in the result.) In such a case $4^h/_a = 4 \times 10^6$, and the logarithm of it is between 15 and 16. Hence the charged area may be 15 or 16 square metres without bringing down the square root term below 1 per metre; so the resistance of the whole flash may in that case be anything below 300,000 ohms without checking the oscillations. The discharged area may indeed be as much as 1,500 square metres without bringing the critical resistance which the flash must not exceed below 30,000 ohms, or 30 ohms per metre. (Understand that "resistance" here does not mean impedance. It means true dissipation of energy-resistance (sec. 25), and the current squared is so enormous (sec. 29) that the resistance-coefficient may be quite small.)

In another place (Chapter XIII.) I have shown reason for believing that the area of cloud discharged at any one flash is usually very moderate; and hence on the whole I consider it proved, so far as elementary theory can do it, that the lightning flash usually takes place under conditions favourable to oscillation.

24. And these oscillations are extremely rapid. The rapidity depends on the inverse geometric mean of L and S, and this is practically almost independent of h. Referring back to their values we see that

$$LS = \frac{\mu K A}{2\pi} \log \frac{4h}{a};$$

and so the number of complete alternations per second is

$$\frac{\text{"}v\text{"}}{\sqrt{\left(2\pi A \log \frac{4h}{a}\right)}},$$

which, in the second example of section 23, with A as 100 square metres, becomes

3 million per second.

If the discharged area were as great as 10,000 square metres, the rate of alternation would still be a third of a million per second.

Material of the Lightning Conductor.

25. A few words may suffice to explain the nature of the impedance which alternating currents meet with in passing through a conductor, and of the reason why iron is as good as, or even better than, copper for the purpose of conveying currents alternating with extreme rapidity.

A rising current has to magnetize the space all around it, and the production of this magnetization delays and impedes the rise of the current to its maximum value. A falling current permits the magnetization of the space all round it to decay, and the dying out of this magnetization delays and impedes the fall of the current to its minimum. The more rapidly the current changes, the more powerfully felt is the influence of the accompanying magnetizations and demagnetizations. Now the total magnetization produced by a current—its total number of lines of force, or its total "magnetic *induction*," as it is often called—is proportional to the current strength: equal to it multiplied by some constant, which we may call L, and write

$$I = LC,$$

where I is the total induction produced by the current C. L is a coefficient characteristic of the circuit, its value being defined by this equation, and is called the coefficient of induction excited by the current's own self, or the coefficient of self-induction.

If the current goes through $p/2\pi$ complete alternations in a second, it can be shown that the impedance it meets with, due to the reversals and re-reversals of its own magnetic field, is pL.

This is not the whole obstruction it meets with, but it is the only part which does not dissipate energy and cause its vibrations to decay. It may be called the *inertia* part of the impedance. The remaining part of the total impedance is resistance, R, the dissipation of energy coefficient, defined by

$$\text{heat per second} = RC^2;$$

and the total is the resultant of these two as if they were at right angles to each other. So that, calling P the total impedance,

$$P^2 = (pL)^2 + R^2.$$

Now in respect of the R term, iron is much worse than copper, not only seven times worse, hundreds of times worse; but then for very rapid alternations the R term is altogether insignificant compared with the pL term.

26. In respect of the pL term does the material of the conductor matter?

Well, in so far as the magnetization spoken of is that of the space surrounding the conductor, of course the substance of the conductor itself matters nothing. But in so far as the conductor itself gets magnetized, the material of which it is made does matter. Now a linear current magnetizes at right angles to itself everything surrounding it—most intensely the things close to it. A hollow cylindrical current magnetizes everything outside itself, but nothing inside. If the current were to distribute itself uniformly through the section of the wire, the outside of the wire would get magnetized in concentric cylinders; but if the current were to confine itself to the outer surface, and flow as a hollow cylinder, it would escape the necessity of magnetizing the wire at all. In cases where the pL term is much more important than the R term this is precisely what it does therefore. It always flows so as to meet with the least possible total obstruction; and it finds less total obstruction by cramping itself into the periphery of the wire than it would find if it utilized the whole section. The cramping into the periphery increases R, but it decreases pL; and on the whole with rapidly alternating currents this is an advantage, and gives a smaller value to P.

Slowly changing currents think most about R, and use the biggest cross-section they can find. Rapidly changing currents think most about pL, and avoid having to magnetize more than they need. Especially must they avoid having to magnetize the conductor if it consists of iron: hence in that case they cramp themselves tremendously into its outer skin, and thereby avoid having to overcome much more total impedance than they meet with in the case of copper; but though they thus keep down impedance, they increase their dissipation of energy term, and get their oscillations damped out far more quickly than when they only have to pass through copper. The violence of the flash therefore subsides more quickly in an iron, than it does in a copper, conductor; and so I find it experimentally.

Nothing here said is in the least hypothetical. It is all absolutely clear and certain, and has been abundantly verified. I will not go into the history of the subject. It is well known.

27. Let it be clearly understood once for all, that in comparing copper and iron conductors I never mean comparing them as of unequal thickness, *i.e.*, of equal conductivity. Such a comparison is ridiculous in the present state of knowledge. They are to be compared when of the *same* thickness; and under those circumstances I assert iron to be a trifle better, certainly not a whit worse, than copper; irrespective of all its other

manifest advantages, cheapness, fusing-point, etc., etc. Want of flexibility is sometimes urged against iron, but stout telegraph wire is flexible enough; and, until experience decides to the contrary, I feel sure it is thick enough for lightning conductors.

The one and only thing on which anything can be said against iron is on the subject of its durability; and this being a chemical question, I offer no positive opinion; at the same time I am absolutely certain that any slight disadvantage in that respect is a hundredfold compensated in most localities by its other superlative advantages.

Current and Potentials during Discharge.

28. In cases where a conductor is pretty thick, say anything like a quarter inch diameter, or even a tenth, and of any moderate length, such as 100 yards or less (not many miles), the two terms of impedance are so unequal that it is for many purposes needless to think about R at all, the impedance is practically pL simply; and this, as I have shown in Chapter XIII., is under any given circumstances half the critical *resistance* which determines whether the discharge shall be oscillatory or not under the same circumstances. The total impedance is commonly to be reckoned in hundreds or even thousands of ohms.

29. The strength of current passing in the alternations can be estimated by considering that the whole quantity stored up in the discharged body has to be transmitted in a quarter of one oscillation period—say, for instance, in the millionth of a second. If the quantity discharged were one coulomb this would mean a current of a million ampères. In any case the current must be hundreds or thousands of ampères.

30. The difference of potential needed to drive so strong a current through so great an obstruction is enormous, being equal to the product *PC*, and may be reckoned in millions or hundreds of millions of volts; hence it is that lightning conductors afford no easy path for lightning, but that it tends to spit off in all directions, even to what would seem, and indeed are, very inferior conductors. It will spit off from a well-earthed stout copper rod to bits of wood and to perfectly insulated bodies.

Effects in the Neighbourhood of a Discharge.

31. Putting together the facts of the enormous electrostatic potential existing at the different points of a conductor conveying a discharge, and the violent oscillation to which so strong a current is subject, we perceive how great must be both the electrostatic and the electromagnetic induction in all space anywhere near it.

These disturbances, thus rendered certain and accounted for, can easily be experienced experimentally. I proceed to relate a few experiments out of a multitude which I have made.

32. I took a considerable length of highest conductivity No. 0 electrolytic copper wire, kindly lent me by Messrs. Thos. Bolton and Sons, and arranging one end so as to be accessible to Leyden jars, etc., carried the wire up to a high gallery and then down to earth. Earth was made in several ways on different occasions: water-pipes, gas-pipes, hot-water pipes which ramified the whole building, outside gas-mains, bars buried in ground; but the most effective and indeed perfect return circuit could be had when wanted by connecting the end of the wire metallically with the outside coat of the jar, without

any earth contact at all. When this direct contact was not made, the outer coat of jar had of course to be connected to earth also, in order to complete the circuit. Very often they were both connected to each other and to earth as well. It makes no essential difference; whatever may be considered the most satisfactory method, that may be adopted, and the phenomena will go on just as well. They may be briefly summarized as follows:

(1) If the conductor pass within an inch or two of any uninsulated piece of metal, it gives off a violent side-flash to it.

If the far end of the conductor is neither earthed nor connected to jar, but is left insulated in air, side-flashing from it occurs still more easily, but not very markedly so.

(2) If the conductor pass within, say, half an inch of an insulated conductor, it gives off a side-flash to it; the strength of the flash depends on the capacity of the insulated conductor, being considerable if it be large; but some side-spark occurs to an absurdly small body perfectly insulated, *e.g.*, such a thing as a coin on a stick of sealing-wax; and this when the conductor is absolutely well earthed at its far end.

(3) Sparks can be obtained from everything, even from quite uninsulated things, connected to the conductor: for instance, if it be connected to the gas-pipes, small sparks will fly to the finger or to an insulated body from all the gas-brackets about, and these sparks are sufficient to ignite gas.

(4) Sparks can be obtained between the ends of any long curved conductor, be it insulated or uninsulated, if they are brought close enough together to form a nearly closed circuit.

(5) Sparks can be obtained from or between insulated bodies, in the neighbourhood of the conductor but not connected to it at all, every time a flash occurs.

For instance, a large piece of wire gauze connected to nothing gave off sparks to a gas-bracket and ignited the gas. Moreover, one piece of wire gauze sent small sparks into another, neither connected with anything. [*In illustration of this, some gilt key-pattern high up on the wall of the hall, 25, Great George Street, was seen to be sparking while Leyden jar discharges were going on through a wire lying on the floor. They were not so bright as the sparkings in the Royal Institution wall paper* ("Nature," vol. xlix. p. 473), *but the gilding was further from the wire, and I believe is quite far from any wire. As illustrating the electric currents produced in conductors during the act of reflecting electro-magnetic waves, they were therefore still more satisfactory.*]

(6) Sparks can be obtained from quite uninsulated bodies, even when not connected with the conductor, *e.g.*, from hot-water pipes, from gas-pipes, from water-pipes, from strips of brass let into the table, from gas-brackets in other rooms, from a wire lying on the floor of a distant corridor ; and, in fact, all over the building, with few exceptions. The sparks can be taken by a penknife held in the hand; sometimes they can be seen going to the knuckle or finger-tip even when pressed against blackened or painted metal.

(7) Sparks can be got to pass between two totally uninsulated things, neither of which have any connection with the lightning conductor. For instance, let the conductor be thoroughly earthed in some outdoor and distant manner, and under favourable circumstances a bright short spark can be seen passing at every discharge between a gas-tap and a water-tap of my lecture table

which happen to approach each other closely. Let the gas escape near these sparks and it ignites.

(8) Sparks can be got between two thinly insulated electric light wires if they lie close enough together, and if a storage battery be connected to them an arc will be started, destroying the insulation and burning the wires.

(9) If at any distant place, or out of doors in daytime, the sparks are too feeble to be seen, disturbances can still be easily detected by means of a telephone; connecting one terminal to the thing—say, the roof of a shed, or a wire fence—and holding the other end in one's hand. Or, of course, by connecting the two terminals to two different things, or to different parts of one thing.

(10) Arrange Abel's fuses between gas and water-pipes, between pieces of wire gauze and gas-pipe, between hot-water pipe and a bit of sheet metal, between the lightning conductor and a 6-inch metal sphere on long glass stem, between two large insulated bodies, between the gas-bracket of another room and an empty Leyden jar; in short, in almost any place, likely or unlikely. Then take a few strong discharges through the thick copper rod with end well earthed, and the fuses will pop off, some at one discharge, some at another. Very visible sparks can sometimes be seen passing into the fuses, if they be not closely connected, without exploding them. A certain energy of spark is necessary to ignite the composition: much more than is sufficient to excite the retina.

Lightning Protectors.

33. Now let me say a word about the possibility of protecting telegraph and other instruments from damage by lightning currents which may have entered the aerial lines.

There is indeed no guarantee that burying wires beneath a pavement, or even beneath water, will effectually secure them from lightning disturbance, as I have now fully illustrated, but certainly overhead wires are more exposed.

The ordinary and well-known form of lightning protecting arrangement is to attach a pair of plates, or a double set of points, or a pair of points in a vacuum, or some other small air space, as a shunt to the instrument or coil of wire to be protected. Here are arrangements of the kind. Now it is perfectly easy to see that the protection such things afford is of the most utterly imperfect kind.[1]

Take a coil of tangled silk-covered wire, or any other coil that you don't mind damaging, and attach it as a shunt to one of these protectors. On discharging a jar through it, the insulation of the coil is pierced in heaps of places. Or one may use a short fine wire, and see it deflagrated by the branch discharge.

34. It may be said that my air space is too wide. Very well, then, abolish it altogether. Bring the plates of your protector into direct metallic contact. By so doing, the coil is indeed shunted out of the circuit, and if it were a telegraph instrument no signal could be given, for no appreciable fraction of the current takes that route; but with a Leyden-jar flash it is otherwise. Al-

[1] This was first shown in 1865, by Messrs. Hughes and Guillemin, in papers the gist of which has been already quoted. See Chap. XII.

though the plates of the protector are in contact, soldered together if one pleases, and led up to by stout wire or rod, a branch flash still breaks through the insulation of our wire tangle, or deflagrates our little bit of thin wire.

35. It will be said the joints are bad. Well, then, do without joints; take a stout rod of highest conductivity copper bent in an arc, say, 2 feet long, and, bridging it across with the wire tangle, discharge a jar round it. Still

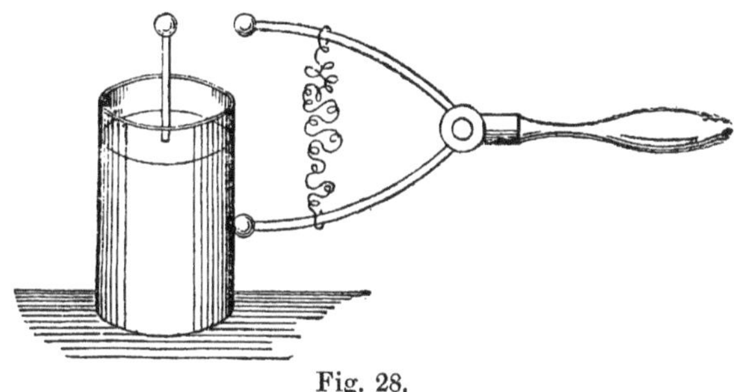

Fig. 28.

a portion takes the thin wire, even though it offer a path yards in length.

Take a straight bar of copper an inch thick, and arrange an invisibly fine Wollaston wire of greater length as a tapping circuit. Some of the discharge shall leave the bar, and spark across a minute air gap at each end, in order to make use of the hair-like platinum wire.

Go further still than this. Connect a coil of thin-covered wire *by one end only* to a wire conveying a discharge, standing the reel upon a block of paraffin or other good insulator, and connect the other end to any little thing of any capacity at all—say, a bullet lying on the block of paraffin: you can see sparkings through the insulation of the wire on the reel at every discharge.

These experiments render manifest the hopelessness of any simple shunt arrangement as a lightning protector.

36. The easiest mode of exhibiting the essentials of these experiments—one that can be tried by any one possessing a Leyden jar, a pair of discharging tongs, and a yard or two of fine silk-covered wire—is to hang a tangle of the wire loosely on to the tongs, not necessarily making any sort of good contact, and then use them to discharge a jar in the very ordinary way. Some of the flash will take the thin wire, and will spark through the insulation at a number of points (Fig. 28).

37. It may be very well objected to me that it is pretty useless if I only point out the imperfection of present methods, and offer no suggestion as to a proper lightning protector. Well, this struck me too, and the result is that I have devised and made what I think I may call an absolutely perfect protector—one into which great flashes may be sent, and yet the galvanometer or instrument intended to be protected shall not wink, nor shall the slightest palpable or visible disturbance be discernible, notwithstanding that complete metallic contact is maintained all the time, and not a trace of the signalling or useful current wasted.

I will explain this at a later stage.

38. The next part of this paper is largely controversial. I do not take much pleasure in this portion, and wish it were unnecessary. But it is a purely impersonal controversy; and so long as the old-fashioned views are in existence, one way of arriving at the truth is to try and thrash them out of existence. If there is any real vitality in them, the attack will fail. It will be well believed that I have no feeling of hostility to the Lightning Rod Conference, when my best scientific friend, Professor Carey Foster, was one of its members. Had

I been one myself, I no doubt should at that time have signed the very documents which now, in a few places, I criticise. Had I indeed so signed, I would abuse what I now see to be its erroneous portions with still more vigour than I now permit myself to employ.

Summary of Points of Difference and Controversy.

39. It may be convenient here to summarize a few of the points wherein the doctrines which I advocate differ from the views held by the older electricians. And the summary will give me an opportunity of emphasizing the incorrectness of the older views in many instances.

I quote a few statements prefixed by capital letters from an abstract I made for the "Electrician" after the Bath meeting of the British Association. (See "Electrician," Sept. 21 and 28, 1888.)

A. Rods as at present constructed, though frequently successful, may and do sometimes fail, even though their earth is thoroughly good; the reason being that they offer to a flash a much greater obstruction—a much worse path—than is usually supposed: an obstruction to be reckoned in hundreds or thousands of ohms, even for a very thick copper rod. N.B.—This is not resistance proper, but impedance.

I may be permitted here to repudiate the doctrine which has several times been attributed to me since the Bath meeting, that it is safest to be without lightning conductors altogether. The long experience of persons learned in this art is by no means to be despised, and until an agreement as to improvements has been arrived at, the safest plan for ordinary persons is to adhere to existing practice. Nevertheless, that conductors sometimes fail, is as certain as that they often succeed. My statement is that customary arrangements are not perfect and are susceptible of improvement. The statement of the Lightning Rod Conference is that "there is no authentic case on record where a properly constructed conductor failed to do its duty." Mr. Preece calls the statement "most decisive." It is certainly *decided*, and in the light of other matter contained between the same red covers I assert it to be in its natural and intended signification decidedly false. The only signification which makes it true makes it also senseless; as if one should record the statement that white things are white. In my Society of Arts lectures, I said nothing against the report of the Lightning Rod Conference, because the work done by that body, in collecting information, abstracting papers, and recording instances, was obviously very valuable, and much of the report itself is correct; while as for the occasional rash statements in that document I imagined the signatories would wish them to sink into oblivion in silence. But Mr. Preece has revivified them, and conspicuously made himself afresh responsible for them; accordingly I now extract one or two more sufficiently dogmatic and unfortunate statements from the same source.

"A man may with perfect impunity clasp a copper rod an inch in diameter, the bottom of which is well connected with moist earth, while the top of it receives a violent flash of lightning."

"If all these conditions be fulfilled; if the point be high enough to be the most salient feature of the building, no matter from what direction the storm may come, be of ample dimensions, and in thoroughly perfect electrical connection with the earth, the edifice with all it contains will be safe, and the conductor might even be surrounded by gunpowder in the heaviest storm without risk or danger."

To adhere to such views as these now, with tenacity sufficient to cause them to be promulgated as authoritative scientific statements, would be, in my opinion, little less than criminal.

"All accidents may be said to be due to a neglect of these simple elementary principles." Certainly this "*may be said*," because it already has been said over and over again; but it cannot be said with truth.

Whenever an accident happens, a believer in the modern exponents of the Lightning Rod Conference who could not point out a flaw, or a bad joint, or a bad earth; or a possible flaw, or a possible bad joint, or a possible bad earth; would feel himself disgraced as a practical man. An instance occurred in quite a recent number of "Nature." The writer of "Electrical Notes" records that a number of fish had been killed in a pond into which the earth end of a conductor had been led, and concludes with the ejaculation, "When will people learn to make proper earth connections?" I select this instance as typical of the extraordinarily contradictory advice often bestowed on that long-suffering body the British Public. Before an accident, the pond would be pointed out as an excellent earth, as indeed it most likely was. After the slaughter of the fish, the erector of the conductor is impersonally ridiculed for having utilized it. So with any struck building. After an accident defects *must* be forthcoming, because else there would be "an authentic case on record where a properly constructed conductor failed to do its duty," which *ex hypothesi* is absurd and impossible; therefore it was not constructed in accordance with the directions of the Lightning Rod Conference, therefore it was defective. Q.E.D.

B. When a Leyden jar is charged it corresponds to a bent spring, and its discharge corresponds to the release of the spring. Its discharge current alternates, therefore, in the same way and for much the same reason as a twitched reed or tuning-fork vibrates. The vibrations decay in either case because of frictional heat production, and because of the emission of waves into the surrounding medium. A single spark of a Leyden jar, examined in an exceedingly fast revolving mirror, is visibly drawn out into a close succession of oppositely-directed discharges, although its whole duration is so excessively minute.

It is very likely that this statement will now no longer be denied. So I pass to the next.

C. A lightning flash is a spark between cloud and earth, which are two oppositely electrified flat surfaces, and the flash corresponds therefore to the internal sparking between the two plates of a great air condenser. All the conditions which apply to a Leyden jar under these circumstances are liable to be true for lightning. Sometimes the resistance met with, either in the cloud itself or in the discharger, may be so great that the spark ceases to be oscillatory, and degenerates into a fizz or rapid leak; but there can be no guarantee that it shall always take this easily manageable form; and it is necessary in erecting protectors to be prepared for the worst and most dangerous form of sudden discharge. The apparent duration of some lightning flashes is due to their frequently multiple character, and indicates successive discharges, not one long-drawn-out one. Nothing that lightning has been found to do disproves its oscillatory character; because Leyden jar discharges, which are certainly oscillatory, can do precisely the same.

(This was in answer to Mr. Preece's contention that lightning could not be oscillatory, because it magnetized steel bars and deflected ships' compasses. The next is an antidote to the continually made statement that the one thing needful for an efficient lightning conductor is *conductivity.*)

D. Although some conductivity is necessary for a lightning conductor, its amount is of far less consequence than might be expected. The obstruction met with by an alternating or rapidly varying discharge depends much more on electro-magnetic inertia or self-induction than upon common resistance. So much obstruction is due to this inertia, that a trifle more or less of frictional resistance, in addition, matters practically not at all. It is very desirable to have a good and deep earth in order to protect foundations and gas and water-mains from damage, and in order to keep total impedance as low as possible.

I find it sometimes thought that I have argued against the need of a good earth. This is not so. I have argued against the exclusive and exaggerated attention that has been paid to this need.

Some Incorrect Statements.

In opposition to the following statements, *e, f, g, h, i,* the substance of which may be considered as hitherto orthodox, I make the statements subsequent, labelled *E, F, G, H, I:*

e. No danger is to be feared from a lightning conductor if only it be well earthed and be sufficiently massive not to be melted by a discharge. All masses of metal should be connected to it, that they may be electrically drained to earth.

f. The shape of the sectional area of a conductor is quite immaterial; its carrying power has nothing to do with extent of surface; nothing matters in the rod itself but sectional area or weight per foot run, and conductivity.

g. Points, if sharp, should constitute so great a protection that violent flashes to them ought never to occur.

h. Lightning conductors, if frequently tested for continuity and low resistance by ordinary galvanic currents, are bound to carry off any charge likely to strike them, and are absolutely to be depended upon. The *easiest* path protects all other possible paths.

i. A certain space contiguous to a lightning rod is completely protected by it, so that if the rod be raised high enough a building in this protected region is perfectly safe.

These statements I say are erroneous. The following (correspondingly lettered) I believe to be correct:

E. The obstruction offered by a lightning rod to a discharge being so great, and the current passing through it at the instant of a flash being enormous, a very high difference of potential exists between every point of the conductor and the earth, however well the two are connected; hence the neighbourhood of a lightning conductor is always dangerous during a storm, and great circumspection must be exercised as to what metallic conductors are wittingly or unwittingly brought near or into contact with it. When a building is struck, the oscillations and surgings all through its neighbourhood are so violent that every piece of metal is liable to give off sparks, and gas may be lighted even in neighbouring houses. If one end of a rain-water gutter is attached to a struck lightning conductor, the other end is almost certain to spit off a long spark, unless it also is metallically connected. Electric charges splash about in a struck mass of metal; and even a small spark near combustible substances is to be dreaded.

F. The electrical disturbance is conveyed to a conductor through the æther or space surrounding it; expressed more simply, lightning currents make use of the periphery of a conductor only, and so the more surface it exposes the better. Better than a single rod or tape is a number of separate lengths of wire, each thick enough not to be easily melted, and well separated so as not to interfere with each other by mutual induction.

The liability of rods to be melted by a flash can easily be over-estimated. A rod usually fails by reason of its inertia-like obstruction, and consequent inability to carry off the charge without spittings and side-flashes; it very seldom fails by reason of being melted. In cases where a thin wire has got melted, the energy has been largely dissipated in the effort, and it has acted as an efficient protector; though, of course, for that time only. (See pp. 122, 197.) Large sectional area offers very little advantage over moderately small sectional area, such as No. 5 B.W.G.

G. Points, if numerous enough, serve a very useful purpose in neutralizing the charge of a thunder-cloud hovering over them, and thus often prevent a flash; but there are occasions, easily imitated in the laboratory, when they are of no avail: for instance, when one upper cloud sparks into a lower one, which then suddenly overflows to earth. In the case of these sudden rushes, there is no time for a path to be prepared by induction—no time for points to exert any protective influence—and points then get struck by a violent flash just as if they were knobs. Discharges of this kind are the only ones likely to occur during a violent shower, because all leisurely effects would be neutralized by the rain-drops better than by a forest of points.

H. The path chosen by a galvanic current is no secure indication of the course which will be taken by a lightning flash. The course of a trickle down a hill-side does not determine the path of an avalanche. Lightning will not select the easiest path alone; it can distribute itself among any number of possible paths, and can make paths for itself. Ordinary testing of conductors is therefore no guarantee of safety, and may be misleading. At the same time it is quite right to have some system of testing and of inspection, else rust and building alterations may render any protector useless.

I. There is no space near a rod which can be definitely styled an area of protection, for it is possible to receive violent sparks or shocks from the conductor itself; not to speak of the innumerable secondary discharges which, by reason of electro-kinetic momentum and induction, by reason of electro-magnetic waves, and of the curious recently discovered effect of the ultra-violet light of a spark, are liable to occur as secondary effects in the wake of the main flash.

CHAPTER XVII.

INSTRUCTIVE EXTRACTS FROM REPORTS OF DAMAGE BY LIGHTNING.

40. Reading between the lines of existing reports on damage done, one can frequently find evidence of many of the phenomena to which I have now called attention. Of course they are not recorded in any prominent manner, because they are to the observers ill-understood and puzzling facts: it is very difficult to note what has exactly happened in any given case unless some clue or expectation has been formed beforehand. The bad conductivity clue, which was the only one prominently available to the skilled recorders in the following cases, is a very partial one, and in many cases is quite insufficient to account for the facts without undue pressure being put upon it.

Under such circumstances the record of the facts is of course far more valuable than the comments made upon that record; and in the following extracts the theoretical remarks should be eliminated or slurred over.

One minor imperfection, common to many accounts, is that they are not sufficiently alive to the possibility of all manner of branching discharges: so that the discharge is said to leave a conductor and go to something else, when the truer statement would be that some *portion* of it branched off at such and such a point.

In the following quotations the references are usually

to the pages of the volume of the Lightning Rod Conference, published by Spon in 1882:

41. Illustrating Side Flashes and Surging Circuits.

"*L.R.C.*,[1] p. 39.—*J. Murgatroyd. St. Mary's, Crumpsall, near Manchester.*—A lightning conductor from spire touched the eaves gutter, and a gas-pipe touched the end of this gutter. The lightning passed from the conductor along the gutter to the gas-pipe, melted it, and set the church on fire by igniting the gas."

"*L.R.C.*, p. 39.—*Wyatt Papworth.*—Tall spire struck. The church stands in an open position with no large trees near. It was provided with an iron lightning conductor $\frac{3}{4}$ in. diam., fixed with iron holdfasts, and carried down inside the spire and tower into ground; the top of it was said to be attached to a bold copper finial on the spire about 150 feet from the ground, and 50 feet above ridge of roof; the lightning is supposed to have first struck the finial, it slightly deranged some beds of masonry in upper part of spire, then descended by iron rod to belfry, melted a gas-tube in the floor, and set fire to the belfry by igniting the gas."

From an Abstract of Report on the Destruction by Lightning of a Gunpowder Store at Bruntcliffe, Yorkshire. By Major V. D. Majendie, R.A. (*L.R.C.*, p. 77.)

"The gunpowder exploded at 4.30 p.m. on August 6, 1878, during the greatest intensity of a violent thunderstorm. The building was brick, with brick arched roof, length 9 feet, width 5 feet, height 6 feet (internal dimensions). The store had a uniform thickness of three bricks, and was furnished at one end with an iron door, at the other end with a lightning conductor. The conductor consisted of a copper wire rope, 10 gauge copper wire, the

[1] L.R.C. stands for Report of Lightning Rod Conference.

rope being $\frac{7}{16}$ inch thick, having four points at the top (one large one in the centre, and three smaller ones round it); it extended to about 13 feet above the top of the building, and about the same length was carried into the ground and terminated in a drain. The conductor had been erected in 1876 by Mr. John Bisby, of Leeds, and was fixed to a pole distant about 2 inches from the end of the building opposite to that in which the iron door was fixed (it was not connected with the iron door in any way). No one was near the store when the powder exploded, and it seems probable that [the earth connection of the conductor was bad, that] the mass of iron in the door offered at least an equally good path—and that the gunpowder was ignited by a flash passing between the two imperfect conductors."

To make this report more completely scientific, it would be well to omit the words I have put in square brackets.

Extract from a reply of Mr. Baldwin Latham, C.E.

"It is no uncommon thing for buildings provided with what are called lightning conductors to be damaged by lightning, and the cause is due to the inadequacy of the conductor to carry the electric fluid, which will leave the conductor for a better or a larger conductor." (See also sec. 43.)

42. As illustrating that a good conductor affords no absolute security, the following document is worthy of reproduction in full, as given in *L.R.C.*, p. 115:

"*Ueber Blitzableiter und Blitzschläge in Gebäude welche mit Blitzableitern versehen waren. Von G. Karsten. Kiel,* 8vo., 1887. (*Abstracted by R. Van der Broek.*)

"In this pamphlet, Dr. Karsten gives an account of

two cases in which buildings that were provided with lightning conductors were damaged by lightning. The author states that the statistics for the year 1873 show that in Schleswig-Holstein twenty-six per cent. of all the cases of fire were caused by lightning; $\frac{1}{130}$ part of these cases occurred in the towns and the remainder in the country.

"Do lightning conductors guarantee absolute protection? The author answers this question as follows:— There is no absolute certainty in empirical matters; each new case may direct our attention to circumstances that had been overlooked. If lightning conductors cannot be said to insure perfect safety, they certainly afford a very high degree of protection.

"The flash of lightning which struck the church at Garding on the 18th of May, 1877, fractured the conductor in fifteen places, and pierced the wall of the steeple in two places. The inefficiency of the conductor resulted from the carelessness with which it was fixed; the line was laid down the north side of the steeple and fastened with twenty-five wall eyes; these wall eyes were hammered too deep into the wall, thus damaging the line and forming a short and sharp bend in each case, besides also unduly straining the wire. The damage to the steeple was the consequence of a neglected secondary circuit. There are an excessively large number of tie-rods in the steeple; the heads of these rods are not connected together, neither are they, except in one case, in close proximity to any of the larger masses of metal that are about the building. The conductor passed close to one of those heads; the south side of the steeple, where the opposite head is, becoming wet through the rain, a secondary circuit was formed, and a return shock followed; the damage to the steeple was trifling.

"The rod was provided with a conical point, rather blunt, but surmounted by a short platinum point. The copper line-wire was of good material—not of a uniform thickness, but at the weakest places not weighing less than 240 grammes per lineal metre (8 oz. per yard, or rather less than $\frac{1}{4}$ inch diameter if solid). The earth-plate was sunk into a well 10 metres deep, and tested faultless after the discharge."

43. The following series of notes are better not separated, though they bear upon different points.

L.R.C., p. 39 *et seq.—From Abstract of Statistics of Buildings and Ships struck by Lightning. By F. Duprez, Member of the Academy.*

Illustrating occult dangers from secondary disturbances in the neighbourhood of conductors.

"The author cites three cases of buildings set on fire, though protected by lightning rods. But the precise cause of the fire was not ascertained."

Illustrating multiple flashes.

"In each of two cases the lightning struck at once the three rods fixed to a building."

Illustrating that a poor earth is not necessarily fatal to the efficiency of the conductor, except as to the protection of the soil itself and of things buried in it.

"Out of fifteen cases of lightning rods struck, in which the conductors were simply buried more or less in the soil, they carried off the strokes in eleven without the buildings being injured or any trace being left of it, except that the ground was upheaved where the latter was too dry."

Illustrating a matter important, if true: suggested by Professor Fitzgerald theoretically at Bath: not yet verified by me.

"In two cases the stroke broke the conductor at points where its direction was abruptly changed."

Illustrating side-flash.

"In two other cases the lightning left the conductors struck, and fell upon buildings near, without causing damage to those on which the rods were fixed."

Illustrating brush discharge from conductor, and surging circuits in its neighbourhood.

"Two electrical phenomena are to be noted as sometimes occurring when a lightning rod is struck. First, when a conductor is formed of metallic plates a peculiar noise is heard like water pouring on a fire. Second (independently of the form of the conductor), electric sparks are emitted from bodies near. The author cites examples at Berne, 1815."

44. The following illustrate the carrying power of fairly thin wire, and the fact that thin wires may protect although themselves deflagrated.

From Mr. Preece's paper on "Lightning" to Society of Telegraph Engineers, November, 1872 (*L.R.C.*, p. 101):— "There were only two cases in the past season where line wires (No. 8 iron, 0·17 in. diam.) were absolutely fused."

From letter by Admiral Sullivan (*L.R.C.*, p. 195) :— "You will like to know a case in which a copper wire acted as a perfect conductor, *though fused throughout its length.* It was at Monte Video, in the house of the English Consul; a flagstaff was struck, and conducted the lightning through a flat roof near the bell-wire of a suite of rooms (the wire ran in sight near the cornice) through a hole in each dividing wall, and then down to the bell in the basement: the wire was melted into drops like shot, which burnt a row of small holes in the carpet of each room. A dark mark on the cornice above showed where the wire had been. At the bell there was a slight explosion and some little damage, but I do not recollect whether anything acted partially as a conductor from that point and so carried off that part of the charge.

"This, I think, shows that even an ordinary bell-wire will act as a conductor for a rather strong stroke of lightning, as the large flagstaff was shattered."

45. As illustrating that experience has led to the perception of the value of large surface to a rod, I may quote the standard American work, Spang's "Practical Treatise on Lightning Conductors." Philadelphia, 1887 (*L.R.C.*, p. 113):

"A conductor of large surface exercises a much greater protective action than the same quantity of metal in the form of a wire or solid rod.

"Not because electricity in motion resides on the surface, but that the expansive action of a discharge may have a wider scope through the metal."

[The incoherence of the reason does not destroy the correctness of the stated fact.]

Messrs. D. Munson and Co., of Indianapolis, Indiana, have sent me specimens of their rods, which have flanges and sharp edges, and in various ways aim at large surface. Their rods are composed of copper and iron mixed, which is curious, and their mode of attainment of large surface is needlessly complex.

The following illustrates the contradictory views about manner of conduction by lightning rods. *Letter from Admiral Sullivan* (*L.R.C.*, p. 199):

"I firmly believe in the surface theory of Harris. I had been with him often when he made experiments nearly fifty years since, and witnessed a strip of tinfoil of the thinnest kind, and about $\frac{1}{4}$ inch wide, protect a model mast of about 6 inches in diameter from electric shock, that without it split the mast to pieces, aided by a small hole through its centre filled with gunpowder. And I always thought that the surface-conducting theory of Harris was indisputable. But about twenty years since, having to approve a proposal of the Trinity House for a new conductor of a lighthouse, which, like previous ones, was an inch in diameter copper rod called 'Faraday's Plan,' I thought I would go up to the Royal Institution and ask him why he did not use a copper tube instead, giving much greater conducting power with less copper. I did so, and he asserted positively that the conducting power depended entirely on the volume of copper in the section of the conductor, no matter whether it was in a bolt, plates, or tube; and that if Harris said differently, 'he knows nothing whatever about it'; of course I approved the rod-conductor. But singularly enough, though I had not seen Harris for years, he came to town a few days after, and came to the Board of Trade to see me, and bring me a piece of his large tube conductor, with a connection that he was fitting to the Houses of Parliament. When I told him what Faraday's opinion was, he answered, 'Then he knows nothing about it.'"

46. I may call attention to some apparently sound doctrines in a paper by Capt. Bucknill, R.E., abstracted in *L.R.C.*, p. 242. (See also Appendix.)

That lightning can penetrate into collieries is proved by an account of a meeting of Mining Engineers (*L.R.C.*, p. 237).

The same sort of thing is illustrated by the accident at Bootham Bar, York (*L.R.C.*, p. 219), when the lightning struck down into a cavity surrounded by high buildings with lead roofs, iron rain-water pipe, and an iron portcullis, to get at a street bracket 11 feet 6 inches above the pavement, melting its pipe, and setting a house on fire by the large flame produced (*L.R.C.*, p. 219).

47. That the existence of gas-pipes make houses more difficult to protect, and in fact causes dwelling-houses to require almost as much attention as powder magazines, can be illustrated by a good number of extracts. The following may serve (*L.R.C.*, p. 239):

"On July 13, 1880, during a thunderstorm, the large 400-light gas-meter of this mill, though locked up in a cellar, and with no light near it, exploded, and the gas, which is supplied through a 4-inch main, was ignited. This was repaired, but on July 5, 1881, during another thunderstorm, precisely the same accident occurred."

From an interesting report, signed J. Gavey, on damage to a church at Cardiff (*L.R.C.*, p. 218), I make the following extract:

"On examining more closely the surroundings of the lightning conductor, I observed that the church gas-pipe, an iron one, about $1\frac{1}{4}$ inches in diameter, passed through the wall of the building about 6 feet from the conductor, and was carried in a direction corresponding with the hole caused by the explosion. I immediately concluded that this explosion was due to the current breaking across from the conductor to the gas-pipe, and on opening up the hole I found this to be the fact. The conductor crossed the gas-pipe at nearly a right angle, being about a foot above it. The under portion of the conductor bore evident marks of fusion, and, more interesting still, the gas-pipe was slightly coated with a very thin deposit of copper, so thin that it perished in my attempt to remove it; but still there was an undoubted coating at one spot."

Continuing the account of this kind of damage, the following interesting remarks by Professor Kirchhoff have a bearing on the important practical question whether gas-pipes should or should not be utilized for earth; but their main utility lies in the proof afforded of the unique kind of danger which gas-pipes introduce:

Injury to Gas and Water Pipes by Lightning.[1]

The city gas company of Berlin, having expressed the fear that gas-pipes may be injured by lightning passing down a rod that is connected with the pipes, Professor Kirchhoff has published the following reply:

"As the erection of lightning rods is older than the system of gas and water pipes as they now exist in nearly all large cities, we find scarcely anything in early literature in regard to connecting the earth end of lightning rods with these metallic pipes, and in modern times most manufacturers of lightning rods, when putting them up, pay no attention to pipes in or near the building that is to be protected. Kirchhoff is of the opinion, supported by the views of a series of professional authorities, that the frequent recent cases of injury from lightning to buildings that had been protected for years by their rods, are due to a neglect of these large masses of metal. The Nicolai Church, in Griefswald, has been frequently struck by lightning, but was protected from injury by its rods. In 1876, however, lightning struck the tower

[1] See also on this subject two Abstracts in "Journal of Institution of Electrical Engineers," No. 77, vol. xviii., 1889, of papers by Prof. W. Kohlrausch and A. Voller respectively.

and set it on fire. A few weeks before, the church had had gas-pipes put in it. No one seems to have thought that the new masses of metal which had been brought into the church could have any effect on the course of the lightning, otherwise the lightning rods would have been connected with the gas-pipes, or the earth connection been prolonged to proximity with the pipe. A similar circumstance occurred in the Nicolai Church in Stralsund. The lightning destroyed the rod in many places, although it received several strokes in 1856, and conducted them safely to the earth. Here, too, the cause of injury was in the neglect of the gas-pipes, which were first laid in the neighbourhood of the church in 1856, shortly before the lightning struck it. The injury done to the school-house in Elmshern, in 1876, and to the St. Lawrence' Church, at Itzehoe, in 1877, both buildings being provided with rods, could have been avoided if the rods had been connected with the adjacent gas-pipes."

"If it were possible," says Kirchhoff, "to make the earth connection so large that the resistance which the electric current meets with when it leaves the metallic conducting surface of the rod to enter the moist earth, or earth water, would be zero, then it would be unnecessary to connect the rods with the gas and water pipes. We are not able, even at immense expense, to make the earth connections so large as to compete with the conducting power of metallic gas and water pipes, the total length of which is frequently many miles, and the surface in contact with the moist earth is thousands of square miles. Hence the electric current prefers for its discharge the extensive net of the system of pipes to that of the earth connection of the rods, and this alone is the cause of the lightning leaving its own conductor."

[This last, like many another casual remark, is not to be supposed true.—O. L.]

Regarding the fear that gas and water pipes could be injured, he further says: "I know of no case where lightning has destroyed a gas or water pipe which was connected with the lightning-rod, but I do know cases already in which the pipes were destroyed by lightning because they were not connected with it. In May, 1809, lightning struck the rod on Count von Seefeld's castle, and sprang from it to a small water-pipe, which was about 80 metres from the end of the rod, and burst it. Another case happened in Basel, July 9, 1849. In a violent shower one stroke of lightning followed the rod on a house down into the earth, then jumped from it to a city water-pipe, a metre distant, made of cast iron. It destroyed several lengths of pipe, which were packed at the joints with pitch and hemp. A third case, which was related to me by Professor Helmholtz, occurred last year in Gratz. Then, too, the lightning left the rod and sprang over to the city gas-pipes; even a gas explosion is said to have resulted. In all three cases the rods were not connected with the pipes. If they had been connected the mechanical effect of lightning on the metallic pipes would have been null in the first and third cases, and in the second the damage would have been slight. If the water-pipes in Basel had been joined with lead instead of pitch, no mechanical effect could have been produced. The mechanical effect of an electrical discharge is greatest where the electric fluid springs from one body to another. The wider this jump the more powerful is the mechanical effect. The electrical discharge of a thunder-cloud upon the point of a lightning rod may melt or bend it, while the rod itself remains uninjured. If the conductor, however, is insufficient to receive and carry off the charge of electricity, it will leap from the conductor to another body. Where the lightning leaves the conductor, its mechanical effect is again exerted, so that the rod is torn, melted, or bent. So, too, is that spot of the body on which it leaps. In the examples above given it was a lead pipe in the first place, a gas-pipe in the last place, to which the lightning leaped when it left the rod, and which were destroyed. Such injuries to water and gas pipes near lightning rods must certainly be quite frequent. It would be desirable to bring them to light, so as to obtain proof that it is more advantageous, both for the rods and the building which it protects, as well as for the gas and water pipes, to have both intimately connected. Finally I would mention two cases of lightning striking rods closely united with the gas and water pipes. The first happened in Düsseldorf, July 23, 1878, on the new Art Academy; the other August 19, last year, at Steglitz. In both cases the lightning rod, the buildings, and the pipes were uninjured."—*Deutschen Bauzeitung.* (Quoted in "The Building News," September 10, 1880.)

48. That lightning sometimes does things which is on the hitherto available views inexplicable, is evidenced by the following record by one who, of all others, made lightning conductors his hobby, who acquired an immense amount of experience concerning them, and to whom the carrying out of probably the most perfect system of protection in actual use is largely due—I mean the late M. Melsens:

Note on the Lightning Flash at Antwerp Railway Station, July 10th, 1865. By M. Melsens.[1]

"I have described this lightning stroke, which was very harmless, since all the damage was confined to the breaking of a square of glass in the roof of the covered station of the railway at Antwerp; but in connection with it I found myself confronted by so extraordinary a phenomenon that I had to search through all the descriptions of lightning strokes which might offer a certain analogy to this with which I have to do. It was only on looking through my notes again, and returning several times to the place, that I have ventured to describe this very extraordinary stroke, and I have only published it after a long time, when it seemed to me that I could establish it upon careful observations and experiments which seemed of a nature to corroborate my conclusions. [Mention made of 19 lightning flashes, which present some analogy to that which struck Antwerp Station.]

"A few words are sufficient to show the peculiarity of the phenomenon. The lightning crossed a square of glass and produced in it a hole similar to that which would be produced by a projectile thrown upwards from below at a slow rate, say from 30 to 50 metres per second, and travelling from earth towards the sky; the edges of the hole were melted. I have given a description of the experiments I made, together with Ruhmkorff, to prove that in reality lightning does travel from the earth towards the sky; the proof which I give of it seems to me decisive.

"It is extraordinary for lightning to pass, without any

[1] "Paratonnerres: Notes et Commentaires," par M. Melsens. Brussels, 1882. Extracted from Reports of Belgian delegates to the Paris Exhibition, 1881.

conductor, through a square of glass 4 millimetres in thickness; (forming a parallelogram of $0^{m.}35 \times 0^{m.}38$, having angles of 83° and 97°).

"The opening produced was at a distance of some centimetres from iron and lead conductors, which were in perfect metallic communication with all the iron of the station. The weight of this latter exceeds 120 tons. But the anomaly does not stop there: to the right and left of the glass roof in which the broken pane was, the covered platform of the station has a roof of zinc No. 13, presenting to the lightning a surface of 3,000 square yards; the weight of this zinc is not less than 15 tons; the three tall rods of the lightning conductor are in immediate metallic contact with this zinc, and with the conductor, and the whole is connected to 28 hollow columns, serving to carry off rain-water. The collection of these metals would allow one to suppose that they were well adapted to carry off lightning or any such disturbance easily; but in this instance it despised them, and chose a path entirely unexpected. Let me add that there was a shed some 62 metres from the station built on hollow columns and iron framework, and the greater part of it roofed with zinc; moreover, about 40 metres on the opposite side were more sheds roofed with zinc; the metallic surface was not less than 2,000 square metres. Moreover, in my investigation, I might mention plenty of other metals in communication more or less perfect with the earth—grids, sconces, gas-pipes, telegraph wires, rails, etc. Besides, the whole of the platform may be considered as one large lightning conductor in perfect communication with a very damp earth, and offering hardly any resistance to the passage of the current from the building."

49. Of all the buildings in the world not wholly made

of metal, the Hôtel de Ville at Brussels was and is the most perfectly and elaborately protected. No electrician exists who would not a year ago have asserted, had he gone over it, that it was absurdly and exaggeratedly safe from damage by lightning. Last July it was struck and set on fire.

The case has been investigated and published in the "Bulletin de la Société Belge d'Electricians" for September and October, 1888.

It seems to have been owing to secondary or induced electric surgings in a horizontal bar of metal totally unconnected with anything; not pretending to lead toward earth, and, being some little distance below a stretch of lightning conductor, not offering itself as an object to be struck. On all the old views it was utterly insignificant. However, as a matter of fact, although not struck, although it did not (probably) even receive a side-flash, yet the induced surgings set up in it, induced by Maxwell and Heaviside's electro-magnetic waves, were so violent as to ignite some gas and cause a small fire. Had it been connected to the conductor, the sparking from it would probably have been still stronger.

This occurrence, and certain sparkings in the wall-paper of the Royal Institution (see "Electrician" or "Nature," March, 1889),—the sparkings, too, in the gilding of this present hall (sec. 32),—are plain and straightforward intimations that the old views on the subject of electric conduction are hopelessly, and absurdly, and dangerously inadequate.

CHAPTER XVIII.

PRACTICAL QUESTIONS

50. There remains to consider what is to be the practical outcome of all this. What improvements in the erection and testing of lightning conductors are possible. This is a matter well worthy of discussion, and eminently suited to it. It is a matter on which I have not the slightest wish to be dogmatic; and if I make a few apparently definite assertions, it is only by way of expressing such judgment as I have been able to form at present, and they are to be taken as intended more by way of suggestion and question than anything else.

With this proviso, I should be disposed to make some such statements as the following:

51. All parts of a lightning conductor, from points to roots, should be of one and the same metal, to avoid voltaic action.

52. Joints should be avoided when possible, and should be made substantially when necessary. Allowance for expansion and contraction must not be forgotten.

53. Sharp bends, and corners, and curves, and roundabout paths to earth should be avoided as far as possible.

54. The use of copper for lightning conductors is a needless extravagance.

55. Iron has advantages over every other metal.

56. The shape of cross-section is but little matter.

Flat ribbon has a slight advantage over round rod, but not enough to override questions of convenience.

57. Liability to be deflagrated by a powerful flash determines the minimum allowance for size of cross-section. No consideration of conductivity and greater ease of path has the least weight in this connection.

58. It is hopeless to pretend to be able to make the lightning conductor so much the easiest path that all others are protected. All possible paths will share the discharge between them, and a lot of apparently impossible ones.

59. A good and deep earth should in general be provided, independent of water and gas mains.

60. If the conductor at any part of its course goes near water or gas mains, it is best to connect it to them.

61. If the place to be protected has water or gas pipes inside it, the conductor should be connected to their mains underground.

62. At all places where water and gas pipes come near each other, and, in general, wherever one metal ramification approaches another, it is best to connect them metallically.

63. The neighbourhood of small-bore fusible gas-pipes, and indoor gas-pipes in general, should be avoided in erecting a lightning conductor.

64. Into powder magazines and such like places no gas or water pipes should be permitted to enter, unless the whole building is made of metal, and they are elaborately connected to it at the point where they enter.

65. It is not wise to erect very tall pointed rods above the roof of a building.

66. A number of points all along the ridge of a roof is better than only a few.

67. Any part of a building is liable to be struck, and,

to make quite secure, every prominent part of the outside should have a rod running along it.

68. Earth-connected as well as insulated bodies are liable to spit off sparks.

69. No complete security can be attained unless the whole building be metal-lined, floor and all.

70. In ordinary houses it may be well to try and insulate the lightning conductor from the walls so as to lessen the chance of side-flash to metal stoves and things inside.

71. In chimneys it may be well to use insulators to protect the bricks from concussion.

72. The cheapest way of protecting an ordinary house is to run common galvanized iron telegraph wire up all the corners, along all the ridges and eaves, and over all the chimneys; taking them down to the earth in several places, and at each place burying a load of coke. Rain-water spouts and other outside metal, if all well connected together, may likewise be utilized.

73. Connecting a lead roof or other such expanse with a lightning conductor is not an unmixed good, for it virtually increases the dangerous proximity of the lightning conductor, and may inadvertently bring it near to many objects which else might have escaped.

74. One of the most difficult things is to know what to connect and what to avoid.

75. The orthodox rule, "Connect all pieces of metal to the lightning conductor," requires modification thus: —Connect all pieces of metal to each other and to the earth, but not to the lightning conductor. (?)

76. It may be always reckoned safe to earth things independently. It is often not safe to connect them to the lightning conductor: *e.g.*, an inside lining of a chimney should be well earthed, but should not be used

as lightning conductor nor connected with it. The same with rain-water pipes and gutters. The same also, probably, with lead roofs. (?)

77. In connecting pieces of metal to each other, if they happen to form a nearly closed circuit, the circuit should be metallically completed.

78. Over the top of tall chimneys it is well to take a loop or arch of the lightning conductor, made of any stout and durable metal.

79. Lightning conductors should be always outside and easily visible.

80. A conductor detached from the building to be protected is safer than one in close contact with it.

81. For powder magazines and such like, an outer cage surrounding the building, with sky points and earth roots, and an inner cage on the building, with independent earth terminals only, is the safest plan.

82. If under these circumstances there be no gas-pipes nor much ramifying metal work inside the building, and no metal at all going near either cage, the interior may be considered perfectly safe.

83. The inner cage may often be conveniently made of continuous sheet iron. The outer cage need not then be at all small-meshed; in fact, it need be little more than a dozen vertical conductors.

84. The resistance of an earth may be tested in the customary way to guard against actual breaches of contact by rust or workmen; but no overweening confidence must be felt, however small the resistance turn out to be.

85. A Wimshurst machine and couple of Leyden jars afford a convenient mode of testing a conductor for flagrant defects. The testing should be done in the dusk or in moonlight, so that there may be light enough to work by and yet sparks be visible.

86. Telephones are the handiest things to detect electric surgings in conductors inside the house while discharges are being made to the conductor. But vacuum tubes, gas leaks, Abel's fuses, etc., etc., can also be employed.

87. Another mode of testing can be carried out on an insulated rod, with an induction coil and spark gap after the manner of a great Hertz oscillator. This is probably the most searching plan. Sparks will then be probably obtained from all the gas-brackets, water-taps, and picture-rails in the house.

88. Telegraph stations, and houses supplied with electricity from overhead wires, should have an efficient lightning protector at the place where the wires enter the house.

89. When a number of houses are wired up together for lighting, even by underground wires, it may be well to disconnect them by means of intervening lightning protectors: on the principle of fire-proof doors.

90. A central lighting station, having a tall chimney connected to its boilers and dynamos, should be, by a lightning protector, disconnected from the leads which carry the current from it; because a small fraction of a stroke getting into the leads might destroy a number of lamps, especially if they were already working at something like their full power.

91. Telephone arrangements, and any long length of close-packed insulated wires (even underground), should be protected by lightning protectors, else the insulation is apt to be sparked through and spoiled.

CHAPTER XIX.

DISCUSSION.

A FULL report of the discussion on this paper appears in the " Journal of the Institute of Electrical Engineers " for 1889, and I am permitted to quote here the remarks of Sir Wm. Thomson and some other speakers; reproducing also such portions of my own contribution as usefully amplify what has been already said, or such as fill up lacunæ in the main communication.

Many remarks were made—and some weighty ones —on the subject of the probable absence of conducting power in clouds, and the consequent difficulty in satisfying the conditions of the *B* flash (*i.e.*, the impulsive rush, or spark between bodies initially at the same potential). I admit at once that obtaining *B* flashes from metal sheets proves nothing whatever concerning the possibility of obtaining them from clouds. But my argument is rather converse to this. I argue (whether rightly or wrongly) that flashes often occur from clouds under circumstances which, under the *a* or steady strain, or high potential condition, scarcely seem natural. Thus points are sometimes struck and melted by flashes; and it is not easy for points to be struck in case *a*. Moreover, from a cloud violently raining flashes occur; whereas one would expect rain to lower the potential gradually. Again, from a cloud resting on a hill-top flashes occur;

and sparks from a badly insulated body are at once suggestive of impulsive rush conditions—that is, sparks from a body of zero potential. Similarly, I understand that a cloud perforated by the Eiffel Tower has carried on a thunderstorm to people below. I doubt if any of these things could happen under the conditions of case *a*.

But, it will be said, surely some of the facts you adduce establish the bad conducting power of clouds. To a great extent the Eiffel Tower case (if a fact) does. The hill-top case merely proves that the *hill* was a poor conductor. I can easily get long sparks from metals roughly uninsulated, as by wood, water, or soil.

Suppose, however, it admitted (not as proved, but as probable) that clouds are poor conductors, what then? All that we can assert is that the whole of a cloud, or even a large portion of a cloud, is unlikely to discharge at once. I quite think that that is so. A calculation of energy shows that a violent flash need only discharge a very small portion—a few square metres—of a charged cloud, and that the same cloud may therefore go on sparking for a long while, as, indeed, it appears to do. Now no great conducting power is needed for a discharge from a small area at a great elevation: the lateral component of rush is in that case negligible.

Of course it *may* be asserted that clouds conduct so badly that no "flash" in the proper sense can ever occur. An answer to that assertion is the existence of lightning.

All that poor conductivity in clouds has to say concerning the impulsive rush seems to me this: that when any part of the cloud receives a violent disturbance from some *A* flash in its neighbourhood, that same part which receives it is most likely to spit off the consequent *B* flash. Whereas, with a perfect conductor, any other portion would be almost equally liable. Poor conductivity goes, indeed, to *help* the violence of the impulsive rush; for the essence of it is that a conductor of small capacity at zero potential shall be suddenly overloaded. Now, if a charge suddenly communicated to a portion of a large cloud could instantaneously be shared with the whole, the potential would be reduced, and nothing very violent need occur.

This may sound like special pleading, but it is only recording the circumstances that have to be attended to. What the *facts* are must be determined by direct observation on flashes and clouds.

But let me here beseech meteorologists to remember that establishing the condition for some one flash or class of flashes does not establish the impossibility or improbability of very different conditions obtaining elsewhere or at other times. There are varieties of thunderstorms, varieties of clouds, and varieties of flashes. Every good and accurate observation will be a help to fuller knowledge, but it will take years of enlightened experience and observation before all possible varieties and circumstances of discharge can be supposed exhausted.

Before leaving this subject I should like to remind the Institution that I have never hesitated to contemplate the imperfect conductivity of clouds, whatever the consequences of that imperfect conductivity might be; and, in proof of this, I enclose an extract from the report of my remarks at Bath, as published in the scientific journals at the time:

Extract from British Association Discussion.—"There was one point where Mr. Preece might have attacked him, but where he did not think Mr. Preece had made out the full strength of his case, namely, the question—What are the conditions of a flash? He (Professor Lodge) had assumed that a flash behaves, or may behave, like the discharge of condensers in a laboratory; but it was a question whether a cloud discharge was of this kind. A cloud is not a good conductor; it consists of globules of water separated from one another by inter-spaces of air; it may be compared, therefore, to a kind of spangled jar; when a spangled jar discharges there is no guarantee that the whole of it discharges, it may discharge out in a slowish manner; it may be that you have first a bit of discharge, then another bit, and so on, so that you may have a kind of dribbling of the charge out of it, and you may thus fail to get these oscillatory and sudden rushes. At the same time he did not think that they could always guarantee doing this with cloud discharges; and it would not be safe in arranging protectors to protect for only one case, and that the easiest; they must provide for the possibility of a sudden and violent discharge. Still, the conditions of actual lightning were to be ascertained by observing lightning, and not by experiments in the laboratory."

Proceeding now to the experiments shown by Mr. Wimshurst.

I feel obliged to him for exhibiting and emphasizing several points which, for want of time or otherwise, I had rather slurred over; also for recalling to my memory a little point which I had forgotten, though it is in my assistant's note-book; and, lastly, for detecting an interesting matter which had escaped me.

To take these successively. 1. He seems to have exhibited side-flashes, from badly-earthed conductors and to well-earthed bodies, more satisfactorily than in my hurry during the paper I managed to do. It is not likely that he exhibited them so strongly as I have obtained them in the laboratory, because the violence of many of these side-flashes is a thing that strikes observers with astonishment; and with a *long* conductor, as stout as you please, it makes but little difference whether its far end is "to earth" or not. It does undoubtedly make some. Professor Threlfall, and Mr. J. Brown of Belfast, have both seen these effects at Liverpool, and neither, I fancy, would contemplate with equanimity the idea of bringing their knuckles near a conductor when struck, however well earthed it was.

At the same time it is perfectly true, and I must have often recorded the fact, that from a well-earthed conductor the sparks will not charge a Leyden jar. They jump in and out again. Sometimes, indeed, there is a residue of charge, but it is accidental what sign it is, and it is always merely the tail end of a series of oscillations cut off by resistance at some arbitrary point.

A more striking experiment is to connect a gold-leaf electroscope to the conductor. If the connection is metallically perfect the gold leaves are nearly quiescent, with imperfect connection they may diverge; they always slightly kick. The experiment is a little rough on the electroscope, for it stains the leaves downwards, and blows fragments off sometimes; but it is a striking thing to be able to take a half-inch or even a one-inch spark, of considerable power and noise, from the cap of an electroscope whose leaves hang stiffly down all the time and barely twinkle.

These side-flashes are not very painful, they look worse than they feel: the charge hops in and out of you without going through you much or disturbing the nerves seriously. I by no means assert that a man would *necessarily* be killed if touching a conductor struck by lightning; but it would surely be a position of considerable danger.

2. The fact which Mr. Wimshurst has observed, but

which I had missed, is this: that when side-flashes are tried for at different points in the length of a wire joining the outer coats of two equally-insulated symmetrical jars, they are obtained more strongly towards either end of the wire, and are not obtainable at all at the middle. The middle is, in fact, a node. There are stationary waves set up in the wire, whose ends are now at high potential and now at low potential alternately, just like a long bath which has been tipped and set down sharply. To and fro the water splashes, and the ends are now at high and now at low level alternately, but the middle is a node and remains of average level all the time: it is at zero potential, and no spark is obtainable from it. It is an interesting fact, and one that it would have been a pity to miss. We may congratulate Mr. Wimshurst on discovering it.

3. Lastly. The little point Mr. Wimshurst has recalled to my memory is this: that when a cloud or top-plate is negative, a small terminal or point gets struck rather more easily—*i.e.*, at a lower elevation—than a big terminal or dome, even under the circumstance of the impulsive rush.

The fact is often so; but Mr. Wimshurst's account of it may lead persons who have not tried the experiment to over-estimate the magnitude of the difference, which is frequently quite inappreciable: being, indeed, often non-existent.

If the impulsive rush is violent—*i.e.*, if it proceeds from a pair of large jars, highly charged, into a plate of moderate size—the difference is non-existent. Careful measurement fails to show that the things equally struck are not all the same height; and this whether the "cloud" be negative or positive. This was the case I exhibited at the meeting, and there was no need to notice of what sign the top-plate was. But if the rush be made less violent, either by using small jars or by charging them feebly, a difference is observable. A point then gets struck, as Mr. Wimshurst showed, at a distinctly lower elevation than a knob does, whenever the overhead flash is negative—not when it is positive.[1]

Those are the facts, and we are indebted to Mr. Wimshurst for calling attention to them, but, as to what the moral and practical bearing of them is, opinions may differ.

Certainly it in no wise upholds the statement that points always discharge silently and cannot get struck, which is what has always been meant by their "protective virtue." Rather it would seem that they get struck in an impulsive rush always as easily as anything else, and sometimes, as Mr. Wimshurst shows, still more easily. For remember that the "striking" is not a fizz or leak of a gentle and protective kind, but is a violent and destructive flash.

All the experiments of Mr. Wimshurst on the alternative path or "bye-pass" are completely in accord with my theory;[2] and also that the remark of Professor Hughes concerning the probable delay of the *B* spark behind the *A* spark is borne out by theory. The lag should be, in fact, a quarter period of the oscillation.

Referring to another remark, I may also say that at Liverpool we have recently obtained excellent photographs of a slowly oscillating spark on a rotating sensitive disc, and that the constituent oscillations are not only conspicuous, but well spread out and accurately measurable, and in agreement with theory within one half per cent. I am proceeding similarly to examine the *B* spark.

[1] For measurements see Chap. XXIII.

[2] See Chap. XIII.

I should like to remark here, what after all is fairly obvious, how great service has been done by the developers of the modern influence machine, from Nicholson, Varley, and Thomson, to Holtz, who did so much, and on to Mr. Wimshurst himself.

Professor Hughes' experiments on iron *versus* copper no doubt agree with mine whenever he uses alternating currents of the same frequency, but disagree when he uses alternating currents of much lower frequency. For telephonic frequencies iron has much greater impedance than copper. Oliver Heaviside does not for an instant deny this, but he says that there may be circumstances in which this extra inertia is an *advantage*, in that it helps to preserve the character, or quality, or shape, of electric waves, although it admittedly transmits more slowly and weakens them. If Mr. Preece and observers in America find in practice that copper wire is better for telephony, as they apparently do, then that means that the character of the vibrations is *sufficiently* preserved in copper, and the comparative absence of retardation is all to the good. It may, however, still happen that in submarine cables iron will show an advantage.[1]

But the question for ordinary Leyden jar frequencies is much simpler. For them the impedance of iron and copper of the same diameter is practically the same, unless the wire is very long or very thin. The *resistance* of the iron is much greater than that of the copper. All this can be expressed, and has been expressed quantitatively, with, in my opinion, complete certainty, and

[1] This is certainly false, and was never suggested by Mr. Heaviside. He wished for inertia for the reason stated, but not at the expense of the violent throttling resistance of iron. The thing he did propose was conduction through copper *surrounded* by iron. I am glad of this opportunity of correcting my remarks. [1891.]

beyond the reach of any but revolutionary doubt to which all scientific doctrines are liable.

And on the practical side one may say this: The circumstances of a telephone wire and of a lightning rod are not only different, they are, in some respects, opposite. The object of a telephone wire is to convey electric waves, unaltered and unweakened, to a distance. One object of a lightning conductor is to wipe them out and dissipate their energy as soon as possible. The very properties which are detrimental in one case may be desirable in the other. It is no doubt a quantitative question how far it is wise to wipe out energy in the lightning conductor itself, and Major Cardew thinks it is unwise to do so at all. Possibly; but at present I hold that so long as total impedance is not appreciably increased, and so long as the margin of melting is not too closely approached, so long it is desirable to dissipate energy wherever you can, and to check the violence of the oscillations as rapidly as possible; and hence I hold that a moderate amount of true *resistance* is no defect in a lightning conductor.

Everyone must admire the beautiful method by which Professor Hughes tests his wires for circular or cylindrical magnetization.

In Mr. Symons' objection to laboratory experiments being regarded as at all analogous to lightning, and still more clearly in Captain Cardew's solemn protest in favour of the dignity of a thunderstorm and the absence of dignity from experiments conducted with tinplate, I seem to hear echoes of some fine old crusted objections which were current in the time of Franklin, and which were, perhaps, somewhat more in harmony with that time than with the present. Now that the subject has been mooted, however, I may be permitted to assert my conviction that the intrinsic dignity and solemnity of nature is as present in a spark one inch, as in a spark one mile, long; that, looked at with insight, a drop of ink hanging from a funnel[1] may be as inspiring an object of contemplation as a cataract; and that to explicitly claim special dignity for the one is implicitly to reject it from the other. True, one's subjective feelings of awe are not aroused in the one case as in the other, but that has to do with the relative size of the human body; and so far as an observer is overwhelmed or liable to have his nerves shattered out of existence by the phenomenon he is witnessing, just so far he is not in a perfectly collected and scientific frame of mind. Moreover, experiment under modifiable circumstances has enormous advantages over mere observation, especially observation which is only occasionally possible. Hence experiments in a laboratory, and a thorough understanding of what occurs on a small scale, are a very good introduction to the enlightened study of atmospheric electricity, though they are by no means to be regarded as a substitute for that direct study. So let me here emphatically admit and insist, in full agreement with what I suppose was the intention of these speakers, and with the more direct assertion of Colonel Armstrong, that experiments on actual lightning are highly desirable. Such experiments, as a sort of practical outcome of the Mann lectures, are, I hope, in course of establishment, by means of the bright idea of the editor of the "Electrician," and by the co-operation of the Eastern Telegraph Company and Sir James Anderson. At foreign stations storms are frequent, and, with suitable appliances, I trust a record of valuable observations may be forthcoming in, say, five

[1] Sir W. Thomson's "Popular Lectures and Addresses," p. 48.

or ten years—without, let us hope, the "expenditure of any observers." The large number of photographic records of lightning which are now being obtained all over the country are likewise very valuable aids to progress.

Mr. Symons is a determined and consistent advocate of large cross-section for conductors, maintaining that they are liable to be fused; and as this is a question of observation and experience of damage, I should be disposed to allow much weight to his opinion. Unfortunately, however, the two instances he adduces in support of his contention are not such as will bear serious examination;[1] since in one it is the links of a chain which are melted, and in the other the conductor is not fused, but merely "burnt by use."

Mr. Symons beautifully illustrates my remark, that whenever a building is damaged, it is always because the infallible rules of the Lightning Rod Conference had not been followed. The church at Garding was damaged because "they took the conductor down the *north* side of the steeple, not down the *wet* side, as we advised"!

Let it be understood that I have never either said or implied that a well-erected lightning conductor is other than a source of safety as far as it goes. I said, in my first Mann lecture, that the neighbourhood of a factory chimney is "a source of mild danger." And so I believe it is, even when possessing a good conductor. But most distinctly without a conductor it would be a source of danger very much other than "mild." I never contemplated such a case, nor supposed that anyone would endeavour to increase their safety by pulling down lightning conductors!

[1] See Chap. XXII.

Colonel Armstrong quotes several cases of damage,[1] wherein the earth resistance was found to be from 100 to 200 ohms. With all deference to his experience, I feel very doubtful if this amount of resistance is sufficient to account for the damage. I have admitted all along that the better the "earth" the better for everybody; but I

[1] Lieut.-Col. R. Y. Armstrong, R.E., among other things gave the following facts:

"At Slough Fort, on the Thames, about eight years ago a small tower, to which a conductor was attached, was struck and cracked by lightning. How the portion of the discharge which struck the tower got to earth was not clear. The conductor was not injured. Its earth resistance was about 200 ohms, otherwise it was in accord with the rules of the Lightning Rod Conference.

"Again, about five years ago the contents of a small civil magazine or shifting-room in the Lake District were exploded by lightning. The earth of the conductor had been dug up before I had the opportunity of having it tested. The resistance of a similar earth on an adjacent magazine was, however, found to be between 100 and 200 ohms, and the point of the conductor of the magazine where the accident occurred did not test good continuity with low volts.

"At the Chichester Cathedral the earth resistance was over 100 ohms, and part of the lightning left it for another earth of between 100 and 200 ohms in connection with a large metallic surface. In another case which I examined, where lightning went down the chimney of a dwelling-house, the earth resistance of an adjacent conductor was over 100 ohms.

"No doubt many members of the Institution will be acquainted with cases where lightning conductors have been burnt up by lightning at riveted joints. Such a case occurred to a conductor on the citadel at Malta, about seven years ago, but the lightning did no other damage in this case.

"Another instance of this sort occurred more recently, also abroad, the conductor being fused BELOW the point at which part of the current left it through a staple in the wall.

"The portion of the current which left the conductor in this case ignited some small-arm cartridges, and these cartridges were in *metal cases*, so that apparently the current, or electrical action, was not entirely on the "differential outer skin."

have also pointed out numerous other reasons for failure and damage beside a bad earth.

In his most interesting observation of the cartridges exploded in a sealed metal case, we ought to be sure that the flash did not pierce, or melt, or ignite to redness, the case. Any violence of that sort might explode things in a very commonplace and unelectrical fashion. So also the violent shock due to expansion of air, or what Sir W. Thomson at Bath called the sound-wave, may be expected to have an effect on detonators.

Mr. Spagnoletti's statements are most interesting.[1] Mr. W. Groves, of Bolsover Street, tells me he has seen

[1] Mr. Spagnoletti spoke as follows:

"I might mention a case or two bearing on this subject. Having to maintain very many thousands of miles of telegraph wires, I have a good opportunity of learning the effects of lightning upon them. The wires are struck very frequently, but are seldom fused. The instruments attached to them, on the other hand, suffer to some extent. An instance occurred in North Wales, where the wire attached to a bell was struck and the coil of the bell was split like a cross, in four, and the whole of the wooden case was lined like a wire brush, by the small pieces of wire, showing that, although the flash must have been a very strong one, the line-wire was not at all affected, which was a galvanized wire of No. 8 or 11 gauge. These line-wires are capable of taking large currents. Another case occurred at Shrewsbury, where a man was up among the wires on a pole one afternoon and was struck by lightning severely. The lightning struck him, and apparently entered his body underneath one arm and passed out of his leg. I inquired why a man of his experience should have been up among the wires during a thunderstorm, and the reply was that there was no storm at Shrewsbury, it being a mild and calm evening, but I learned that there was a heavy thunderstorm at Hereford at the time, and that the current, which was ultimately the cause of the man's death, was carried from Hereford to Shrewsbury, a distance of fifty miles, by the line-wire—No. 8 galvanized iron. On another occasion, at Reading, a No. 4 gauge galvanized iron wire was fused (and that is the only case where I have known of a No. 4 wire having been fused), and the No. 16 copper wires, covered gutta percha, attached to the alphabetical step-by-step machine worked two or three letters forward by atmospheric electricity of some kind on a wire between his place and Sir Charles Wheatstone's.

Mr. Evershed's observations on clouds go to support the conclusion that the well-known "return stroke," and such like observations, prove the conducting nature of clouds—of some clouds at any rate.

He is quite right in pointing out that all oscillatory character is liable to be wiped out of a discharge which has had to travel a great length of thin wire, and that the finding of a quiet tail of current leaking away in some obscure corner of a telegraph office is no criterion as to the vigour or character of the main flash whence it arose.

When Major Cardew says with respect to No. 30: "We all know there are millions of volts," my point is misapprehended. It is familiar that there are millions of volts between cloud and earth; it is not familiar that there may be millions of volts between the top of a well-earthed and stout copper lightning conductor and the earth. When Major Cardew says that a conductor of small

it were melted, and the gutta percha ran over the ends and sealed them.

"I should like to ask whether a flash of lightning has been proved to be oscillatory. From what Professor Hughes stated to-night it does not appear to be so. Mr. Preece gave several examples, one showing that the action of a flash upon a polarized relay was a continuous line on the printing instrument, therefore it does not look as if it were oscillatory. I have several times tried with hand magneto machines to test the currents they give with a galvanometer, but could get scarcely any motion of the needle; and the faster I turned the less was the motion. Therefore, I think, if lightning is oscillatory the polarized relay would have either caused a dotted line, or, if the vibrations were so rapid as to prevent the tongue not touching the contact pins, no mark would have been made."

impedance is desirable, everyone must agree with him; but when he goes on to state that such a conductor is obtained by following the rules of the Lightning Rod Conference (or any other rules for that matter), it is necessary to disagree with him. The impedance could not be considered in any sense "small," even if a column of pure copper, a foot in diameter, was employed. The impedance of such a column, 100 metres high, to a current of frequency one million per second, is nearly 900 ohms.

The experience which Colonel Bucknill has had in connection with the War Office conductors, and the attention he has for many years given to the protection of powder magazines, render his practical remarks very weighty. I regret they are at present so brief.[1]

[1] I am permitted to quote his draft rules for Army lightning conductors in an appendix.

His remarks at the meeting were as follows:

"Lieut. Col. J. T. Bucknill, late R.E.: This is the most interesting and suggestive paper I have ever read on the subject of lightning and lightning rods.

"Section 5 is very important, as it seems to indicate that 'the violence of the spark is lessened' by an increase of ohmic resistance in the conductor, but that the conductor gathers the stroke as effectively, so far as striking or sparking distance is concerned, as with a much lower ohmic resistance.

"That present practice gives conductors a much higher conductivity than is absolutely necessary has been held before. Thus, Mr. R. S. Brough, in a communication to the Asiatic Society of Bengal, February, 1877, gave scientific reasons in harmony with the more convincing arguments and mathematics now published by Professor Lodge; and I myself suggested in 1881, to a War Office authority, that a large telegraph wire will always carry off a stroke of lightning innocuously.

"The cloud to cloud, or condensers in series action, has been ably examined by the lecturer, but the possibility of subterraneous condensers in series acting similarly is not suggested, and this appears to me to be a more probable explanation of the phenomena

And now I come to the remarks of the President himself. There is one point—that with reference to article 56—where I wish to explain my meaning more fully.

My statement runs, "Flat ribbon has a slight advan-

noted in the Tanfield Moor Colliery than the one given on p. 237 Lightning Rod Conference, and adopted by the lecturer (see section 46).

"The coal strata separated by strata of very low conducting power and connected by the galleries, shafts, and winding-gear, and tramways of the mine, would spark from one to the other through these imperfect connections.

"And this leads to a very important matter, which I think has not received sufficient attention from Dr. Lodge, viz., the *position* of the main induced terrestrial charge. Where water and gas pipes exist, I believe that they become highly charged by induction before the flash, and that the flash follows the route of minimum impedance that exists between the charged cloud and the earth system of conductors which is inductively charged. It is therefore useless to provide "a good earth independent of the water and gas pipes," as proposed in section 59; on the contrary, it would evidently be preferable to connect the highest portions of the water and gas supply pipes to the conductors, and thus get to the induced charge by the *shortest* route.

"I am convinced that this word *shortest* is one that should never be lost sight of in lightning-rod practice. For similar reasons I would add the words, *but where they cannot be avoided they should be connected*, to section 63, as *disruptive* is far more dangerous than *conductive* discharge; and I am utterly sceptical as to a flash melting even a small gas-pipe, or igniting the gas, except by disruptive discharge. Hence, large cast-iron gas-pipes with oakum packing at the sockets are more dangerous conductors than small gas-pipes with threaded connections.

"I should like to ask the Professor how he would deal with the great mass of metal now frequently stored in magazines—(*a*) by metal powder cases, which have replaced powder barrels; (*b*) by live shell in the expense magazines.

"Would he connect them? I say No.

"With reference to section 65. There are notable exceptions—tall rods being absolutely necessary over powder mills, petroleum oil wells, etc."

tage over round rod, but not enough to override questions of convenience."[1] Now it is of course perfectly true that extent of surface diminishes impedance, that

[1] The following are Sir William Thomson's remarks:

"The PRESIDENT: I think we must now consider the discussion as closed. I am quite sure we all agree that Dr. Lodge has done exceedingly good service in having raised the question in the manner in which he has raised it, and in having brought into the discussion of the theory and practice of lightning conductors some very important scientific principles that had not been fully taken into account by those who preceded him in the subject. I think we all admit that the principle of self-induction had not been sufficiently taken into account in connection with the theory of lightning conductors and practical rules for safety in their use. I do not know whether Franklin had any consciousness whatever that there was such a question as the mutual influence of currents in neighbouring conductors, or in different parts of one conductor, in respect to the facility afforded for carrying away the electricity by the conductors. It is quite clear that Snow Harris *had* some correct views on the subject: we must not accept all his views of electricity as correct; but many of us must now feel that in some respects in which we thought him wrong—in which, forty years ago, I, among many others, thought him wrong—he was quite right. There is one thing in Dr. Lodge's summary of results (article 56) that I confess I cannot understand at all: 'the shape of the cross-section is not of much importance.' This seems to be altogether at variance with his own teaching on the subject. Snow Harris thought a great deal of surface and shape of cross-section. In speaking on the subject at the British Association at Bath, I referred to the lightning conductor set up fifty years ago on the tower of the old Glasgow University buildings, under the recommendation of Snow Harris. It was a large *tube* of copper, and I well remember being taught to consider that that had been a mistake, and that the same quantity of copper in a solid rod or a wire rope would have been cheaper and just equally effective. We then thought Snow Harris wrong, and I believe that Faraday himself did not perceive that Snow Harris was right in that matter. We now know that he *was* right, and that spreading copper over a wide area is even better than rolling it up the same breadth in the form of a tube. A sheet of copper, we now know, constitutes a conductive path for the discharge from a lightning stroke much less impeded by self-induction than the same

Snow Harris's hollow tubes were better than Faraday's solid rods, and that if only one single stout conductor is to be used, then tape is distinctly its best, as indeed it is then also its most convenient, form. But I wished to obtain

quantity of copper in a more condensed form, whether tubular or solid.

"Now, as to the 'practical questions' put forth by Dr. Lodge, I think there are some valuable suggestions. No. 72 seems to me important: 'The cheapest way of protecting an ordinary house is to run common galvanized iron telegraph wire up all the corners, along all the ridges and eaves, and over all the chimneys, taking them down to the earth in several places, and at each place burying a load of coke.' The burying of the load of coke is the heaviest part of the business, but the multiplying the mains by connecting a large number of comparatively small wires instead of one close conductor does seem to me an important practical suggestion. On the other hand, he says it is no use connecting them to water-pipes. That I cannot agree with at all. On the contrary, I would take these galvanized iron wires described by Dr. Lodge, and the more of them the better, down all the corners and wherever you can get them, and connect every one of them to a water-pipe. I would far rather do that than to a load of coke, it is more easily done; and I think that that is the best way of doing it for the protection of an ordinary dwelling-house having water supplied to it in many-branched metal pipes. An ordinary house can, I believe, be made exceedingly safe by the water-pipes.

"'Connecting a lead roof or other such expanse with a lightning conductor is not an unmixed good, for it virtually increases the dangerous proximity of the lightning conductor.' Well, I would say connect all pieces of metal to each other, and to the earth if you can, but if you cannot connect each of them to an earth, connect them to the lightning conductor, and give *it* a good earth. I think, on the whole, that the spark coming from a lightning conductor is not one of the main sources of danger, although there is no doubt that Dr. Lodge is perfectly right in saying that there is a liability that it may light gas or other combustible substance. There is no doubt whatever but that the more completely the house can be caged in the better; and for powder magazines I believe that it is perfectly true what Dr. Lodge says (and what I have said myself), that the way to make a powder magazine perfectly safe is to completely enclose small self-induction by splitting up the conductor into detached portions, making each portion fairly thin. For these small conductors also, no doubt ribbon is electrically better than wire. But will it last as long? Is iron

it in iron. Make a complete iron house of a powder magazine: line the floors with wood or soft material to prevent ignition of stray powder by persons walking on the floor; but let a powder magazine be an iron building with an iron floor and then you do not need an earth. The powder should be kept well in, far enough from iron walls, floor, and roof, that no etheric spark can ignite it. Whether on a granite rock or in a swamp it would be equally safe: the need for the earth is absolutely done away with if the magazine is completely enclosed by metal. In that case I suppose iron would be the best metal, although it would be rash to say, seeing how very difficult is the subject of the impulsive current in iron. Remembering Professor Hughes' experiments and illustrations, and the mathematical theory worked out so magnificently by Heaviside, we are not allowed to overlook the impedance due to the magnetization of the iron itself under the influence of a sudden current. I may be wrong in this, but my impression is that this very impedance would help to make the interior of an iron shell freer from electric disturbance than it would be with a mass of equal conductivity of copper, or other metal having equal conductivity.

"The subject is so tremendously interesting that I do hope this is only the beginning of it, and that we shall have a great deal more of it. Colonel Armstrong spoke of the ignition of ammunition completely encased in metal. I hope he will experiment in that direction. The metal was not soldered all round I presume."

"Lieut.-Col. Armstrong: I think it was. The ammunition is made damp proof, and therefore the case must be complete all round."

"The President: I hope that Colonel Armstrong will be able to take up the matter experimentally as a scientific question; to see, for instance, if thin steel instead of copper would make any difference. Besides that, I think that on a larger scale something should be done. We all know how Faraday made himself a cage, six feet in diameter, hung it up in mid-air in the theatre of the Royal Institution, went into it, and, as he said, lived in it and made experiments. It was a cage with tinfoil hanging all round it; it was not a complete metallic enclosing shell. Faraday had a powerful machine working ribbon easy to obtain? So long as common galvanized-iron telegraph wire is so easy to procure, it seemed a pity to insist on any other shape of cross-section, especially since a ribbon of corresponding cross-section would have

in the neighbourhood, giving all varieties of gradual working up and discharges by 'impulsive rush'; and whether it was a sudden discharge of ordinary insulated conductors, or of Leyden jars in the neighbourhood outside the cage, or electrification and discharge of the cage itself, he saw no effects on his most delicate gold leaf electroscopes in the interior. His attention was not directed to look for Hertz sparks, or probably he might have found them in the interior. Edison seems to have noticed something of the kind in what he called the etheric force. His name 'etheric' may, thirteen years ago, have seemed to many people absurd. But now we are all beginning to call these inductive phenomena 'etheric.'

"I cannot sit down without expressing in the name of the Institution our most cordial thanks to Dr. Lodge for having taken all the trouble he took to bring this subject before us, with the beautiful experiments he has shown, and for having stimulated so many minds, whether to defend or oppose his views. I am sure you must all feel grateful to Mr. Wimshurst also for the potent assistance he gave to Dr. Lodge to prove his case, and for the potent application of his splendid apparatus this evening to further illustrate and to criticise some parts of Dr. Lodge's case. The discussion has been sometimes warm, and has been carried on with a considerable degree of humour; but I am perfectly sure that we all feel exceedingly obliged, not only to Dr. Lodge, but to all who have spoken on the subject, whether they have attacked Dr. Lodge wholly, or agreed with him wholly or in part. He has pointed out some flaws in the Lightning Rod Conference Report, but I do think that this book continues practically to hold the field, by its practical rules and recommendations for the rendering of buildings and telegraphic apparatus safe against lightning. We may admit the validity of some, or perhaps even of all his criticisms of the orthodox dogma. We must admire the vigour of his attack; and in the brilliancy of his own exposition we cannot but see much that is instructive and suggestive.

"But, after all, the conclusions adopted by the Lightning Rod Conference do afford us very strong reason to feel that there is a very comfortable degree of security, if not of absolute safety, given to us by lightning conductors made according to the present and to be so thin as to be very liable to rust away. All this I had in my mind in writing section 56. I had so frequently insisted on the advantage of large surface in my theoretical papers, that I thought it permissible to throw it over in the practical portion for solely practical reasons, *i.e.*, because to insist on it to the bitter end seemed to entail trouble and expense.

But, it may be objected, why then did I say that tape had only a *slight* advantage over rod? Well, it is a matter of arithmetic to reckon how much better a given tape is than a given rod. If I make no mistake this is the result.

The self-induction of a rod of sectional radius, r, is to that of a strip of breadth, b, both being of same length, l, very nearly in the ratio

$$\frac{\log 2\,l - \log r - 1}{\log 2\,l - \log \tfrac{1}{4}\,b - 1},$$

the currents in each case being of such rapid frequency as to keep to the outer surface.

Now, unless the rods are very short, or unless the breadth of the tape is enormous—its thinness being likewise excessive, if it is to consist of the same amount of metal as the rod—this ratio is not much greater than unity; and the same will be the ratio of their impedances.

Similarly the difference between hollow tube and solid rod is not of any *great* practical moment in lightning-rod circumstances.

With far lower frequencies, such as 100 per second,

ORTHODOX rules as actually laid down in this book. I am quite sure that the authors of this book will be exceedingly glad to modify their views in any practical way whatever, when cause is shown and proof given that such modification will improve the practical result."

when frictional or dissipation resistance is the important part of total impedance, and when currents penetrate a certain depth into the substance of a conductor, it is an altogether different matter, and the advantage of tube or plate over rod is then enormous; as Sir William Thomson has so thoroughly brought home to everybody.

Suppose, as rather an extreme case, the ratio of self-inductions for tape and rod were as great as 2, then the tape would have half the impedance of the rod for currents of the same frequency. Such a case I have experimented on; but I should not like to insist even then on the use of the tape in preference to the rod, if there were serious practical objections on the score of cost, unsightliness, want of durability, etc., to be made against it.

If there are no such objections, then tape by all means, and the thinner and broader the better.[1]

It may be just borne in mind that decreasing the self-induction goes to increase the frequency, and hence that if ever the conductor forms a large portion of the entire path of discharge, the advantage of reducing its inertia is still less marked, for the impedance depends only on the square root of L in that case.

The President misunderstands me in one place, where he thinks I have said that it is no use connecting conductors to water-mains. I do not know whence this misunderstanding can have arisen; possibly from section 59, where I say, "A good and deep earth should in general be provided, independent of water and gas mains." This may not be perfectly clear, but my meaning was as follows:

Have at least one independent earth, made by a well or other suitable means, in addition to water-main

[1] See Chap. XXI.

connections. In other words, do not depend *solely* on water-main connections.

Probably this is a counsel of perfection for the case of ordinary dwelling-houses, but for an important building I think it may be wise, for these reasons. Mains are near the surface, and in some weathers the soil near them may have become dry. Also they ramify into the house and into other people's houses, and will therefore conduct any violent charge communicated to them partly into these places, where, by a branch flash to a gas-pipe, damage may be done and gas ignited.

I have shown that well-earthed mains can thus give off unexpected sparks at a fair distance, even when only a Leyden jar discharge is run into them; hence I feel sure that some cases of damage result from lightning being thus brought underground into houses.[1]

Having a good independent earth in addition to water-mains is not indeed a *security* against this source of danger, but it is a step towards it. I do not propose to avoid the mains altogether, because in so many places it is not practicable. Whether you connect to them or not, the lightning will go to them if it chooses, unless they are far away; and it is better to give it an easy path rather than let it fly through air or soil, and knock, or melt, or burn a hole in them. It may sound absurd to talk of lightning knocking a hole; but the concussion

[1] In the basement lavatory of the hall, 25, Great George Street, the porter noticed the gas and water pipes sparking loudly into each other during the course of my experiments, and the same thing is often noticed at Liverpool, even when neither gas or water mains are being used as earth. If either system is used as earth the sparks are stronger. It should be remembered in repetitions of these experiments that risk of fire growing from an unnoticed gas-leak ignited by one of these sparks is not negligible; and suitable precautions should be taken.

of air is so great as to produce all the effects of an explosion. I entirely agree with Colonel Bucknill, that damage is most usually done wherever an air-gap is jumped. I think compo pipes are mostly melted where a flash jumps to or from them than where it simply passes along them.

With reference to the load of coke, I was under the impression that it was cheap and easy. It is not novel, and there are dozens of other well-known plans, if any are handier.

Lastly, I come to the most interesting topic of all—the cartridges exploded in metal cases mentioned by Colonel Armstrong (always provided that they were not merely ignited by heat), and the President's remarks thereupon.

Experiments on the effect of screens have gone on at intervals for some time in my laboratory. We can suspend a little electrometer-like needle, charged positive at one end and negative at the other, inside a tinfoil-coated glass box, and can deflect it by moving towards it a charged ebonite rod. But in order to succeed, the lid of the box must be so put on that a Léclanché cell shall not be able to ring a bell by conduction along the box. In other words, there must be a breach of continuity, or at least a very high resistance in the circuit. So soon as a Léclanché current can pass, no practicable motion of the ebonite rod can disturb the needle in the slightest degree. But there must be some limit to this. A stronger charge moved more quickly might do something, so we have taken to firing charged bullets out of a miniature cannon towards the box; or more simply, to give the case sudden sparks. But not a wink does the needle show. That only means that the tinfoil coating is too thick. We are going on to gold leaf, or a silver film, and so gradually thinning down till an effect is obtained. An effect *must* be forthcoming with a thin enough conductor, because one can go by gradual degrees to none at all. Liquid screens can, of course, also be employed, and probably quite a decent thickness of these will be fairly transparent. I would suggest, principally by way of query, that the action will be as follows:

Let the resistance of a metal box to a current along it be R, then when a steady current (C) flows, a difference of potential ($R\ C$) will exist between its ends, whence electrostatic lines of force will radiate both inside and outside, and an electrometer needle inside will feel them. Now, instead of passing a current through the box, move an electrostatic charge, Q, with velocity, v, towards it. An electric displacement occurs which results in a momentary current, proportional to $Q\ v$, in the metal walls of the box, and to a slope of potential some specifiable fraction of $R\ Q\ v$ which the needle may feel.

When a spark strikes the box, a momentary current similarly exists in its coating.

Now, if the momentary current has no time to penetrate the entire thickness of the metal so as to flow in its innermost layers, then none of the slope of potential due to it can be felt inside the box, though outside it would be mixed up with the much greater direct action of the electrostatic charge. But if the covering is thin enough for some portion of the current to travel by its innermost layer, then an electrostatic disturbance will occur inside, which the needle, or a frog's leg, or a vacuum tube, or a microscopic spark gap, may be competent to feel. I may say, however, that frogs' legs do not appear very sensitive to this class of effects. A zinc copper contact disturbs them vastly more.

Now, if the metal be iron, the depth to which the transient currents penetrate is very much less than it is in the case of non-magnetic metals; hence a superficial layer, thick enough to make an effective screen if made of iron, might be a very imperfect screen if made of any non-magnetic metal. On the other hand, the resistance of iron is so immensely greater than that of non-magnetic metals to these transient currents, that if the layer were thin enough to permit an effect to be appreciated at all, the slope of potential to be felt might be greater than with copper, or even with tin or lead.[1]

And now with respect to the "Practical Suggestions," which I provisionally made at the end of my paper in order that they might receive the benefit of criticism; and between which and the main body of the paper I have always drawn a clear distinction. Several have been criticised, and some have been shaken. May I quickly run over the list (p. 206), indicating those which I still strongly uphold and those which I regard as doubtful?

Nos. 51, 52, and 53, I suppose Mr. Symons would say, are "reprinted from the Lightning Rod Conference." They have, certainly, a fine ancient flavour of orthodoxy about them. But he would not have me throw over everything, both bad and good! They seem to me good.

Nos. 54 and 55 I strongly uphold.

No. 56 I have indicated my reasons for provisionally maintaining.

No. 57 I regard as very important, especially its latter sentence. It is just one of the points wherein the rules of the future will differ from the rules of the past.

Nos. 59, 60, 61, 62 are very much open to discussion.

[1] See "Philosophical Magazine" for June, 1889. "Electrostatic field produced by varying magnetic induction."

Nos. 63 and 64, I think, are sound. But very likely Colonel Bucknill's addition to 63 is an improvement.

No. 65 is very doubtful. There are, as Colonel Bucknill points out, very serious exceptions to it, even if it can ever be regarded as a rule.

Nos. 66 and 67 are sound, I think.

No. 68 is a fact.

No. 69 is a counsel of perfection: intended for powder magazines, not for dwelling-houses. Sir W. Thomson said it, or something like it, at Bath. It must be remembered, however, that "gasometers" are damaged when struck, according to reports in newspapers.

Nos. 70 and 71 are very doubtful. I throw them out as suggestions which experience must settle.

No. 72 is, I think, all right, but after the words "a load of coke" one may add, *or any of the well-known earth contact arrangements.*

No. 73 has been wholesomely criticised. I think I am safe in still saying "it is not an unmixed good." But very likely the gain outweighs the loss. In fact, I have in the Mann lectures advocated the proceeding as good on the whole.

No. 74 I should be glad to be able to omit, but see no present chance of it.

Nos. 75 and 76 have been well criticised. I quite feel the force of the criticisms, and am glad to take refuge in No. 74. At the same time a righteous substitute for No. 76, if it be wrong, is very desirable. The middle part of No. 76 (a chimney with inside metal shaft not reaching to the top) is a frequent and very difficult case. It embodies the advice which at present, for want of better, I give. Boiler firemen, engine tenders, and dynamos, would be apt to be damaged, I fear, if contrary advice were followed.

No. 77 is, I think, generally true, for such things as rain-water conduits under eaves, for picture-rods, etc.; not, of course, for a miscellaneous collection of metal objects.

No. 78 is, I think, right, if not too troublesome in practice. A crown of long points leaning well over into the smoke may do as well.

No. 79 probably belongs to Mr. Symons and the Lightning Rod Conference.

Nos. 80, 81, 82, 83 are intended to apply only to desperately important places: dynamite factories, petroleum tanks, and such like. They are of course perfectly open to criticism.

No. 84 is correct.

Nos. 85, 86, 87 are hints towards more elaborate methods of testing than the out-of-date plan at present in use. I call it out of date because it is based upon the untruth of No. 57, and upon entire ignorance (very natural a few years back) of the great obstruction offered by good conductors. It is better than no testing at all, but it is extremely inadequate, in that it detects only one, and that a comparatively unimportant, kind of flaw.

Nos. 88, 89, 90, 91 have to do with lightning "protectors," and, I suppose, are orthodox and indubitable.

CHAPTER XX.

THEORY OF *B* CIRCUITS, OF "ALTERNATIVE PATH" EXPERIMENTS, AND OF SIDE-FLASH.

CONSIDER a couple of jars connected to the terminals of a machine by their inner coats and to a wire circuit by their outer coats (Fig. 1, p. 33, or Fig. 29).

They form an ordinary circuit with a capacity inserted equal to the semi-harmonic mean of the two jars separately, and an air gap of adjustable width at *A*; and the maximum difference of potential producible in it is determined by the distance of the *A* knobs. When the discharge occurs, a current flows of course equally round the whole circuit, but the peculiarity is that up to the instant of discharge the *B* portion of the circuit is at a uniform potential. If a gap exists in *B* also, as it well may, the terminals of the gap may likewise be at the same potential up to the instant when the rush occurs. The discharge will, as usual, be oscillatory unless the resistance of the whole circuit be too great; and the period of oscillation will be approximately $2\pi\surd(LS)$, where S is the capacity of the two jars in series.

Now number the coatings of the two jars as shown in the diagram (Fig. 29), and write down their electrical condition before and during the discharge spark at *A*:

	Plate 1.		Plate 2.		Plate 3.		Plate 4.	
	Charge.	Potential.	Charge.	Potential.	Charge.	Potential.	Charge.	Potential.
Before discharge	+Q	+V	-Q	-V	-Q	0	+Q	0
After ¼ period	0	0	0	0	0	-V	0	+V
After ½ period	-Q	-V	+Q	+V	+Q	0	-Q	0
After ¾ period	0	0	0	0	0	+V	0	-V
After a whole period	+Q	+V	-Q	-V	-Q	0	+Q	0

and so on, with gradual damping (the damping being omitted in the table for simplicity).

Thus, then, between the ends of the *B* wire exists at regular intervals almost the whole difference of potential which is able to jump the air gap at *A*. Strictly speaking, the difference of potential is rather less than that corresponding to the *A* gap, thus:

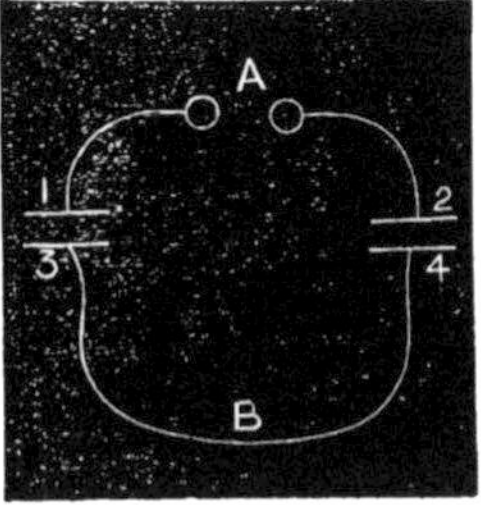

Fig. 29.

The equation to the current at any instant is accurately

$$C = \frac{V_0}{pL} e^{-\frac{R}{2L}t} \sin pt,$$

where V_0 is the initial difference of potential corresponding to the *A* spark, and where

$$p = \sqrt{\left(\frac{1}{LS} - \frac{R^2}{4L^2}\right)}.$$

Now if L_1 is the portion of the whole self-induction which corresponds to the *B* length of wire (*i.e.*, subtracting from the whole L the part belonging to the A wire), and if R_1 is the resistance of the B wire, its impedance is $\surd((pL_1)^2+R_1^2)$; and while a current, C, is flowing through it, the difference of potential between its ends is therefore $\surd((pL_1)^2+R_1^2)\,C$.

Now the current flowing through attains its maximum value one-quarter period after the A spark has commenced. *i.e.*, in a time

$$\frac{\pi}{2p};\left(\text{more exactly, in a time } \frac{1}{p}\tan^{-1}\frac{2pL}{R}\right);$$

and inserting this in C we get the maximum possible strength of current, viz.:

$$C_1=\frac{V_o}{pL}\,e^{-\frac{\pi R}{4pL}}.$$

Hence the maximum possible difference of potential between the ends of the B wire is

$$V=V_o\frac{\surd((p\,L_1)^2+R_1^2)}{p\,L}\,e^{-\frac{\pi R}{4pL}};$$

that is, a certain fraction of V_o; the fraction being

$$\frac{\text{total impedance of } B \text{ wire}}{\text{inertia impedance of whole circuit}}\times\left\{\begin{array}{l}\text{damping during}\\ \text{a quarter period.}\end{array}\right.$$

Very often a sufficient approximation to this is

$$\frac{L_1}{L}\,e^{-\frac{\pi R}{4pL}};$$

and if the wires are thick and short, or non-magnetic, and the capacity big, the damping during the first quarter of a period is often so small that merely the fraction $\frac{L_1}{L}$ will do sufficiently well.

So then, if a supplementary pair of tapping knobs be connected to the ends of the B wire, as shown in Fig. 30, and if their distance be adjusted to be $\frac{L_1}{L}$ of the A distance, a spark is liable to pass at these knobs.

This is what I call a B spark, and the spark gap affords an alternative path to the B wire, or *vice versa*.

There is no need to tap off the *whole* of the B wire. Any portion however small will serve, provided the appropriate value of L_1 is used. The length of the B spark measures the difference of potential needed to propel the current through the portion of wire which is thus tapped. Of course, if a B spark actually *occurs*, it introduces disturbance; the knobs should be set so that it just fails.

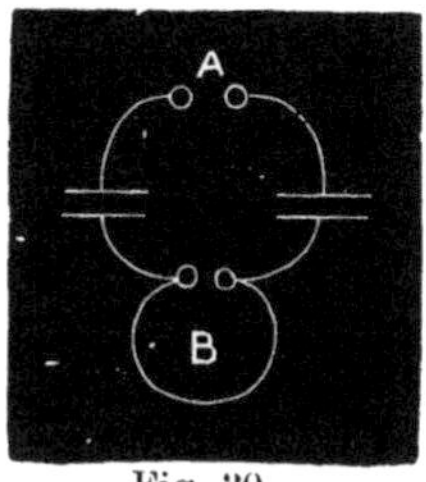

Fig. 30.

There is one thing not here explicitly stated, but which has to be taken into account in calculating the value of R, and that is the loss of energy by radiation. With small jars and circuits this loss is very great, and it increases the value of R enormously. See a paper of mine in the "Philosophical Magazine" for July, 1889, or Chapter XXIV. below. With big jars and circuits it may be safely omitted; the experimentally observed B spark will agree with calculation. But with small jars, if it be omitted, the observed B spark will be always less than the calculated.[1]

In this way a measure of the damping due to radiative dissipation of energy can immediately be made.

[1] See the "Electrician" for 21st June, 1889.

The observation of Mr. Wimshurst about the neutral point, indicates at once that this theory also gives the length of side-flash obtainable from the wire. Let any part of the B wire be put to earth, or let its natural neutral point be found, then the V calculated as above for any other point gives the length of side-flash obtainable from that point to earth.

Side-flash is in fact a special case of the alternative-path experiment. With a symmetrical wire like this, everything insulated and the jars equally charged, the neutral point is naturally the middle. But with a lightning conductor the lower end is to earth more or less completely, hence from the actual bottom of the wire no side-flash should be obtainable. One always will be obtainable, however, owing to the impossibility of making a non-resisting earth of infinite capacity. Higher up, the length of side-flash obtainable must be its length at the bottom plus the V corresponding to height of point tried. The maximum side-flash is obtainable from the top of the wire. The strength or energy of the spark depends, of course, on the capacity of the body receiving it (if insulated); being $\frac{1}{2}SV^2$, when V is calculated as already said. If it be an earthed body, then the whole discharge divides itself between the two paths, according to the laws of divided current appropriate to these conditions.

In testing a conductor, a spark should be given to the top, and the length of side spark obtainable at the bottom should be observed. All else can be calculated, except in so far as there may be defects in the visible portion of the rod.

CHAPTER XXI.

RESISTANCE AND IMPEDANCE FOR FREQUENCIES COMPARABLE TO A MILLION PER SECOND.

If $\frac{p}{2\pi}$ is the frequency of current conveyed by a wire of length l, and of ordinary resistance r, made of a substance of permeability μ; then its resistance to currents of excessively high frequency is

$$R = \sqrt{(\tfrac{1}{2} p l \mu r)},$$

hence the resistance of soft iron is immensely higher than that of any non-magnetic metal.

The self-induction under the same circumstances is

$$L_0 + \frac{R}{p},$$

where L_0 refers solely to the space surrounding the conductor.

The inertia portion of the impedance is[1]

$$pL_0 + R,$$

[1] At first sight it may seem as if I were making a mistake in having an R term in the purely inertia part of the obstruction, but it is quite right. This term R happens to represent exactly the magnetization of the substance of the wire, so far as its outer skin is magnetized.

of which the first term is far the bigger at high frequencies, even for iron, unless the wire is very thin.

The total impedance is

$$\sqrt{(p^2L_0^2+2pL_0R+2R^2)},$$

of which, again, the first term usually far eclipses the others.

Numerical Examples.—1. Let the length, l, of conducting rod be 10 metres, its diameter 1 centimetre, and let it be bent into the form of a circle (if it be straight, there will be but little difference); take $\mu=1$ for copper, or 900 for iron; specific resistance, 1,600 square centimetres per second for copper, or 7 times this for iron; and let $p=2\pi\times 10^6$ per second.

Then, whatever the substance of the conductor,

$$L_0=12{,}000 \text{ centimetres};$$

while, for the ordinary resistance,

$$r=\begin{cases}\cdot 002 \text{ ohm for copper.}\\ \cdot 014 \text{ ohm for iron.}\end{cases}$$

Hence the effective resistance is

$$R=\begin{cases}\cdot 08 \text{ ohm for copper.}\\ 6{\cdot}3 \text{ ohms for iron.}\end{cases}$$

The inertia portion of the impedance is,

$$pL_0+R=\begin{cases}75{\cdot}4+\cdot 08=75{\cdot}5 \text{ ohms for copper.}\\ 75{\cdot}4+6{\cdot}3=82{\cdot}0 \text{ ohms for iron.}\end{cases}$$

The total impedances are practically the same—viz.,

$$\begin{cases}75{\cdot}5 \text{ ohms for copper.}\\ 82{\cdot}0 \text{ ohms for iron.}\end{cases}$$

2. If, instead of taking a rod 10 metres long, we consider a length 100 metres long, of the same thickness, these quantities become:

$$L_0=162{,}000 \text{ centimetres.}$$

$$r = \begin{cases} \cdot 02 \text{ ohm for copper.} \\ \cdot 14 \text{ ohm for iron.} \end{cases}$$

$$R = \begin{cases} \cdot 8 \text{ ohm for copper.} \\ 63 \cdot 0 \text{ ohms for iron.} \end{cases}$$

Inertia part of impedance $\begin{cases} 1{,}003 \text{ ohms for copper.} \\ 1{,}066 \text{ ohms for iron.} \end{cases}$

Total impedance $\begin{cases} 1{,}003 \text{ ohms for copper.} \\ 1{,}067 \text{ ohms for iron.} \end{cases}$

3. Lastly, for a wire 100 metres long, but 1 *millimetre* in diameter, the values would be

$$L_0 = 208{,}000.$$

$$r = \begin{cases} 2 \text{ ohms for copper.} \\ 14 \text{ ohms for iron.} \end{cases}$$

$$R = \begin{cases} 8 \text{ ohms for copper.} \\ 630 \text{ ohms for iron.} \end{cases}$$

Inertia part of impedance,

$$pL_0 + R = \begin{cases} 1310 \times \quad 8 = 1{,}318 \text{ ohms for copper.} \\ 1310 \times 630 = 1{,}940 \text{ ohms for iron.} \end{cases}$$

Total impedance $\begin{cases} 1{,}318 \text{ ohms for copper.} \\ 2{,}040 \text{ ohms for iron.} \end{cases}$

All this supposes the frequency to be determined independently of the conductor considered, and to remain the same; but as the conductor increases in length it has a tendency to decrease the frequency; and that is the meaning of my sentence in section 28, to which Major Cardew objects, "and of any moderate length, such as 100 yards or less (not many miles)."

I ought to say that the here calculated values for R do not take into account at all the loss of energy by radiation. This will always go to increase R, often very perceptibly, sometimes enormously. I will go into this further in some other place.

These examples illustrate sufficiently well the comparative behaviour of iron and copper under well-marked and frequently occurring conditions. I have chosen the frequency of a million a second, because I have shown reason for believing that it is not at all unlikely to apply to the circumstances of lightning; the capacity discharged per flash, and the self-induction of its path, being neither of them very big.

But while we are about it, it is instructive and quite easy to write down the values for some considerably lower frequencies: not for slow frequencies such as alternating machines give, the theory for them is more complicated, but the simple theory will do for, say, 10,000 complete periods per second. The result will be distinctly different. No longer does inertia constitute the whole of the obstruction for iron, though it still does for copper; and for iron it constitutes the largest part.

Frequency, 10,000 per second.		Re istance, R.	Inertia part of Impedance ($pLo+R$).	Total Impedance.
		Ohms.	Ohms.	Ohms.
10-metre rod 1 cm. thick	Copper	·008	·762	·762
	Iron	·63	1·384	1·52
100-metre rod 1 cm. thick	Copper	·08	10·11	10·11
	Iron	6·3	16·96	18·1
100-metre wire 1 mm. thick	Copper	·8	13·9	13·9
	Iron	63·	76·1	98·8

The depth penetrated by the current into the substance of the wires, is definite at a given frequency—unless the wire is too thin to leave a central margin—and is independent of the diameter of the wire; at least for these high frequencies. It is easily calculated with fair approximation, thus, the sectional radius of the wire being a:

$$\frac{2\pi a da}{\pi a^2} = \frac{r}{R},$$

the ratio of the ordinary resistance, when the current is distributed uniformly through the section, to the throttled resistence, when it is cramped in the periphery. Whence da, the depth effectively penetrated by the current, or the thickness of conductor practically made use of, is—

For the million per second frequency	in copper $\frac{1}{16}$ millimetre; in iron $\frac{1}{180}$,,
For the ten thousand per second frequency	in copper $\frac{1}{16}$ centimetre; in iron $\frac{1}{180}$,,

It may be after all, therefore, that I am wrong in saying that rod is anything approaching as good as tape for conductors. It is nearly as good in respect of mere impedance, but whenever there is any chance of the wire being melted, then tape is far better. Rod ought to be apt to have its skin burnt off it, unless the central core has time to exert any cooling action by sharing the heat.[1] But it is because I doubt whether decently substantial conductors are in any real danger from heat that I have asserted the advantage of greater surface to be but small.

[1] The specimens exhibited by Mr. Preece, of copper wire incipiently fused by lightning internally, are interesting. They may have been fused by the dead-beat tail of a current; the outside cooling most rapidly. They *look* as if they had been hottest inside, and if so an explanation is needed; but they are not likely to upheave the foundations of electro-magnetism.

CHAPTER XXII.

ON THE MELTING OF CONDUCTORS.

THE list of fused conductors at the end of the Lightning Rod Conference Report, Appendix J, is very short, but short as it is it includes things not quite free from serious misleading. Over and over again it has been truly asserted that wherever there is an arc or a flash to a conductor damage is likely to be done. Terminals which have to receive the flash should always be thicker than the wire which has only to conduct it. This must be regarded as very ancient and orthodox, as well as very true. I now run through the short list of damage, and analyze it. The table is headed, "LIST OF METALS MELTED."

1. "Copper rod, ·35 inch diameter." This was an upper terminal, tapering from one-third of an inch diameter at the *base* to a point, and only 9½ inches long altogether. This terminal was "nearly all melted."

2. "Copper rope, ·31 inch diameter, at Nantes." Callaud, "Traité," page 89.

3. Rope, *said* to be ·7 inch diameter, at Carcassone. Callaud, "Traité," page 89.

These I will refer to directly. They were not fused, but broken, or eaten into, or otherwise "burnt by use."

4. "Iron rod, ·2 inch diameter." This was a few

inches melted from the *point* of an upper terminal, and some of the links of a *chain.*

5. "Brass rod, ·2 inch diameter." This was a tapering terminal, ten inches long, of the given diameter at the *base*, and it was only melted for one-fourth of its length.

The implied *statement* in the report is, therefore, that a brass rod $\frac{1}{5}$ of an inch in diameter was melted. The *fact* is, that $2\frac{1}{2}$ inches was melted off a sharp brass point! Fortunately in this case, and in case 1 also, the body of the report itself contains the material capable of overthrowing this misrepresentation.

6. "Copper rod, *perhaps* ·13 inch diameter." This was a common bell-wire, and it was legitimately destroyed, but still it protected.

That is the whole list, and it amounts to nothing more than a bell-wire, and to cases 2 and 3, the account of which I now proceed to translate from the treatise of M. Callaud. The Carcassone case is one of the two Mr. Symons quotes in his remarks (the other is case No. 4, above). It is the only one that sounds improbable, and the evidence for it seems to me weak; but I leave readers to judge. The evidence for the French cases, such as they are, is here reproduced:

Extract from "Traité des Paratonnerres," par A. Callaud, p. 89:—"The conductor of the Church Sainte-Croix at Nantes was a cable of red copper, a centimetre in diameter; it was formed of seven strands, each consisting of seven wires, the wires being one millimetre thick. I was witness of a storm and of violent flashes which traversed it, and it showed no trace of deterioration. This size can therefore be permitted, though it seems to me slight. The cable which existed before that of which I speak, also of red copper, was found broken by a flash and damaged over a part of its length; il avait 8 millimètres. I know of conducting bars, 5 millimetres, which a single storm has deteriorated and eaten into in a way that ten years of rust would hardly accomplish.

"M. Viollet-le-Duc, whose words I have had the honour of quoting, has seen at Carcassone some cables of lightning conductors burnt by use. Ils avaient 18 millimètres. 'In this town,' he tells me, 'storms are frequent—daily, in certain seasons.' In such a case the size of 18 millimetres will be then insufficient."

This last is a most vague account. The material is not specified, nor is it perfectly certain whether the 18 millimetres refer to the diameter, or whether it means that it consisted of 18 wires, each a millimetre thick. Evidently, however, M. Callaud supposes it to mean the diameter, and most likely it does. But why in the plural? And does "burnt by use" mean anything more than that some of the thin wires were burnt or fused together, or that the cable was oxidized superficially?

Considering the exceptional character of the testimony, if understood in the Lightning Rod Conference sense, it is a pity it is second-hand.

CHAPTER XXIII.

ON CONDITIONS UNDER WHICH POINTS CAN BE PREFERENTIALLY STRUCK IN CASE B.

Referring to Mr. Wimshurst's observation of the effect of the sign of top-plate (p. 216), the following is an extract from an April note-book kept by my assistant :

" Large sphere (or dome), knobs, and point, arranged between two plates so as to be equally struck by a *B* spark. The plates are connected to the outer coats of the two small or pint jars, whose inner coats are connected to the machine, between whose terminals occurs a moderate *A* spark.

1st. *With the top-plate positive.*

Distances of	Dome	2·5 centimetres.
	Large knob...	3·6 ,,
	Small knob...	3·8 ,,
	Point	3 8 ,,

2nd. *Top-plate negative.*

Distances	Dome	2·5 centimetres.
	Large knob...	3·0 ,,
	Small knob...	3·7 ,,
	Point	8·0 ,,

Lengthening the *A* spark makes the distance at which the point is struck less."

The following measurements have been made quite recently, large jars being used, but the vigour of the rush being diminished in some cases by making the *A* spark (*i.e.*, the distance between the machine terminals) quite short.

Two gallon jars similarly connected, instead of the pint jars. Objects arranged between plates to be easily and about equally struck, as before. First, with the *A* spark 1 centimetre long.

Top-plate negative.

Distances from top-plate of	Large knob ...	1·2	centimetres.
	Small knob ...	1·4	,,
	Point	2·4	,,

Top-plate positive.

Distances of	Large knob	1.5	centimetres.
	Small knob	2·2	,,
	Point	2·0	,,

Lengthen *A* spark to 5 centimetres—

Top-plate negative.

Distances of	Large knob	3·4	centimetres.
	Small knob	3·5	,,
	Point..............	3·9	,,

Top-plate positive.

Large knob.........	4·0	centimetres.
Small knob.........	4·2	,,
Point	3·9	,,

Repeat with *A* spark *about* 5 centimetres, but the *B* distances made greater.

XVIII.

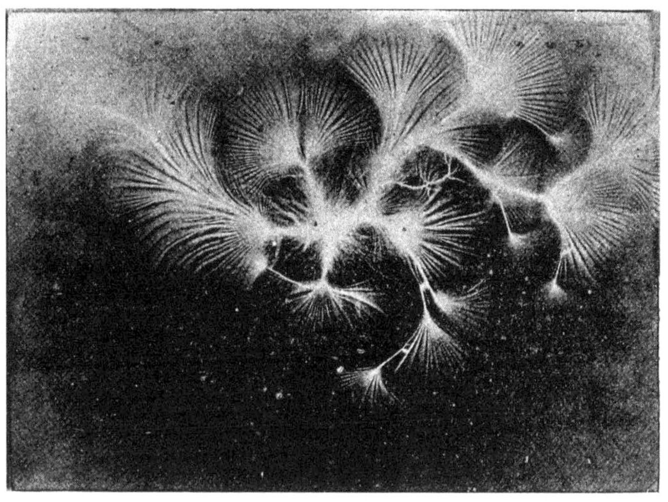

NEGATIVE SPARK TO A DRY-PLATE [J. BROWN].

POSITIVE SPARK TO A DRY-PLATE [J. BROWN].

To face p. 254.

CHAPTER XXIV.

ELECTRIC RADIATION.[1]

As illustrating the far-spreading effects of a lightning discharge, even into regions whither no conductors lead, and the disturbances that can be set up in distant insulated conductors, I have made experiments on Leyden jar discharges in which the Leyden jar coatings were represented by large insulated plates connected by a straight rod after the manner of Hertz : the whole being called a Hertz oscillator. Each plate is connected to the terminal of an ordinary large Ruhmkorff coil, so that the spark occurs between the knobs. At each discharge electricity rushes from one plate into the other, and then surges to and fro, emitting large waves into the ether, until the original energy stored up electrostatically on the plates is dissipated in radiation. There is then an interval of quiet until the next spark occurs, when the whole oscillatory disturbance begins again.

The sparks may succeed one another at the rate of, say, 100 a second, but the disturbance caused by each spark has entirely subsided, and the two or three waves excited by it have travelled a thousand miles away, before the next spark occurs. The size of the waves emitted depends on the size of the plates and on their distance

[1] Being extracts from a paper by Professor Lodge, in the "Phil. Mag.," July, 1889.

apart. Full details for different sized oscillators are given below. The waves emitted are essentially light, though so much larger than the waves of what ordinarily goes by that name. Physiologically speaking they are not light, because they do not affect the retina; physically, they have every one of the attributes of ordinary light, and all the usual optical experiments can be performed with them.

It is much easier to work with a large oscillator than a small one, because the same extraordinary suddenness in starting the oscillations is not then essential; only with large waves, mirrors and everything have to be

Fig. 31.—Large Oscillator used for violent and distant effects Scale $\frac{1}{80}$.

Plates 120 centim. square. Knobs 3·2 centim. diameter. Each rod 230 centim. long and 8 millim. diameter. Spark-gap about 1·5 centim.

$$\text{Static capacity, } \frac{S}{K} = 25 \text{ centim.}$$

$$\text{Self-induction, } \frac{L}{\mu} = 8{,}320 \text{ ,,}$$

$$\text{Characteristic factor, } \log \frac{4l}{d} = 7{\cdot}9.$$

Rate of vibration, 10 million per second.
Wave-length, 29 metres.
Dissipation-resistance, 22,500 ohms.
Initial stock of energy, about 300,000 ergs.
Power of initial radiation, 128 horse-power.
Number of vibrations before energy would be *at this rate* dissipated, about 3.

heroic to match, and our laboratory was not big enough for optical experiments on gigantic waves. Electrical experiments on such waves I have made in large numbers, obtaining them originally by means of discharging Leyden jars, but recently sometimes by a gigantic Hertz oscillator consisting of a pair of copper plates, each consisting of a couple of commercial sheets soldered together and rimmed round with wire, connected by a length of No. 0 copper wire interrupted in the middle by a couple of large knobs. The plates and connecting rod are hung from a high gallery, so that everything occupies one plane, their distance and dimensions being here shown.

The electrical surgings obtained while the Hertz oscillator is working are of just the same character as are noticed when a Leyden jar is discharging round an extensive circuit; but whereas from a closed circuit the intensity of the radiation will vary as the inverse cube of the distance as soon as the circuit subtends a small angle, the radiation from a linear or axial oscillator varies in its equatorial plane only as the inverse distance, as Hertz showed.

Hence, for obtaining distant effects the linear oscillator is vastly superior. Its emission of plane-polarized, instead of circularly-polarized, radiation is also convenient.

(I may mention that a thunder-cloud and earth joined by a lightning rod or by a disruptive path constitute a linear oscillator; and hence radiation effects and induced surgings may be expected to occur at very considerable distances from a lightning flash.)

Exciting this oscillator by a very large induction-coil, extraordinary surgings are experienced in all parts of the building, and sparks can be drawn from any hot-water-pipe or other long conductor, whether insulated or otherwise, and from most of the gas-brackets and water-taps in the building, by simply holding a penknife or other point close to them. From conductors anywhere near the source of disturbance the knuckle easily draws sparks.

Out of doors some wire fencing gave off sparks, and an iron-roofed shed experienced disturbances which were easily detected when a telephone terminal was joined to it, the other terminal being lightly earthed. [Sometimes I utilized the wire fencing as one of the plates of the oscillator, and thus got still bigger and further spreading waves.]

The waves thus excited are from 30 to 100 yards long, and optical experiments with them would be as difficult and vague as are experiments on sound-waves of corresponding length. Small oscillators can, however, easily be employed which shall give waves from a foot to a yard in length.

[Some optical and other details are here omitted.]

The particular form of receiver is a comparatively unimportant matter, but I prefer linear ones to circular or nearly closed circuits as being more sensitive at great distances, for much the same reason as has been stated for oscillators.

Exact timing of the receiver is unessential. If resonance occurred to any extent, so that the combined influences of a large number of vibrations were really accumulated, the effects might doubtless be great; but hitherto I have seen no evidence of this with linear oscillators; the reason being, I suppose, that the damping out of the vibrations is so vigorous that all oscillations after the first one or two are comparatively insignificant; and very bad adjustment, or no adjustment at all, will give you the benefit of all the resonance you can get from such rapidly decaying amplitudes. The main reason of the rapid damping is loss of energy by radiation. The "power" of the radiation while it lasts is enormous, and the stock of energy in a linear oscillator is but small.

Leyden jar discharges in closed circuits only die away after many more oscillations, and for them some approach to exact timing is essential, if a neighbouring circuit is to respond easily.

In working with small oscillators it is essential that the spark-knobs shall be in a state of high polish, else the sparks will not be sufficiently sudden to give the necessary impetus to the electrification of the conductors.

Any hesitation or delay about the spark permits the potentials of the knobs to be equalized by a gradual subsidence which is followed by no recoil, just as a tilted beer-barrel may be let down gently without stirring up the sediment by waves. The period of a natural vibration is comparable to the time taken by light to travel a small multiple of the length of the oscillator, and hence not a trace of delay is permissible in the discharge of a small conductor if any oscillations are to be excited by means of it. Thus if an electrostatic charge on a conducting sphere be disturbed in any sudden way, it can oscillate to and fro in the time taken by light to travel 1·4 times the diameter of the sphere, as calculated by Prof. J. J. Thomson; and hence it is by no means easy to disturb a charge on a sphere of moderate size except in what it is able to treat as a very leisurely manner. Even on large spheres the oscillations cannot be considered slow: thus an electrostatic charge on the whole earth would surge to and fro 17 times a second. On the sun an electric swing lasts $6\frac{1}{2}$ seconds. Such a swing as this would emit waves 19×10^5 kilometres or twelve hundred thousand miles long, which, travelling with the velocity of light, could easily disturb magnetic needles[1] and produce auroral effects, just as smaller waves produce sparks in gilt wall-paper, or as the still smaller waves of Hertz produce sparks in his little resonators, or, once more, as the waves emitted by electrostatically charged vibrating atoms excite corresponding vibrations in our retina. It may be worth while to suspend at Kew a compass-needle with a natural period of swing of 6·6 seconds, and see whether it resounds to solar impulses. Another, but almost microscopic, recording needle with a period of $\frac{1}{17}$ second might also be suspended.

The charge on the oscillator used in the present set of experiments vibrates 300 million times a second, which,

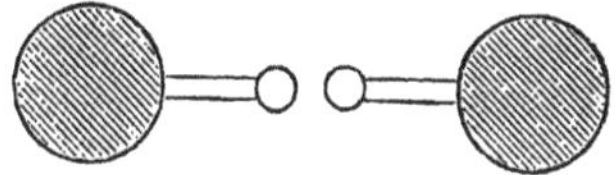

Fig. 32.—Small Oscillator used for optical experiments. Scale $\frac{1}{8}$.

Plates 8 centim. diameter.
Knobs 2 centim. diameter.
Each rod 6 centim. long and 1 centim. diameter.
Spark-gap about 8 millim.

Static capacity, $\frac{S}{K} = 1{\cdot}4$ centim.

Self-induction, $\frac{L}{\mu} = 190$ „

Characteristic factor, $\log \frac{4l}{d} = 4{\cdot}5$.

Rate of vibration, 300 million per second.
Wave-length, 1 metre.
Dissipation-resistance, 7,250 ohms.
Initial stock of energy, about 5,400 ergs.
Power of initial radiation, 128 horse-power.
Number of vibrations before energy would be *at this rate* dissipated, about $1\frac{1}{2}$.

Cf. Mr. Oliver Heaviside, "Phil. Mag.," February, 1888, p. 152.

though slower than the electric quiverings on, say, a three-inch ball, is yet quick enough to demand care and attention.

With very large oscillators, such as that described at the beginning of this paper, no such minute precautions need be taken.

My oscillator is a good deal dumpier, and its ends have more capacity, than those of corresponding wave-length used by Hertz; the reason being that I prefer to make the electrostatic capacity bear a fair relation to the electro-magnetic inertia, so as to gain a reasonable supply of initial energy for radiation. The store of energy is proportional to the capacity; the rate at which it is radiated per second is independent of it. Large terminal capacity helps to preserve a high potential longer, and so prolongs the duration of the discharge.

The wave-length of the emitted radiation is easily calculated approximately from the expression

$$\lambda = 2\pi\sqrt{\left(\frac{L}{\mu}\cdot\frac{S}{K}\right)},$$

where $\dfrac{L}{\mu} = 2l \log \dfrac{4l}{d}$; l being the length of the entire rod portion of the oscillator, and d its diameter. The measurement of l is the most unsatisfactory part. It is best to include the knobs and spark-gap as part of the whole length; the constriction at the spark will increase that part of the self-induction, but the expanse of the knobs will diminish another part. A trifle extra length should be allowed for the currents in the discs or balls at the end; but to measure l from centre to centre is rather too much allowance. From centre of one to nearest point of the other is a fair compromise.

As to S, it will be practically half[1] the static capacity of the sphere or plate at either end of the oscillator, especially if these are pretty big compared with the size of the rod. Strictly speaking they are not isolated, even when far from other conductors, because they are in presence of each other, but the correction is usually small. For instance, for two oppositely charged spheres of radius r, at a considerable distance l from centre to centre, the capacity is about

$$\frac{\frac{1}{2}lr}{l-r} = \frac{1}{2}r\left(1+\frac{r}{l}\right).$$

Hence the ordinary value of the capacity, as recorded for convenience below, is always a minimum which circumstances may increase but hardly diminish.

Values of $\frac{S}{K}$ for Isolated Bodies.

For a globe,	its radius.
For a thin circular disc,	$\frac{2}{\pi}$ times its radius.
For a thin square disc,	1·13 times inscribed circular disc, or ·36 times a side of the square.
For a thin oblong disc,	a trifle greater than a square of the same area.

Intensity of the Radiation.—Hertz has shown[2] that the amount of energy lost per half swing, by a radiator of length l charged with quantities $+Q$ and $-Q$ at its ends respectively, is

$$\frac{\pi^4 Q^2 l^2}{3K(\frac{1}{2}\lambda)^3}.$$

[1] *Half*, because the two spheres are technically "in series."

[2] Wied. "Ann.," January, 1889; or "Nature," vol. xxxix. p. 452.

He omits the dielectric constant K, because he supposes Q expressed in electrostatic units, but it is better to make expressions independent of arbitrary conventions.

So the loss of energy per second, being $\frac{v}{\frac{1}{2}\lambda}$ times the above, is

$$H = \frac{16\pi^4 (Ql)^2 v}{3K\lambda^4};$$

and this therefore is the radiation power.

For a given electric moment, Ql, the radiation intensity varies therefore as the fourth power of the frequency, *i.e.*, inversely as the fourth power of the linear dimensions of the oscillator, as Fitzgerald some time ago pointed out.

But inasmuch as different oscillators will not naturally be charged to the same electric moment, but will rather be charged to something like the same initial difference of potential, as fixed by the sparking interval between their knobs, it will be better to write $Q = SV$, and to insert the full expression for λ.

Doing so, we get for the radiation activity at any instant when the maximum difference of potentials at the terminals is V,

$$H = \frac{\pi^4 S^2 V^2 l^2 v}{3\pi^4 K S^2 L^2 v^4} = \frac{V^2}{3K\mu^2 v^3 \left(2 \log \frac{4l}{d}\right)^2}$$

$$= \frac{V^2 K v}{12 \left(\log \frac{4l}{d}\right)^2} = \frac{V^2}{12\mu v \left(\log \frac{4l}{d}\right)^2};$$

an expression roughly almost independent of the size of the oscillator. Quite independent of it if the length and thickness of its rod portion are increased proportionately.

(The factor μv may always be interpreted as 30 ohms whenever convenient.)

Thus all oscillators, large and small, started at the same potential, radiate energy at approximately the same rate; short stout ones a little the fastest.

But the initial energy of small oscillators being small, of course a much greater proportional effect is produced in them, and the radiation ceases almost instantaneously, their energy being dissipated in a very few vibrations. On the other hand, oscillators of considerable capacity keep on much longer; and with very large ends, as in Leyden jars, the loss of energy by radiation is often but a small fraction of that turned into heat by the frictional resistance of the circuit.

The expression for the radiating power may be compared either with the form $\frac{1}{2}SV^2$ or with the form $\frac{V^2}{R}$; and the loss of energy may be said to be like a static capacity of

$$\frac{30 \text{ earth quadrants}}{6\left(\log \frac{4l}{d}\right)^2}, \text{ or } \frac{5{,}556 \text{ microfarads}}{\left(\log \frac{4l}{d}\right)^2},$$

charged to the potential V, being discharged once a second; or like the heat produced per second in a wire of resistance $360 \left(\log \frac{4l}{d}\right)^2$ ohms, having a difference of potential V between its ends. The duration of the discharge must therefore be exactly comparable to the time a wire of this resistance would take to equalize the potential of the oscillator-ends initially charged to the same difference of potential.

For the small oscillator used in the optical experiments here recorded, the value of $log \frac{4l}{d}$ is approximately $4\frac{1}{2}$; hence the equivalent resistance is 7,250 ohms. And, since the initial difference of potential is, say, 26,400 volts, the power of the initial radiation is 96,000 watts or 128 horse-power.

At this rate the whole original stock of energy (5,400 ergs) would be gone in the two-hundred millionth of a second, *i.e.* in the time of $1\frac{1}{2}$ vibration; but of course the energy really decays logarithmically. The difference of potential at any instant being given by

$$\frac{d\,(\frac{1}{2}SV^2)}{dt} = \frac{V^2}{R}, \text{ that is, } V = V_0 e^{-\frac{t}{RS}};$$

where R is the above 7,250 ohms plus the resistance of the spark and of the oscillator itself to these currents. The resistance of the spark is probably but a dozen, or perhaps a hundred, ohms; that of the small oscillator is about $\sqrt{(lr)}$ ohms, where r is its ordinary resistance to steady currents expressed in ohms, and l is its length in centimetres. This, therefore, is utterly negligible; practically the whole of its energy goes in radiation. For the big oscillator the resistance is about $\sqrt{(\frac{1}{30}\,lr)}$; and so for a linear oscillator in general the dissipation resistance may be considered as simply

$$R = 360\left(log\,\frac{4l}{d}\right)^2 \text{ ohms.}$$

Nothing approaching continuous radiation can be maintained at this enormous intensity without the expenditure of great power, a hundred and thirty horse-power if my calculation is right. Under ordinary circumstances of excitation the intervals of darkness are enormous; if they could be dispensed with, some singular effects must occur. To try and make the radiation more continuous a large induction-coil excited by an alternating machine of very high frequency, or by a shrill spring-break, might be tried. But even if sparks were made to succeed one another at the rate of 1,000 per second, the effect of each would have died out long before the next one came. It would be something like plucking a wooden spring, which, after making 3 or 4 vibrations, should come to rest in about two seconds; and repeating the operation of plucking regularly once every two days.

CHAPTER XXV.

ON THE INFLUENCE OF SELF-INDUCTION ON THE RATE OF DISCHARGE OF A CONDENSER OR CLOUD.

A LETTER by Dr. Sumpner on page 761 of the "Electrician" for 4th May, 1888, establishes his statement that the time required for practically complete discharge of a condenser can be diminished in certain cases by inserting in its circuit a moderate amount of self-induction, leaving everything else the same. The point is a curious one, and I congratulate Mr. Sumpner on having noticed it. Anyone would have thought that since the time-constant of a condenser circuit is $\frac{L}{2R}$ an increase in self-induction would have retarded its discharge, and a decrease would have accelerated it. And this is what does happen so long as there is sufficient self-induction to make the discharge oscillatory. Nevertheless, if one proceeds to diminish self-induction still further, the time of discharge begins to lengthen in an unexpected manner, until it is ultimately possible exactly to double the time of a discharge by removing all trace of self-induction from the discharging circuit supposing this to be experimentally feasible.

One finds that with $L = \frac{1}{2}SR^2$ the time taken over a complete discharge is just the same as when $L = 0$; also that when $L = \frac{1}{4}SR^2$ the time of discharge is just half the preceding, and is then a minimum. This happens to be just the condition when the character of the discharge changes from oscillatory to continuing. The minimum value of the constant is $\frac{1}{2}SR$, which it has when $L = \frac{1}{4}SR^2$. Altering the self-induction either above or below this value lengthens the time of the discharge, though not in a symmetrical manner; increasing the self-induction above $\frac{1}{4}SR^2$ lengthens the time con-

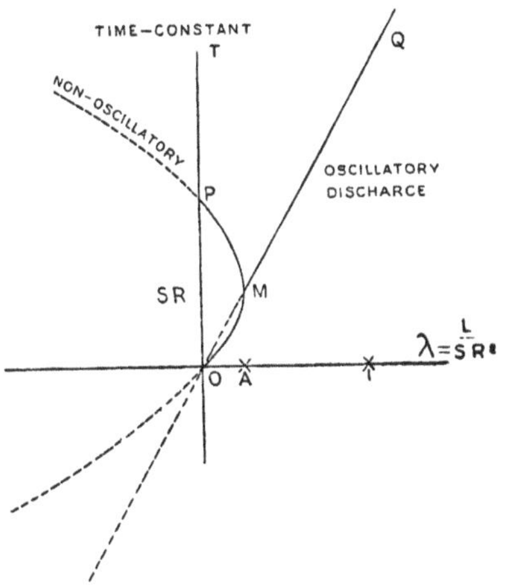

Fig. 33.

stant in a simply proportional manner without limit, but decreasing it below $\frac{1}{4}SR^2$ lengthens the time-constant in a parabolic manner towards an upper limit SR, which it attains when $L = O$.

The above curve shows the whole thing.

Plotted horizontally are successive values of the ratio L to SR^2, a ratio which I call λ, and which may be altered by varying the discharging circuit. Unit length is shown by OI. Plotted vertically are corresponding values of the time-constant T; viz., the time required for the charge of the condenser to sink from any value to $\frac{1}{e}$th of that value. In a time $2T$ the charge is reduced to $\frac{1}{e^2}$, in $3T$ to $\frac{1}{e^3}$ and so on.

Now, as e^3 is $(2{\cdot}71828)^3$, or about 20, and e^4 is about 54, it follows that in a time equal to five or six times T the condenser is, for all practical purposes, completely discharged. In a time $7T$ only one-thousandth of its original charge remains in the condenser; and in $21T$ it is reduced to less than a thousand-millionth.

Hence, if we plot the value of T we represent all we need know as regards the total time of discharge of a condenser or Leyden jar.

Now, the curve consists of two distinct portions. One portion is a straight line, MQ, sloping upwards at the gradient $\frac{1}{2}SR$ vertical to 1 horizontal, and indicating the time-constant of the oscillatory discharge for different values of λ. The dotted part of this line below M has no particular meaning.

The other portion of the curve is a parabola, with axis horizontal and vertex at M, a point characterized by the co-ordinates $\lambda = OA = \frac{1}{4}$ and $T = AM = \frac{1}{2}SR$.

The height AM represents the minimum time-constant spoken of as above, viz., its value when $L = \frac{1}{4}SR^2$. The height OP represents the value of the time-constant SR when L is nothing. The most important part of this parabolic curve is PM; the dotted part does not concern us at all, and the bit MO is of small importance, for there are two time-constants to the non-oscillatory discharge; points on MP indicate values of one of them, and points on MO indicate values of the other. The bigger of the two is not only intrinsically more decisive of the time of the discharge, but its term is multiplied by a larger coefficient than the other enjoys. One coefficient always exceeds the other by the amount SR. The smaller one vanishes for $L = O$. Both become infinite for $\lambda = \frac{1}{4}$.

Rate of Discharge from Instant to Instant.

But it now remains to consider what effect the sum of these coefficients exerts upon the *rate of discharge.* Plotting the time-constant does not tell us everything, for though it is the value of the time-constant that decides the *ultimate time of complete discharge,* yet upon its earlier stages the coefficients can exert a very appreciable effect, and accordingly conditions which cause the rate of discharge to be at first comparatively slow may ultimately make it overtake its competitor.

One can hardly make this fully clear without writing down the equations.

The equation to the current at any instant during the discharge of a condenser of capacity S is of course

$$L\frac{dC}{dt} - RC = V - V^1 = \frac{Q}{S}. \quad . \quad . \quad . \quad (1)$$

where Q is the charge remaining in condenser at any time, and $C = -\frac{dQ}{dt}$. This equation contains the whole theory of a discharging condenser, neglecting the static capacity of the discharging conductors, and ignoring the series of facts experimentally observed with certain dielectrics as "residual charge."

It should be well known (perhaps it is) that Sir Wm.

Thomson originally established this equation and worked out its prime consequences in 1853.

Integrating the equation, it becomes

$$Q = Q_0 e^{-mt}\left(\cos nt + \frac{m}{n}\sin nt\right). \quad . \quad . \quad (2)$$

which in case n is imaginary may be more conveniently written, with $n\sqrt{-1}$ for n, becoming

$$Q = Q_0\, e^{-mt}\left(\cosh nt + \frac{m}{n}\,\mathrm{shin}\, nt\right). \quad . \quad . \quad (3)$$

where

$$m = \frac{R}{2L}; \text{ and } m^2 - n^2 = \frac{1}{LS}.$$

Sometimes it is more convenient to write the equation in the form

$$\frac{Q}{Q_0} = \frac{m+n}{2n} e^{-(m-n)t} - \frac{m-n}{2n} e^{-(m+n)t}. \quad . \quad . \quad (4)$$

It is to be understood that all these four equations are mathematically identical, and express the same series of facts in other words. Sometimes one form is convenient, sometimes another. One can always write the ratio of actual to original potentials, V/V_0, instead of the ratio of the charges, Q/Q_0, if one so prefers.

Now what we are at present interested in is to see how different values of L affect the rate of decay of Q; especially do we want to compare the rate of discharge with any specified small value of L with the corresponding rate when L is completely abolished—supposing this to be experimentally possible. It is not practically possible to dispense with self-induction altogether, but for the case when a jar overflows its edge, sparking direct from one coating to the other, it is as small as practicable. Perhaps, however, in this case R^2 is smaller still, so this overflow discharge may be very oscillatory.

Let us for purposes of comparison write down the form of the discharge equation when L is zero. It is very simple, representing a simple logarithmic curve,

$$Q = Q_0 \, e^{-\frac{t}{RS}} \quad . \quad . \quad . \quad . \quad (5)$$

$\frac{1}{RS}$ may be called the logarithmic decrement; or

$\frac{1}{e^{RS}}$ may be called the common ratio of the decreasing geometrical progression formed by the charge remaining in the jar at equal short intervals of time after the discharge has begun; or RS may be called the "time-constant" of the discharge (the meaning of this term being popularly explained above); and this last plan is commonly the handiest.

CHAPTER XXVI.

THEORY AND RECORD OF THE EXPERIMENT OF THE ALTERNATIVE PATH.[1]

IN the "Philosophical Magazine" for August, 1888, I gave a general statement of the considerations which had to be attended to in a discussion of experiments on the division of a Leyden jar discharged between the two branches of a divided circuit, and pointed out that by making one of the branches an air gap a considerable simplification of the theory would result.

1. The diagram of connections may be drawn in various forms, which, though apparently different, are essentially the same. Figs. 34, 36, 38 are identical, and are the most convenient arrangements in practice. Figs. 35, 37, 39 are really the same thing, but they have the disadvantage of being liable to make the alternative path part of the charged system, so that it is unpleasant to touch; moreover, the effective capacity is more troublesome to reckon. If these arrangements had any advantage, these slight objections could be easily overcome, but I see no advantage in them. I have employed them all, however, at one time or another. Fig. 39 has the additional disadvantage of leaving the length of the *A* spark vague, so that there is no guarantee that the jar shall be charged to the same potential every time.

The electrical machine is not shown in these figures. It can be connected up in any convenient way to the *A* knobs. To prevent the machine itself from becoming an important part of the circuit and having to be taken into account, I often connect it up to the *A* knobs by imperfect conductors,

[1] Reprinted from the "Electrician," vols. xxi., xxii., xxiii.

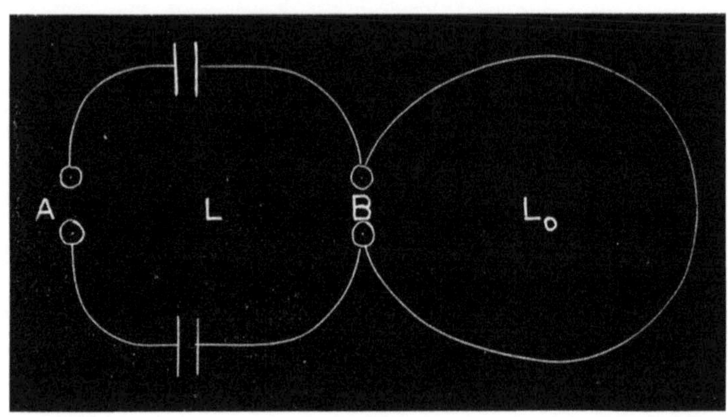

Fig. 34.

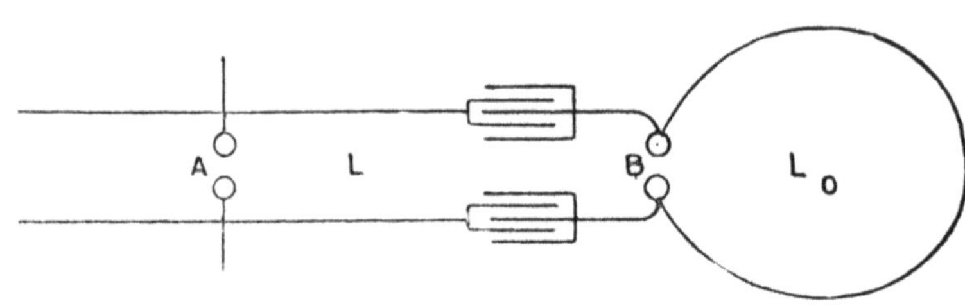

Fig. 36.

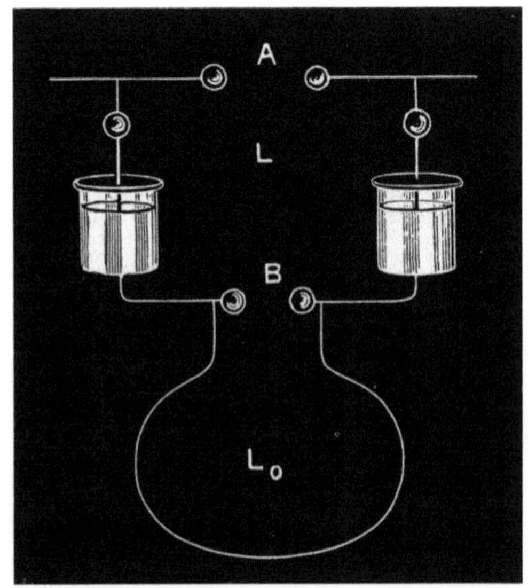

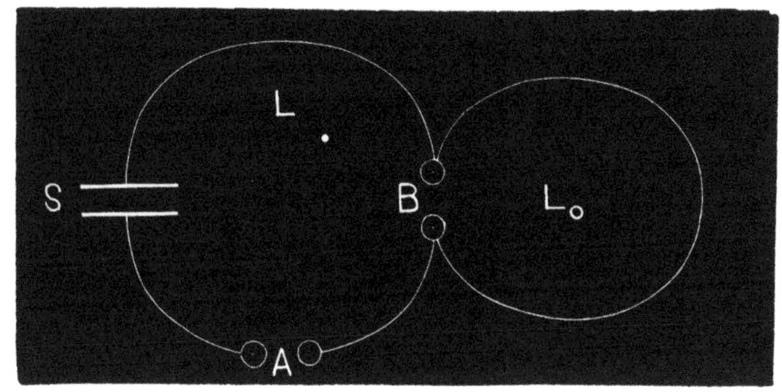

Fig. 35.

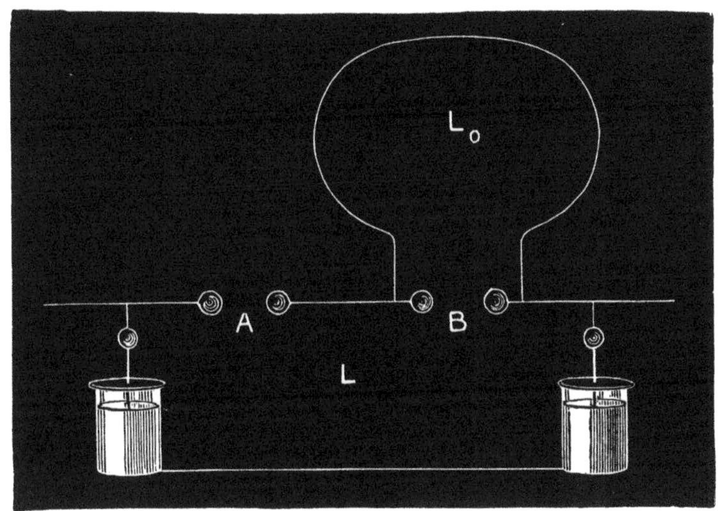

Fig. 37.

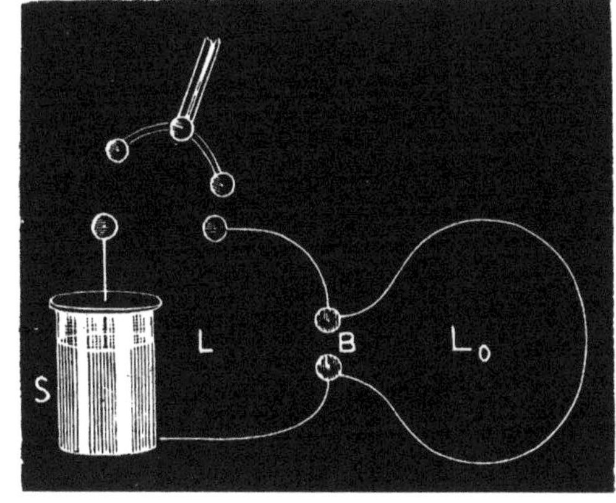

Fig. 39.

To face p. 274.

such as a couple of penholders or lead pencils. They supply the charge fast enough, but they take no appreciable portion of the discharge. Very often, however, the *A* knobs are those of the machine itself, fixed a measured distance apart.

The capacity charged is the *A* knobs and wires and the coats of the jars attached to them. The other two coats of the jars are obviously connected together, and may, for convenience of handling, be connected by some imperfect conductor with the earth. The difference of potential to which the jars are charged is given by the length of the *A* spark and the size of the *A* knobs. The circuit conveying the discharge current varies according to circumstances. It consists of two loops, the portion labelled L and the portion labelled L_0. One might call them the *A* loop and the *B* loop; but perhaps it is less ambiguous to call them the L loop and the L_0 loop, respectively; the latter being the alternative path.

Now the circuit conveying the discharge current consists of the whole of L and a variable portion of L_0. On arriving at the *B* knobs the current divides, part going on round L_0, part jumping across the air gap. It is not possible without bringing the *B* knobs into contact, nor would it be of any service, to make the discharge circuit consist of L alone; some portion is sure to patronize the L_0 route. Only when the alternative path is removed altogether does the circuit consist of L alone.

It is, however, easy to arrange that nothing jumps across the air gap; and this is the simplest case; for in that case the discharge circuit consists solely of both loops in series, $L+L_0$. But ordinarily there is a spark at *B*; and even if there is no real spark a little brush is visible in the dark between them at every discharge. Perhaps the quantity passing in this brush is negligible; if so, the simplest case is attained by separating the *B* knobs until their spark just fails. Experimentally it is easier and feels more satisfactory, to separate the *B* knobs till about half the sparks fail and half pass; but perhaps the best combination of advantages, convenience of experiment and simplicity of theory, is secured by adjusting the *B* knobs so that some 90 per cent. of the sparks there fail, the occurrence of the remaining tenth showing that they are not separated altogether beyond a possible range. The distance of the *B* knobs is read, with great ease and more than sufficient accuracy, by supporting them on a micrometer screw arrangement with its head graduated to the 400th of a millimetre.

2. Calling the self-inductions of the two loops L and L_0, and their resistances R and R_0 respectively, we have in the simplest case when the B sparks are just failing, the inductance of the discharge circuit equal to $L+L_0$, and its resistance equal to $R+R_0+$ resistance of the A spark.

The capacity of the jars (in cascade, if there are two) and charged portion of the leads being called S, the total quantity discharged each time is SV_0; and the strength of the discharge current at any instant after commencement of discharge is given by the equation

$$C = \frac{V_0}{n\,(L+L_0)} e^{-mt} \sin nt \quad . \quad . \quad . \quad . \quad . \quad (1)$$

where $$m = \frac{R+R_0+A}{2\,(L+L_0)}; \quad n^2 = \frac{1}{S\,(L+L_0)} - m^2;$$

and $$(R+R_0)^2 = \tfrac{1}{2}\,n\,(\mu\,l\,r + \mu_0\,l_0\,r_0) \quad . \quad . \quad . \quad . \quad . \quad . \quad . \quad (2)$$

$$\triangleq \frac{\mu\,l\,r+\mu_0\,l_0\,r_0}{2\sqrt{S(L+L_0)}} - \left\{\frac{\mu\,l\,r+\mu_0\,l_0\,r_0}{8\,(L+L_0)}\right\}^2,$$

where l is the length, and r the ordinary resistance, of the L loop; and l_0, r_0, the same things for the L_0 loop of the circuit.

The impedance of the alternative path is

$$P_0 = \sqrt{\{n^2\,L_0^2 + R_0^2\}},$$

and the difference of potential between the B knobs at any instant is

$$V = P_0 C \quad . \quad . \quad . \quad . \quad . \quad . \quad . \quad . \quad (3)$$

This is the quantity whose maximum value is measured by the length of the B spark. This, therefore, is the quantity which we experimentally observe, and have to analyze the theory for, in its applicability to different conditions.

First Approximation.

3. Now, the complete expression for V being rather long and unwieldy, it is better to begin by making approximations; and as a first approximation we can suppose resistance negligible as compared with inertia, at any rate for the first

few oscillations. Making this simplification, $m=0$, and the current amplitude is—

$$C_1 = V_0\sqrt{\left(\frac{S}{L+L_0}\right)}$$

$$P_0 = n\,L_0 = \frac{L_0}{\sqrt{S(L+L_0)}},$$

and
$$V = \frac{V_0 L_0}{L+L_0}, \quad . \quad . \quad . \quad . \quad . \quad . \quad (4)$$

a very simple expression, independent of the capacity of the jar, and showing that the B spark with the alternative path L_0 is shorter than the exciting spark A in the ratio

$$\frac{V}{V_0} = \frac{L_0}{L+L_0}.$$

It should be easy to verify this under circumstances when the first approximation is applicable; and if it turn out reasonably accurate, it would give an easy means of comparing the self-induction of two short and well-conducting simple circuits; for

$$\frac{L}{L_0} = \frac{V_0}{V} - 1$$

and V_0/V may be taken as the ratio of the B spark length to the A spark length; the knobs being of the same size.

One sees from these equations, that when the fixed portion of the circuit, L, is much bigger than the alternative path, L_0, the B spark is very short, and roughly proportional to L_0; but that when L is much smaller than L_0, the length of the B spark hardly depends upon the alternative path at all, and is much the same whatever the nature of that path, or even if it is absent altogether. For sensitiveness, therefore, it is better to have L rather the bigger of the two. But then remember that all this only applies to the case when the *resistances* concerned are sufficiently small to be neglected. It is not likely to apply, therefore, to long or badly-conducting circuits. For these we must proceed to a further approximation.

Second Approximation.

4. As a second approximation we had better take into account that effect of resistance which is likely to become important first.

Now, resistance has three main effects. First, it diminishes the value of n, thereby lessening the inertia part of the impedance; but since it increases C just as much, this does not affect the value of V. Secondly, it increases the damping coefficient and more rapidly brings down the current amplitude, so that if the B spark does not jump until after the accumulated momentum of several oscillations, or even if it does not jump until after the occurrence of a portion of an oscillation, it is likely, by reason of damping, to be shortened. Its third effect is an increase in the total impedance by reason of the R_0 term.

Of these three effects, the first scarcely affects V, the last tends to increase it; only the second acts so as to make V smaller with increasing resistance. Hence any diminution of V which may be actually observed to accompany an increase of R_0 must be due (so it would appear) to an effect of the damping term, involving the coefficient m.

Let us see what this term becomes in any very short time, say at the end of a complete oscillation period T—

$$m\,T = \frac{2\,\pi\,m}{n} = \frac{\pi\,(R+R_0+A)\sqrt{S}}{\sqrt{\{(L+L_0) - \frac{1}{4}\,S\,(R+R_0+A)^2\}}} \quad . \quad . \quad (5)$$

$$\triangleq \pi\,(R+R_0+A)\sqrt{\left(\frac{S}{L+L_0}\right)}.$$

For this to be of any effective magnitude, S and one of the R's must be great, but the L's have most effect when they are small; in fact, their smallness may easily imperil the oscillatory character of the discharge, and may end by making $m\,T$ infinite or worse.

The effect of high resistance may thus, in a curious way, diminish total impedance, and so tend to make the B spark actually shorter for a resisting material than it is for a better conducting one, especially in the case of a long circuit of not very high conductivity—a circuit, that is, in which resistance is an important portion of total impedance; while its damping effect, by bringing down the violence of the current oscillation, tends still further in the same direction.

Notice further, that the expression for R, besides the length of the conductor and its ordinary resistance, contains the magnetic permeability of its substance; and thus the effective resistance, or throttling part of total impedance, for an iron wire, is considerably greater than for a non-magnetic substance of the same dimensions and specific resistance, while the inertia portion is much the same. Hence it would appear possible that iron wire used as alternative path is liable in certàin cases to give a shorter B spark than such substances as copper on the one hand or lead on the other. (See further development in §§ 34 and 36.)

Third Approximation.

5. If there is the least tendency for m to become comparable in size with n—that is, for the oscillation period to be appreciably lengthened by the resistance, or damping, coefficient—there is a fourth effect which may possibly have to be taken into account, viz., this:—

The equation (2) is by no means a complete expression for the resistance of a conductor to currents of all frequencies of vibration; it is only accurately true for infinitely rapid vibrations. So soon as n falls below a certain magnitude, a much more complicated expression has to be employed, viz., as Lord Rayleigh has shown,

$$\frac{R}{r} = \text{real part of} \frac{J_0\sqrt{\left(-\frac{4nl\mu}{r}\right)}}{J_0'\sqrt{\left(-\frac{4nl\mu}{r}\right)}} \quad . \quad . \quad (6)$$

where R is the resistance of a conductor to currents alternating $\frac{n}{2\pi}$ times a second, and r is its ordinary resistance to steady currents; l is the length of the conductor, and μ its magnetic permeability; J_0 stands for Bessel's function of order zero, and J_0' for its first derived function.

For $n=0$ the value of the right-hand side is 1. For $n=\infty$ it becomes infinite, being equal to

$$\sqrt{\left(\frac{nl\mu}{2r}\right)}.$$

Now, although this last expression may be used as a good approximation to the right value of the resistance ratio for very great values of n (and it is what we have used so far), yet if there be any decrease in the frequency of oscillation below a certain ill-defined value, the more complicated expression must be used to calculate the effective resistance, and it will have a smaller value.

The whole thing could be easily calculated out and exhibited in curves or numbers if only a table of Bessel's functions were available.

I hope the recently appointed British Association Committee on Mathematical Tables will speedily see their way to calculating and issuing these much-needed tables. Meanwhile I do not lay any stress on this, what I have called, "fourth effect" of resistance; I merely note it as *possibly* acting in the direction suggested, and anyhow not to be lost sight of when the second approximation is used.

6. Lastly, there is a fifth effect, which I may as well also note, though it is probably too small to have much practical bearing. It depends on the value of the inductance of a conductor being affected by the mode of distribution of a current throughout its substance. With steady or slowly-changing currents the current is distributed uniformly all over the cross-sectional area of the conductor, and accordingly the self-induction of a single circular loop of length l, made of non-magnetic wire whose circumference is c, is

$$L = 2\,l\left(\log\frac{8\,l}{7788\,c} - 2\right). \quad . \quad . \quad . \quad . \quad (7)$$

But very-rapidly-varying currents are concentrated near the outer surface of their conductor, and hence the constant under the logarithm has to be modified for them, until in the limit it is unity; so, when n is very great,

$$L' = 2l\left(\log\frac{8\,l}{c} - 2\right). \quad . \quad . \quad . \quad . \quad . \quad (8)$$

and this is slightly smaller than the preceding value; but I doubt if the difference is great enough to be noticeable.

I now propose to record and examine in detail some experimental results, with a view to ultimately seeing how far the above attempted theory is really applicable to them.

Record of Experiments on the Alternative Path.

As just explained, I proceed to record the series of experiments I have made on the E.M.F. needed to drive a given discharge through a conductor. These experiments are mixed up in my note-book with a number of others made with slightly different objects, but it will be clearer if I pick out all those of one kind first, and attend to the others afterwards.

I begin with some early experiments—the first tried by me —when the arrangements for measuring spark length were comparatively imperfect (a graduated wedge was inserted between the knobs of the discharger), and when the essential conditions were less understood. They are not so immediately available for testing theory, but there are some instructive points about them nevertheless.

Early Experiments. February, 1888.

7. A copper wire, about No. 22 B.W.G., and an iron wire, about No. 20, were stretched round the lecture theatre (a length of some 35 yards), being suspended by silk threads from four vertical posts a good way off every wall. A large condenser, consisting of 16 pairs of 11in. square tinfoils separated by double thicknesses of window glass each about $\frac{1}{10}$in. thick, the whole soaked and embedded in a mass of paraffin and enclosed in a teak box, was specially made and used. [The capacity of this condenser was subsequently determined as ·028 microfarad.]

The condenser was charged through the long wire, and a choice was offered the discharge, so that it might go either round the wire or leap an air gap, as it chose. The arrangement is shown diagrammatically in Fig 35.

A are the ordinary terminal knobs of the machine where the spark occurs; *B* is the discharge interval alternative to the wire or to other resistance. The spark length *B* was adjusted so that it was an off-chance whether the discharge chose it or the wire. It was noticed that when the discharge chose *B* the *A* spark was strong, but when the discharge chose the wire the *A* spark was weak. The difference appeared to be only in the noise or suddenness of the spark, for when a Reiss's electro-thermometer was inserted in the circuit it indicated about the same in either case. A number of observations were made with this energy-measuring arrangement, but I appear to have no record of them.

The following are the spark-length readings:

Critical *B* Spark Length.	Choice open being
·60 inch . .	Iron wire round room.
·61 „ . .	Copper ditto.
·64 „ . .	Both wires in series (currents in same sense).
·60 „ . .	Ditto (currents in opposite senses).
·97 „ . .	A 6-inch capillary glass tube full of tap water.
·74 „ . .	The same tube full of dilute acid.
·02 „ . .	A short copper wire 2ft. long.
·075 „ . .	Very thick iron rod a yard long.
·075 „ . .	Very thick brass rod a yard long.
·11 „ . .	18 inches of thin brass wire.
·11 „ . .	The same taken once through a large ring of iron wire.
·16 „ . .	Ditto.

This was the first experiment that distinctly suggested that the ordinary magnetic properties of iron were suspended under these circumstances. The iron ring was in the position of the iron core of a transformer, and yet it exerted no appreciable effect on the impedance of the wire round it.[1] The experiment was made because the previous experiments had indicated how equal iron and copper seemed, showing none of the expected great difference between them.

The readings in these experiments are not to be supposed at all accurate, and the second place of decimals means very little. No doubt the actual *readings* are taken correctly enough, but we had as yet by no means hit on the best procedure for determining the critical *B* spark with any nicety, and in this first set of experiments the length of the *A* spark is not recorded.

8. A set of observations was now taken with the object of seeing what effect the length of the *A* spark had upon *B* under these circumstances.

[1] This led on to a series of experiments on the effect of iron cores, and on the magnetization of steel and iron by a discharge. See below, §§ 58, 59, etc., pp. 329 and 331.

Experiments on the Connection between Lengths of A and B Sparks.

(The alternative path being the iron wire round the room all the time.)

A Spark Length.	Critical *B* Spark Length.	$\frac{B}{A}$
·12 inch	·095 inch	·8
·225 ,,	·22 ,,	·98
·28 ,,	·30 ,,	1·07
·31 ,,	·35 ,,	1·13
·36 ,,	·405 ,,	1·10
·40 ,,	·48 ,,	1·20
·46 ,,	·575 ,,	1·25
·537 ,,	·648 ,,	1·21
·578 ,,	·726 ,,	1·26
·587 ,,	·726 ,,	1·24
·587 ,,	·744 ,,	1·27

The ratio of *B* to *A* thus increases with the energy of the discharge, but seems to approach a limit.

9. Now arrange a gallon Leyden jar (of capacity, as subsequently determined, 55 metres electrostatic units, or ·006 microfarad) as a lateral appendage to the *B* terminals, in the same way as the condenser of an induction coil is a lateral appendage to the contact breaker; the idea being that it would reduce the liability to spark across *B*, and help the discharge to choose the wire path.

Alternative Path.	*A* Readings.	Critical *B* Readings.	$\frac{B}{A}$
Iron wire round room	·533 inch	·498 inch	·92
,, ,,	·428 ,,	·397 ,,	·93
,, ,,	·28 ,,	·267 ,,	·95
,, ,,	·145 ,,	·123 ,,	·85
Copper wire round room	·145 ,,	·123 ,,	·85

10. The jar was now removed and the experiments continued.

Wire round room	·295 inch	·339 inch	1·15
Capillary tube containing acid (§ 11)	·240 ,,	·339 ,,	1·41
	·255 ,,		1·37
	·266 ,,		1·27
Acid tube and wire in parallel	·226 ,,	·339 ,,	1·15

Same with wire cut in middle, and ends separated a few inches [1] . .	·286 ,,	. ·339 ,,	. 1·18
Wire and acid tube in series; acid tube in middle of wire, filling up gap	·282 ,,	. ·339 ,,	. 1·2
Ditto, with gap mended; acid tube at one end of wire	·274 ,,	. ·339 ,,	. 1·2
Ditto, acid tube at other end . . .	·266 ,,	. ·339 ,,	. 1·27
Acid tube only, the wire being included in the rest of the circuit .	·266 ,,	. ·339 ,,	. 1·27

When the acid tube was the alternative path, and the *B* knobs so far apart that the discharge was obliged to choose it, the *A* spark was very weak, being reduced to a quiet spit, which could be analyzed by a slowly rotating mirror into several detached sparks.

11. The resistance of the capillary acid tube so frequently mentioned was afterwards estimated as follows:

The tube full of mercury	·437 ohm.
The tube full of zinc sulphate solution, of strength $\frac{ZnSO_4, 5H_2O}{20}$ in 100 cc. with zinc electrodes. No difference on reversing battery	177,000 ohms.
The tube full of extremely dilute acid as generally used	from $\frac{1}{4}$ to $\frac{1}{2}$ a megohm.

Experiments with a Smaller Condenser.

12. Instead of the large condenser, I now tried either a pint or a gallon Leyden jar (capacity ·0016 or ·0061 microfarad) and got much the same results. The jar often overflowed its edge and self-discharged, but a spark still occurred at *B* when this happened, though not at *A*.[2] Hence, an overflow path was provided, at *C* (Fig. 40); and sparks at *B* and *C* could be got, but not at *A*; or at *A* and *B* and not at *C*; or at *C* only.

It was found that connecting the knob of the jar to earth, the jar itself being insulated, increased the strength of the

[1] The brushes and vigorous disturbances subsequently seen at this gap led to a series of experiments on "the recoil kick." See below, p. 365.

[2] This led to a series of experiments on "overflow." See below, p. 346.

B sparks very much, and made them easier to get. Evidently the wire was acting as one coat of a condenser, the wall being the other coat.

13. Putting acid resistance into the circuit at *M* or at *N* weakens but does not stop the *B* sparks, and it has the same effect at *M* as at *N*. But inserting resistance at *Q* does not weaken the *B* spark perceptibly; neither does cutting the wire there, only of course, in order to permit the charging of the jar, the *B* gap has to be bridged by some very imperfect

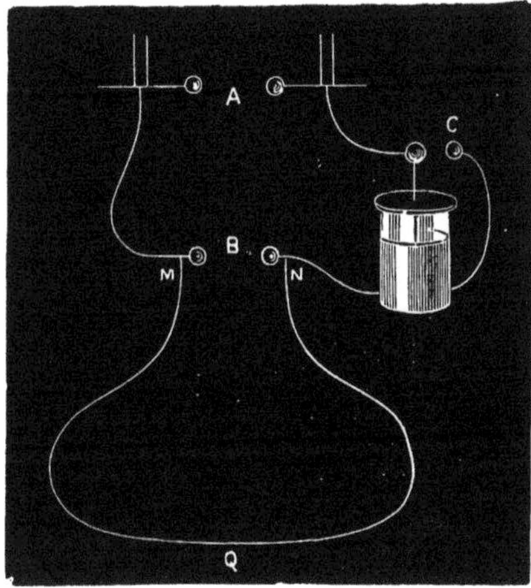

Fig. 40.

conductor, this shunt high resistance having no appreciable effect upon the *B* spark.

14. The jar was now discarded, and the wire alone used, as in Fig. 41. The sparks occurred at *B* perfectly well,[1] and were much the same whether the wire was cut at *Q* or not.

After a great number of experiments in various other directions, "overflow," "recoil kick," "liability to be struck," "side flash," "gauze house," "bye path," etc., some more experiments on alternative path were done, and to these we now proceed.

[1] This led to a series of experiments on the "surging circuit." See below, p. 363.

Further Experiments with Various Long Leads, March 1888.

15. I had now arranged around the theatre a couple of very stout wires, almost rods, supported by silk ribbon, each about 27 metres long, one of No. 1 iron, the other of No. 1 copper; besides these there were a couple of leads of ordinary thickness, each about 30 metres long, viz., No. 18 iron, No. 19 copper, and one lead of fine iron wire, No. 27, also 30 metres in length.

The resistance of these leads measured in the ordinary way,

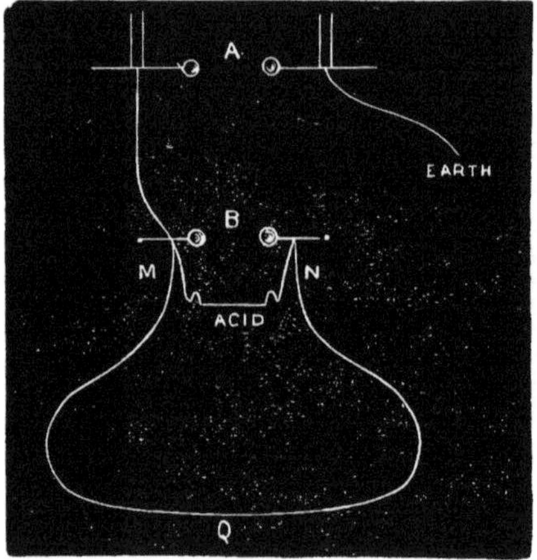

Fig. 41.

Carey Foster's method being used for the short ones, was

Resistance of No.	1 copper	·0254 ohm.
,, ,,	1 iron	·0881 ,,
,, ,,	19 copper	2·72 ,,
,, ,,	18 iron.	3·55 ,,
,, ,,	27 ,, ...	33·3 ,,

The copper is evidently of shockingly bad electrical quality, but so is most of what one buys at an ordinary metal merchant's. I must say, however, that the Birmingham-wire-gauge nomenclature is never intended to be anything more than roughly descriptive of the wire. When one wants to know its diameter really, of course one measures it.

Sectional radius of the No. 1 copper lead	·37	centimetre.
,, ,, ,, 1 iron	·355	,,
,, ,, ,, 19 copper	·0425	,,
,, ,, ,, 18 iron	·060	,,
,, ,, ,, 27 ,,	·0175	,,

Each of the two thick leads formed a rough rectangle about 840 centimetres by 515 centimetres, the entire length of each wire being 2,710 centimetres. The three thinner leads formed rough rectangles of 840 by 675 centimetres, and the length of each of these was 3,030 centimetres.

These dimensions are necessary to be known, in order that we may hereafter calculate their coefficient of self-induction (see § 19).

The static electric capacity of the leads was determined roughly by charging them with 144 small cells, discharging through a ballistic, and comparing the kick with that obtained from a small jar of known capacity treated in the same way. The results (quite rough) are—

Electrostatic capacity of either thick lead	5	metres.
,, ,, either medium lead	$3\frac{1}{2}$	,,
,, ,, thinnest lead	3	,,

16. For the alternative path experiment with these leads, arrangement Fig. 38 was adopted, two gallon jars being used; the *A* knobs being those of the machine, while the *B* knobs belong to a universal discharger. The only divergence from the figure was that the ends of the alternative path were connected direct to the outer coats of the jars instead of quite directly to the *B* knobs. [This plan is not quite so good, and was subsequently modified into agreement with Fig. 38. See § 23.] The length of the *A* spark was kept at 1 in., the *B* knobs were adjusted till their spark just began to go.

Alternative Path.	Length of B Spark.
No. 1 copper round room	1·43 inch.
No. 27 iron ,,	1·03 ,,
No. 1 iron ,,	1·08 ,,
No. 19 copper ,,	1·34 ,,
No. 18 iron ,,	1·08 ,,

At the rooms of the Society of Arts these same, or very similar, leads were rigged up, and the experiment repeated, with much the same result. But instead of going round the room the leads were only taken to the end of the room and back, with about 3 yards interval between going and return wire, and the circuits were not arranged one over the other, but all in the same horizontal plane, side by side.

It will be seen how strikingly shorter the sparks are for the iron alternative paths than for the copper. The *B* knobs were however rather small, and this tends to exaggerate the effect. Their size is given in § 21.

Experiments to Determine the Effect of Self-Induction.

17. To help study how much of the effect was really due to self-induction I procured a piece of thick stranded double-wire cable, each wire being thickly covered in gutta-percha, and then the two enclosed in a gutta-percha sheath. The ends could be joined in various ways, so as to put the wires in series with the currents going in the same direction, *i.e.*, with a maximum of self-induction, or in series with them going opposite ways, so as to have a minimum of self-induction ("anti-inductively," so to speak), or in parallel, or in any other way.

The length of the bit of cable was 770 centimetres, and it was arranged to form a rude circle. The ordinary resistances of its wires were ·0608 and ·0586 ohm respectively.

Keeping exactly the same arrangement as before, with the *A* spark still 1 in., the experiments were as follows:

Alternative Path.	Length of *B* Spark.
The two gutta-percha-covered wires in series, so as to enclose max. area	1·77
The two wires in series, currents enclosing min. area	·91
The two wires in parallel	·91
One wire alone	·91
One wire alone, the other having its ends joined, so as to form a closed secondary	·79

18. A number of experiments were now made with this bit of cable, and with a great thick No. 1 copper open spiral, by inserting large masses of iron, bundles of iron wire, iron bars and rings, into their contour, to see what difference the iron

made. So far as one could tell it never made any. To take definite readings, therefore, I got my assistant to make an induction spiral of tinfoil, consisting of a strip of tinfoil 3 inches wide and 21 feet long, wrapped round a glass tube with sufficient insulation between the turns. Also an anti-induction tinfoil zigzag with the same length of tinfoil.

The resistances of these were ·614 ohm for the spiral and 708 ohm for the zigzag.

The length of the spark *A* being ·73 inch, the following are the readings taken:

Alternative Path.	*B* Spark.
Tinfoil zigzag joined up to outer coats of jars by two leads, each a couple of feet of No 12 copper wire	·23 inch.
Tinfoil spiral ditto ditto	·64 ,,
Spiral with iron wire bundle inserted in its tube	·64 ,,
Leads only (the two No. 12 wires)	·21 ,,
No path at all	1·11 ,,
Zigzag connected direct to jar, without leads .	·06 ,,
Spiral ditto ditto .	·64 ,,
Spiral with iron bundle inside	·64 ,,

Compare also the similar experiments related on pp. 341, 353, and 354.

Calculation of Self-Induction of Long Leads.

19. The long leads round the room really enclose rude rectangles; but inasmuch as I do not know how to calculate the self-induction for a nearly square rectangle, and inasmuch as bending them out into a circle *of the same perimeter* would certainly not have effected their self-induction much, I reckon it by the formula (8) given previously, assuming that the depth to which the currents penetrate into the substance of the conductor is certainly small.

Using the dimension of these leads as given in § 15, one thus reckons in electro-magnetic units,

Self-induction of either No. 1 lead	.	390 metres.
,, ,, ,, 18 ,,	.	550 ,,
,, ,, ,, 19 ,,	.	570 ,,
,, ,, ,, 27 ,,	.	630 ,,

All this time it had not struck me that it was important to

specially know the particulars of the part of the circuit containing the jars, and they were accordingly connected up to the machine by wires in a very ordinary manner; but as far as I can estimate the self-induction of this part of the circuit now I should say it was between 40 and 60 metres usually.

Further Alternative Path Experiments, April, 1888.

20. A spark micrometer was now constructed out of a micrometer screw arrangement obtained from Lutz, of Paris, as part of a "Desains' apparatus" for measuring Newton's rings. It is depicted in various optical books. Insulated knobs (of diameter, one 1·940, the other 1·965 centimetres) were supplied to this, and the distance between them could be varied and read with great ease and accuracy.

The pitch of the screw was one millimetre, and its micrometer head was divided into grades, so that actual readings were in 400ths of a millimetre. Seldom, however, was it necessary to read so close as this.

This instrument was always used henceforth to measure the length of the B spark. Knowing this length, the corresponding potential can be determined as follows:

Connection between Differences of Potential and Length of Spark.

21. The potential needed to give a spark in air between two flat plates d centimetres apart is 110 d electrostatic units, or 33,000 d volts. The corresponding spark length between two large spheres, of radii r and r' respectively, is b, where

$$\frac{1}{d} = \frac{1}{b} + \frac{2}{3r} - \frac{1}{3r'}.$$

(See Maxwell's "Electrical Papers of Cavendish," p. 386.)

In the present case the knobs are not large enough, in comparison with the usual length of spark, for this formula to accurately apply, but the error cannot be great if we take the potential corresponding to any very short spark, expressed in fractions of a centimetre, between the above micrometer knobs (which are each about 1 centimetre radius), as

$$\frac{99000\,b}{3+b}\text{ volts.}$$

For longer sparks or smaller knobs one must make a special series of experiments to compare one with another, and link all the different sized knobs together into one set, so that all their readings can be interpreted. To this end I have recently made a series of experiments, arranging all the pairs of knobs commonly used in parallel with a couple of large spheres, and adjusting them all so that a spark was willing to choose any of the intervals at random. The large spheres were more easily manageable than a couple of flat plates would have been; while calculation from them to flat plates is easy for all the lengths of spark tried. The results are shown in the annexed Table:

Corresponding Lengths of Spark between different sized Spheres, Simultaneously compared (Expressed in centimetres).

Spark Micrometer Knobs, $\left\{\begin{matrix}1{\cdot}97\\1{\cdot}94\end{matrix}\right\}$ cms. diameter. Commonly used for *B* spark.	Knobs of Voss Machine, $\left\{\begin{matrix}2{\cdot}34\\2{\cdot}34\end{matrix}\right\}$ cms. diameter. Often used for *A* spark.	Universal Discharger, $\left\{\begin{matrix}1{\cdot}38\\1{\cdot}41\end{matrix}\right\}$ cms. diameter. Often used.	Large Spheres 12·1 cms. diameter. Used to compare with the others.	Corresponding Spark Length, calculated for flat plates from the preceding column. $\frac{18b}{18+b}$
·843	·87	·92	·86	·82
1·010	1·02	1·11	1·00	·95
1·410	1·38	1·62	1·33	1·24
1·455	1·41	1·74	1·41	1·31
1·783	1·73	2·28	1·60	1·47
2·130	1·93	3·30	1·74	1·59
2·365	2·17	3·70	1·85	1·68
2·638	2·37	4·14	1·97	1·78
2·972	2·57	5·60	2·10	1·88
4·361	3·85	7·92	2·50	2·19

Hence the length of *A* spark, 2·4 centimetres, between the knobs of the universal discharger, corresponds to a length of 1·50 centimetre between flat plates, and therefore to a potential difference of 50,000 volts.

Comparison of Wire, Gauze, and Foil.

22. In continuing the alternative path experiments the first question I wished to examine was how far the distribution of the same amount of substance over different surface would affect the result. I wished, in fact, to compare a round wire, a flat strip, and a bundle of detached wires, all as nearly alike as possible.

Three alternative paths were therefore prepared, each exactly 218 centimetres (about 7ft.) long. One was a piece of No. 12 copper wire, weighing 91·6 grammes. Another was a ribbon of copper foil 6·4 centimetres wide, and weighing 88·7 grammes. The third was a strip of fine regular copper wire gauze of the same width as the foil, consisting of 154 parallel wires, each about No. 34, with a corresponding woof. The weight of the whole gauze is 136·12, but only half this is effective, so its effective weight is 68·1 grammes. The total surface exposed by the foil is 12·8 times its length; the surface exposed by the wire is ·77 times its length; and the surface exposed by the effective wires of the gauze strip is 8·23 times its length.

The resistances, measured in an ordinary way, ought to be inversely as their weights; or approximately the same; but, as so often happens, the copper wire turned out a bad material, and these were the results:

Resistance of copper foil	. . . . .	·01125 ohm.
,, ,, gauze	. . . .	·02931 ,,
,, ,, wire	. . . .	·02446 ,,

23. The two gallon jars in series were used as before. The *B* knobs were those of the spark micrometer. The *A* knobs, however, instead of being those of the machine, as heretofore, were the smaller knobs of the same universal discharger as had hitherto been often used for the *B* spark (see § 16). It was now employed in order to obviate danger of any variation in the length of the *A* interval by reason of shaking the machine.

The alternative path was sometimes connected to the outside of the jars direct, sometimes direct to the knobs of the spark micrometer. Hitherto no special care had been taken about this because the alternative paths used were so long; but now that one began to use much shorter ones it was necessary to be careful where they were connected. Connecting them as directly as possible to the *B* knobs, as in Fig. 38, is plainly the best plan. As this had not been usually done hitherto, it was necessary to examine what sort of difference would be made by the change. The *A* knobs were adjusted to a distance of 2·4 centimetres apart, and the critical length of *B* sparks, when about half of them failed, is recorded below:

Alternative Path.	Length of *B* Spark.
	Cms.
No. 12 copper wire, enclosing maximum area, and connecting outer coats of jars direct	·743
The same, but connecting *B* knobs direct, as in Fig. 38	·552
Ribbon of copper foil connecting *B* knobs and enclosing area	·268
The same, but distorted from circular into long U form, so as to enclose much less area	·242
Ribbon of copper gauze, connecting knobs and enclosing area	·267
The same, but connecting outer coats of jars . . .	·414
No alternative path at all. (Frequent spitting at A)	2·118
No. 19 copper wire round theatre, connecting knobs .	1·649
The same, but connecting jars	1·698
No. 27 iron wire round theatre, connecting knobs . .	1·682
The same, but connecting jars	1·663
High liquid resistance, connecting knobs	2.080
No alternative path	2·080 or 2.00

But when no alternative path is used, the critical *B* length is less definite and sharp; and, instead of a single weak spark as with a high resistance, there is frequent fizzing and spitting at *A*; though a strong spark passes at *A* whenever the *B* spark occurs. For comparison of these experiments with theory, see § 40.

24. Henceforth, the alternative path was connected direct to the knobs of the spark micrometer, so as to have no uncertainty about the effect of the short leads joining the outer coats of jars to these knobs.

In the next set of experiments, the same arrangement was continued precisely, except that sometimes two pint jars were employed instead of the two gallon jars. The *A* spark was 2·4 centimetres, as before.

	Alternative Path.	Length of *B* Spark.
Two pint jars; capacity in series, 733 centimetres.	None	1·906
	Liquid resistance	1·990
	No. 19 copper round room . . .	1·380
	No. 18 iron ,, . . .	1·358
	No. 27 ,, ,, . . .	1·305
	No. 1 copper ,, with short stranded connections .	1·297
	No. 1 ,, connected direct . .	1·280
Two gallon jars; capacity in series, 2,800 centimetres.	The same	1·628
	No. 27 iron round room	1·628
	No. 18 ,, ,, ...	1·675
	No. 19 copper ,, ...	1·675

The big jars thus give a longer *B* spark than do the small ones. A possible reason for this will be suggested later on (§ 48).

It will be further noticed that in these experiments the iron is not maintaining its former position of superiority with respect to copper (see § 16). Thin iron is no longer able to beat the thickest copper, but appears about on an equality with it.

25. Since this seemed possibly due to a difference in the arrangement of the *A* part of the circuit—the part containing the jars, and which in the first of this series of papers we called the *L* loop, a part which hitherto had not been much attended to—the following modification was made:—One of the two gallon jars was connected up to the machine by means of one of the long leads round the theatre, so as to increase the self-induction of this part of the circuit enormously. The experiment was now repeated, with entirely fresh results. Everything else was the same as before.

	Alternative Path.	Length of *B* Spark.
Two gallon jars, but with one of them joined to machine through the No. 19 copper wire round the room.	No. 18 iron . . .	·782 centim.
	None	2·618 ,,
	No. 1 copper . .	·551 ,,
	No. 27 iron . . .	1·074 ,,
	No. 1 iron . . .	·585 ,,

This experiment shows how vital is attention to every part of the circuit if the result is to be thoroughly understood.

Hereafter, therefore, the circuit containing the jars and A knobs was much more carefully constructed, so that its dimensions could be exactly ascertained.

This last experiment is, moreover, the first to which we can apply our theory as set forth at the beginning of this chapter. It does not happen to apply in its simplest form, however, when resistance can be neglected. The resistance of the long, thin leads seems to be an appreciable item in their impedance, and we shall have to take it into account.

Application of Theory to the Last Recorded Experiments.

26. The L loop of the circuit in these last observations consisted of the No. 19 copper wire round the room, and some ordinary connecting wires in addition, the self-induction of which portions may be estimated as about 100 metres. It is fortunately unnecessary to know it exactly, because it is almost lost in the much greater inductance of the No. 19 lead.

Collecting the data required from §§ 15 and 19, and reproducing them here, we have, for the inductance, length, and ordinary resistance of the conductors employed, the following values:

	Self-Induction L	Length l	Resistance r
	Centimetres.	Centimetres.	Ohms.
No. 19 copper + connections	67,000	3,500	3·0
No. 1 copper	39,000	2,700	·0254
No. 1 iron	39,000	2,700	·0881
No. 18 iron	55,000	3,000	3·55
No. 27 iron	63,000	3,000	33·3

The first thing necessary to be calculated is the frequency of vibration; and inasmuch as it is improbable that in any of these cases is the resistance sufficient to make the coefficient m of § 2 comparable in size to n, unless the resistance of the A spark be supposed excessive, I shall proceed to calculate n in the first instance as simply equal to

$$\frac{1}{\sqrt{\{S(L+L_0)\}}}.$$

The electrostatic capacity charged is, the jars, which together equal 28 metres, and the No. 19 lead and connections, which equal four metres more. So the value of S is 3,200 electrostatic units.

When either of the No. 1 leads is the alternative path the total self-induction of the whole circuit is

$L + L_0 = 67000 + 39000 = 106000$ electro-magnetic units.

Before multiplying these quantities together we must reduce them both to the same units, or, what is better, express both fully and without any convention, as $S = 3200\ K$ centimetres, and $L + L_0 = 106000\ \mu$ centimetres.

So the product

$$\sqrt{\{S(L+L_0)\}} = 18400 \sqrt{(\mu K)} = \frac{18400 \text{ centimetres}}{3 \times 10^{10} \text{ centimetres per sec.}}$$

Hence $$n = 1{\cdot}63 \times 10^6 \text{ per second,}$$

and the frequency, being $\frac{n}{2\pi}$, is in this case just about one quarter of a million complete alternations per second.

A similar calculation for the other leads enables us to fill up the column headed n in the subjoined table (p. 298).

27. The next thing is to reckon the effective resistance to currents alternating at this rate. The rapidity being so great, one may use the simple formula of § 2, which makes the effective resistance the geometric mean between r and $\sqrt{(\frac{1}{2} n \mu l)}$; and hence the next column of the table is calculated, and the column of resistances headed R_0 filled up from it. Unfortunately there is great uncertainty as to the proper value of μ, the magnetic permeability which should be used. Prof. Ewing has found values of μ varying from 60 to 130 or so for hard iron or steel; and all kinds of values for very soft iron up to 2,000, and even, when vibrated, up to 20,000. Now the skin of wires is liable to be hard, and, as they tend to get magnetized by the current in concentric rings, the lower values of μ seem most probable, at least for thin wires. None of the wires were softened or annealed in any way—they were put up as bought at an ironmonger's. It will be well to try a few experiments with some specially selected iron wire. Meanwhile, as I am uncertain what value of μ to employ, I propose to try two very different ones, and see what happens in each case. First I will consider $\mu = 100$, and then $\mu = 2500$.

The quantity $\sqrt{(\frac{1}{2} n \mu l)}$ has the dimension of a resistance, and to bring it to ohms it is divided by 10^9. The value of $n L$ can be similarly expressed.

28. The rest is pretty simple. Having filled up the R_0 column as the mean of the preceding and of r, we calculate impedance by reckoning the value of

$$\sqrt{\{R_0^2 + (n L_0)^2\}}.$$

For the No. 1 copper wire this comes out 64 ohms.

One estimates the potential V_0, to which the jars were charged, by the length of the A spark, and thus calculates the current amplitude

$$C_1 = \pm \frac{V_0}{n (L + L_0)}.$$

The A spark being 2·4 centimetres long, between the knobs of the universal discharger, will be found, by means of the information contained in § 21, to correspond with a length of spark 1·5 centimetres between flat plates. This gives as the initial potential, V_0, 50,000 volts; and accordingly the maximum current comes out 284 amperes for the circuit including the No. 1 copper.

Between positive and negative values of this magnitude, the current oscillates 260,000 times a second until damped out by the exponential term and dissipated into heat. This process does not take long. A column shows the time taken for the current amplitude to decay to one-millionth of its initial value, *i.e.*, 14 times what is ordinarily called the "time constant." The next three columns show the proportion of current amplitude remaining after 1 complete vibration, after 10 vibrations, and after $\frac{1}{4}$ of a vibration.

Near the middle of the table is a column showing the estimated potential available for the B spark, allowing nothing for damping. This is got by simply multiplying the impedance of the alternative path and the current amplitude. Dividing this by 33,000, we get the calculated length of B spark between flat plates; and to obtain from this the corresponding length of spark between the actual micrometer knobs we have to make use of information contained in § 21. These lengths are tabulated in the adjoining column, and for comparison with them the experimentally observed numbers are tabulated alongside of those with which they best agree. Considering

TABLE I.

Wire round room. Alternative path.	Number of alternations per second multiplied by 2π.	Assumed magnetic permeability of wire, extreme values.		Throttled resistance of alternative path.	Throttled resistance of whole circuit.	Inertia obstruction in alternative path.	Total Impedance of alternative path.	Amplitude of current. $\frac{V_0}{n(L+L_0)}$	E.M.F. available at B knobs without damping.
	n	μ	$\frac{1}{2} n l \mu$	R_0	$R+R_0$	$n L_0$	P_0	C_1	$P_0 C_1$
	millions per second.		ohms.	ohms.	ohms.	ohms.	ohms.	amperes.	volts.
No. 1 Copper	1·63	1	2.2	·75	3·8	64	64	284	18,300
No. 1 Iron .	1·63	100	220	4·4	7·4	64	64	284	18,300
		2,500	5500	22·0	25	64	67·7	284	19,200
No. 18 Iron .	1·52	100	228	28·4	31·4	83	87·8	267	23,200
		2,500	5700	142·0	145	83	164·5	267	44,000
No. 27 Iron .	1·47	100	221	86·1	89	93	126·6	256	32,600
		2,500	5525	430·5	434	93	440·0	256	112,000

Continued on next page.]

Continued from previous page.]

Wire round room. Alternative path.	Theoretically calculated length of B spark without damping.	Experimentally observed length of B spark.	Damping coefficient $\frac{R+R_0}{2(L+L_0)}$ m	Total duration of discharging spark.[1] $\frac{14}{m}$	Proportion of current amplitude left after one complete vibration. $e^{-\frac{2\pi m}{n}}$	Proportion of charge or of current left after ten complete alternations. $e^{-\frac{20\pi m}{n}}$	Damping ratio for one quarter period. $e^{-\frac{\pi m}{2n}}$	Theoretically calculated B spark with damping.	Experimentally observed B spark.
	centimetres.	centimetres	millions per second.	thousandths of a second.				centimetres	centimetres
No. 1 Copper.	·59	·55	·018	·78	·933	·5	·983	·58	·55
	·59	·58	·035	·40	·87	·25	·966	·57	—
No. 1 Iron.	·62	—	·118	·12	·63	·01	·89	·55	·58
	·76	·78	·128	·11	·59	·005	·88	·67	—
No. 18 Iron.	1·54	—	·595	·023	·09	·00000000002	·54	·78	·78
	1·07	1·07	·340	·041	·23	·0000005	·70	·75	—
No. 27 Iron.	10·0	—	1·650	—	—	—	—	—	1·07

[1] Or time required for the charge to be permanently reduced to one-millionth of its original value.

that the calculation attempts *absolute*, and not merely relative values, the agreement is surprisingly close, for the numbers calculated by means of $\mu = 100$.

29. But then, so far, no allowance for damping has been made. We must next make the supposition that the *B* spark lags a quarter period behind the *A* spark, so that the current amplitude has had time to be slightly reduced. The theoretical *B* spark thus obtained is entered in the last calculated column of the table, and alongside of it the experimental numbers are again quoted, but this time they agree best with the larger value of μ.

As a matter of fact, this lag of a quarter period *must* occur, in order to give time for the current to rise from nothing to its maximum value. The only question is whether the *B* spark does not lag a little longer, and only occur at the second or third oscillation. If so, of course damping becomes much more important; but I see no reason making it probable that if it cannot jump at first it will be able to jump at all. I propose, however, to settle this point by arranging the two sparks *A* and *B* alongside of one another, and examining them both in a very rapidly rotating mirror.

30. There is just the possibility that the *B* spark jumps the gap before the current has had time to be fully excited in the alternative path, and if so, the circumstances regulating its critical value must be different; the hypothesis really requires a fresh theory, in which the electrostatic capacity of the alternative path lead will play an important part. While testing the theory originally suggested above, we must ignore this possibility. Besides, it can be tested experimentally by cutting the wire forming the alternative path at its middle, and seeing whether the length of the *B* spark is or is not affected. This I have done, the numbers will be given later on; suffice it to say now that the *B* spark is immensely lengthened by the change, showing that the wire acts by conduction, not by electrostatic displacement.

31. The column of the table headed "Damping coefficient" gives values of m calculated on the assumption that the resistance of the spark itself is negligible. This may be untrue, but I do not know how to estimate it. If it is more than five or six ohms it ought to be taken into account; though its only effect will be to slightly increase the damping coefficient. Towards the end of the discharge the resistance of the spark

may become very large, but by that time the B spark is over.

The last entry in this damping coefficient column, 1·65 million per second, is the value of m for the No. 27 iron wire on the improbable assumption that $\mu = 2{,}500$. This value of m is actually bigger than n, which is only 1·47 million per second, so the whole second approximation breaks down for this case, for it depends on m^2 being moderately small compared with n^2. The column headed n ought, indeed, strictly speaking, to be headed $\sqrt{(n^2+m^2)}$; and, knowing this, a more correct value of n can easily be calculated by successive approximations, except for the last line of the table. It is because of this failure to satisfy the second approximation conditions that no further entries are made in the last line of the table. It is fully worked out in § 38. The most generally interesting part of the table is included between a pair of double lines.

Application of Theory to the Last Recorded Experiments.

32. The agreement between theory and observation turned out so remarkably good in the last article that it encourages one to go into the matter a little further; for, although some of the precise numerical agreement is no doubt accidental, the numbers being easily changed one or two per cent. either way by differences in calculation, yet the discrepancies are well within errors of experiment, and their smallness seems distinctly to show that we are on the right tack, and have hold of the clue to a thorough understanding of these experiments.

But, concerning the value of μ, I was in considerable uncertainty, and chose two very different values merely to see how the results would come out. That they came out so very comparable with experiment was a matter of surprise to me, and was not expected till quite the end of the calculation, when I found it necessary to make the observations detailed in § 21 in order to interpret the spark-length indications.

In order to gain some information as to what value of μ was probable for the case under consideration, I wrote to our authority in magnetism, Prof. Ewing, asking him what value of μ was probable for an iron wire conveying a rapidly alternating current; and have just obtained from him a reply, which, being of general interest, I quote:

"Next to nothing is known, I believe, about the value of μ for rapidly-alternating magnetic force. There is certainly a true magnetic viscosity; therefore μ for rapid alternation is certainly less than the static μ. How much less I have no definite idea. I think, however, that for such frequency of alternation as occurs in a transformer, say, it is not much less. As to the static value, if your wire is annealed, it may be anything from 3,000 downwards to, say, 100, according to the particular value of $\mathfrak{H}$ at the place considered. If your wire is hard-drawn, it may be anything from about 500 downwards, according to the value of $\mathfrak{H}$. Take an example: a wire four millimetres in diameter, carrying 10 amperes, $\mathfrak{H}$ at the circumference will then be 10 C.-G.-S. This in an annealed wire would make $\mathfrak{B} = 13,000$, giving $\mu = 1,300$, about. In a hard-drawn wire μ for the same force might be about 500. At points nearer the axis, where $\mathfrak{H}$ is less, μ would (in the soft wire) be at first somewhat greater, and then less as you go nearer still to the axis. I should say that for soft wire $\mu = 1,000$ would be a fair average value, and for a hard wire $\mu = 300$ or 400."

33. It will be seen that he suggests a probable value intermediate between the extreme values I had previously made the calculation for; and that for the wires used $\mu = 400$ is a likely value for oscillations such as occur in transformers. But then these oscillations are only a hundred or two per second. Neither Prof. Ewing nor anyone else is able to say what the value of μ will be for currents alternating at the rate of a quarter of a million per second. It would naturally occur to anyone that my experiments themselves should, when calculated out, be able to furnish the appropriate value of μ; and so I hope they will, but they cannot give a value at all closely, because an enormous difference in μ makes but little difference in the length of the *B* spark. The table printed at p. 298 shows that it makes but little difference, at least for cases when the second approximation is easily applicable; and a further theoretical discussion of the point brings out the same thing.

There is some advantage about the fact, however, viz., that it is not necessary to know the value of μ at all closely in order to make calculations. It may be asked why I do not determine μ for the particular wires used by direct experiment. It would be easy to determine the μ appropriate to steady longitudinal magnetization, but this has no connection whatever with the μ appropriate to magnetization in concentric cylinders alternating hundreds of thousands of times a second. *A priori* one could not say that for such a case as this μ is any greater than unity, but I hope that the experiments will enable us to pronounce on this point one way or another.

I may say that in writing these sections I have not done the calculations beforehand. I am making them as I go along, so as to test the theory by the facts whenever we have data sufficient to make an application of theory possible. The facts themselves were obtained before I had worked out the theory. The agreement so far found is sufficiently encouraging to make us willing to write out the theory a little more fully, so as to investigate all the circumstances on which the length of the spark alternative to a given conductor in a given circuit depends.

Second Approximation Theory further Worked Out.

34. In § 4 a general notion of what is necessary for the second approximation to the alternative spark length is suggested, but it is not worked out in that first article. Now that we have a case for which it seems very clearly to apply we must proceed to work it out further.

The principal effect of resistance is the increase of impedance; the effect next in importance is the damping out of the vibrations. It being tolerably certain that the current must rise to a maximum before a critical B spark can occur, a quarter period, or $\frac{\pi}{2n}$, has to elapse between the A and the B spark. Hence the potential available for the B spark is (see § 2)

$$V = P_0 C_1 e^{-\frac{\pi m}{2n}} \quad . \quad . \quad . \quad . \quad (9)$$

Now,
$$C_1 = \frac{V_0}{n(L+L_0)} = \frac{V_0}{p+p_0}, \text{ say}, \quad . \quad . \quad . \quad (10)$$

Hence $$\frac{V}{V_0} = \frac{P_0}{p+p_0} e^{-\frac{\pi m}{2n}} \quad . \quad . \quad . \quad . \quad . \quad . \quad . \quad (11)$$

So, if we neglect damping, the ratio of the B spark to the A spark is as the total impedance of the alternative path to the inertia part of the impedance of the whole circuit.

35. The first approximation, § 3, equation (4), gave us for the ratio of the two sparks

$$\frac{p_0}{p+p_0}.$$

The second approximation, without damping, gives us

$$\frac{P_0}{p+p_0};$$

whose numerator contains not only the inertia part of the impedance of the alternative path, but the resistance part as well.

Suppose we call the inertia part of impedance "inertia" simply; and the resistance part let us call "throttling," so that (impedance)^2 = (inertia)^2 + (throttling)^2.

The throttling of the whole circuit write

$$R_0 + R = \sqrt{n}\,\{\sqrt{(\tfrac{1}{2}\, l_0\, \mu_0\, r_0)} + \sqrt{(\tfrac{1}{2}\, l\, \mu\, r)}\} = (a+b)\sqrt{n}\,. \quad (12)$$

where $a^2 = \frac{1}{2}\, l_0\, \mu_0\, r_0$, and is proportional to both the magnetic permeability and the ordinary resistance of the alternative path; while b^2 represents the same thing for the rest of the circuit.

We have also the equations,

$$P_0^{\,2} = p_0^{\,2} + R_0^{\,2} = n^2 L_0^{\,2} + n\, a^2 \quad . \quad . \quad . \quad (13)$$

$$m^2 + n^2 = \frac{1}{S\,(L+L_0)} \quad . \quad . \quad . \quad . \quad . \quad . \quad (14)$$

$$\frac{m}{n} = \frac{R+R_0}{2(p+p_0)} = \frac{a+b}{2\sqrt{n}(L+L_0)} \quad . \quad . \quad . \quad (15)$$

It is cumbersome and unnecessary to combine all these equations with absolute accuracy; but quite inappreciable error will be made by using the approximations

$$n = \frac{1}{\sqrt{\{S(L+L_0)\}}} - \frac{(a+b)^2}{8(L+L_0)^2} \quad . \quad . \quad (16)$$

$$P_0 = n\sqrt{(L_0^2 + a^2\sqrt{S(L+L_0)})} \quad . \quad . \quad (17)$$

and
$$\frac{m}{n} = \frac{(a+b)S^{\frac{1}{4}}}{2(L+L_0)^{\frac{3}{4}}} \quad . \quad . \quad . \quad . \quad . \quad . \quad . \quad . \quad (18)$$

Hence

$$\frac{V}{V_0} = \frac{L_0}{L+L_0}\sqrt{\left\{1 + \frac{a^2}{L_0^2}\sqrt{S(L+L_0)}\right\}}e^{-\frac{\pi(a+b)S^{\frac{1}{4}}}{4(L+L_0)^{\frac{3}{4}}}} \quad (19)$$

This clearly shows the relation of the first approximation (the term outside the square root) to the second. The smaller the capacity of the condenser (other things being the same), the better serves the first approximation.

36. By means of this formula, it is easy to see how the ratio of the B spark to the A spark varies with the different arbitrary variables, L, L_0, S, a, and b.

Iron or Lead versus *Copper*.

The thing which most interests us is to examine how the B spark depends upon the resistance and magnetic properties of the alternative path—that is, how it varies with a, since $a = \sqrt{(\frac{1}{2}\, l_0\, \mu_0\, r_0)}$.

So take logarithms of (19) and differentiate, abbreviating

$$\sqrt{S(L+L_0)} \text{ to } \frac{1}{n};$$

then
$$\frac{1}{V}\cdot\frac{dV}{da} = \frac{a}{n\,L_0^2 + a^2} - \frac{\pi}{4\,(L+L_0)\sqrt{n}} \quad . \quad . \quad (20)$$

This clearly shows that when the alternative path is short, $\frac{dV}{da}$ is negative, and that hence the B spark is shortened by an increase in either its resistance or magnetic susceptibility; but that, as a increases, $\frac{dV}{da}$ passes through an insensitive stage, and ultimately becomes positive. When this is the case, the length of the B spark is increased by increasing r or μ.

The intermediate insensitive condition $\frac{dV}{da} = 0$ is attained

when
$$a^2 = n\left\{\frac{4}{\pi^2}(L+L_0^2 \cdot\cdot L_0^2\right\} \quad . \quad . \quad . \quad . \quad (21)$$

that is, approximately, when $P_0 = \frac{2}{\pi}(p + p_0)$ . . . (22)

or, in other words, when the undamped potential available for the B spark is about two-thirds that available for the A spark.

Under these circumstances (which are very probable ones) a change in the material of the alternative path from copper to iron or lead makes hardly any difference.

The statement (21) for the insensitive condition may be also written, approximately,

$$R_0^2 = \frac{4}{\pi^2}\left(p_0 + p\right)^2 - p_0^2 \quad . \quad . \quad . \quad . \quad (23)$$

If the throttling of the alternative path is greater than the square root of the difference of the squares of $\frac{2}{\pi}$ times the inertia of the whole circuit and of the inertia of the alternative path, then a further increase in its r or μ will diminish its efficacy as an alternative path; but so long as the throttling is kept less than the above value, then the worse a conductor, or the more magnetic its substance, the better an alternative path will it afford.

Thus, the apparently abnormal fact that under certain circumstances thin iron serves as a better alternative path than thick copper (actually *because* it has a higher resistance and a higher value of μ, and thereby damps out the vibration sooner) is fully explained; and the circumstances under which iron ceases to be better and becomes worse than copper are completely specified.

On the Value of μ which makes Calculation and Experiment best Agree.

37. It is an important matter to know what is the true value of the magnetic permeability of iron applicable to these very rapid alternations, and we will see if the last article can help us to an answer.

First, it will be instructive to calculate out the whole circumstances of the experiment of § 25, assuming the value of μ to be 400 for iron wire under the circumstances. The following table (Table II.) shows the result, and it can be compared with the table in § 28, which gave the results for $\mu = 100$ and $\mu = 2{,}500$.

The meaning of the symbols heading the columns has been

already explained, but it may be well to recapitulate. P is the total impedance of a conductor, p is the inertia part, and R the resistance or throttling part of it; n is 2π times the number of times the current alternates in a second; and b is a constant such that $R = b\sqrt{n}$. Its value is $\sqrt{(\frac{1}{2}\,l\,\mu\,r)}$ where l is the conductor, r its ordinary resistance, and μ its magnetic permeability.

In one of the rows of the table the expressions by which these things are calculated, or references to the equation or section where such expressions can be found, are, for convenience, added.

The No. 19 conductor is given a part of the table to itself because it formed a permanent part of the circuit; the other four conductors are alternative paths, and were successively placed in series with the No. 19 conductor. V_0 being the potential acting in the whole circuit, V is the potential acting in the alternative-path portion. The last column of the table exhibits the ratio of these potentials as obtained by direct measurement of the A and B sparks, for the purpose of comparison with the calculated or theoretical values. I may mention that, by help of § 35, the values of m and n are now calculated more exactly than they were in § 31.

38. Plainly the agreement between theory and experiment is, for this value of μ, not very good. It was much better for $\mu = 100$ if the damping were neglected; but it can hardly be legitimate to neglect the damping.

The agreement was very good with $\mu = 2{,}500$ for the three cases which could be calculated out (see Table I., p. 298). The fourth case was not then worked out, but we ought now to be able to do it by help of § 35. The result is shown in Table III.

Thus the agreement between theory and observation is very close for this fourth case also, if μ be taken as 2,500. This set of experiments seems, therefore, to indicate 2,500 as a probable value for μ. It is surprisingly large.

39. To bring out the effect of different values of μ more clearly I append here a table (Table IV.) showing the calculated ratio of potentials, available for the A and the B spark respectively, with different values of μ, namely, with $\mu = 1$, $\mu = 100$, $\mu = 400$, $\mu = 2{,}500$. This table gives not only the theoretical ratio of potential but also its undamped value in each case, and thus clearly exhibits the effect of damping.

TABLE II.—*Calculated on the Assumption that μ for Iron is 400.*

Conductor.	L	b	n	R	p	P					
			Millions.	Ohms.	Ohms.	Ohms.					
No. 19 Copper .	67,000	$2{\cdot}3\times10^6$	1·63	2·94	109	109					
			1·50	2.82	101	101					
			1·32	2·65	88·5	88·5					
	(§ 19) L_0	$\sqrt{(\frac{1}{2}\mu l r)}$ a	(16) n	$(a\sqrt{n})$ R_0	$(n L_0)$ p_0	$\sqrt{(p_0^2+R_0^2)}$ P_0	$p+p_0$	(18) $\frac{m}{n}$	$e^{-\frac{\pi m}{2n}}$	(11) Calculated. $\frac{V}{V_0}$	Observed. $\frac{V}{V_0}$
			Millions per second.	Ohms.	Ohms.	Ohms.	Ohms.				
No. 1 Copper .	39,000	$\cdot18\times10^6$	1·63	·23	63·7	63·7	173	·00924	·986	·362	·346
No. 1 Iron .	39,000	6·9	1·63	8·81	63·7	64·3	173	·0336	·948	·351	·366
No. 18 Iron .	55,000	46	1·10	56·4	82·5	99·9	183·5	·160	·779	·423	·490
No. 27 Iron .	63,000	142	1·32	163	83·5	183·1	172	·456	·489	·520	·660

TABLE III.—*Case of the No. 27 Iron Wire Worked Out for μ = 2,500.*

Conductor.	a	n	R_0	p_0	P_0	$p+p_0$	$\frac{P_0}{p+p_0}$	$\frac{m}{n}$	$e^{-\frac{\pi m}{2n}}$	Calculated. $\frac{V}{V_0}$	Observed. $\frac{V}{V_0}$
No. 27 Iron	Millions. 354	Million. ·52	Ohms. 256	Ohms. 32·6	Ohms. 258·1	Ohms. 67·4	3·83	1·14	·169	·65	·66

TABLE IV.—*Summarizing the Comparison for Different Values of μ.*

Conductor.	μ=1.		μ=100.		μ=400.		μ=2,500.		Observed.
	$\frac{P}{p+p_0}$	Calculated. $\frac{V}{V_0}$	$\frac{P_0}{p+p_0}$	Calculated. $\frac{V}{V_0}$	$\frac{P_0}{p+p_0}$	Calculated. $\frac{V}{V_0}$	$\frac{P_0}{p+p_0}$	Calculated. $\frac{V}{V_0}$	$\frac{V}{V_0}$
No. 1 Copper	·367	·367	·367	·363	·367	·362	·367	·360	·346
No. 1 Iron	·367	·367	·367	·354	·371	·351	·389	·347	·366
No. 18 Iron	·447	·445	·473	·416	·544	·423	·91	·485	·490
No. 27 Iron	·483	·472	·660	·461	1·06	·520	3·83	·650	·660

There is no doubt but that the largest value of μ gives the best agreement. We must not put too much confidence in the indication of this one set of experiments, but we will remember the hint and proceed to discuss a few earlier observations, recorded before § 25.

We may, however, just notice before leaving that this last table serves to illustrate in a general way the truth of what we reckoned in § 36, viz., that if $\frac{P_0}{p+p_0}$ was less than $\frac{2}{\pi}$ (or ·636), an increase in μ diminishes the ratio $\frac{V}{V_0}$; whereas so soon as $\frac{P_0}{p+p_0}$ gets up to (or apparently near) $\frac{2}{\pi}$, an increase in μ makes $\frac{V}{V_0}$ increase. For instance, in the case of the No. 1 iron an increase in μ steadily decreases the ratio; whereas for the No. 27 iron it gets first decreased and then increased.

Application of Theory to Previously-recorded Experiments.

40. Although in the experiments recorded before § 25 we had not quite complete data for applying theory, inasmuch as the particulars of the jar part of the circuit were not specially noted, yet we can try the estimate of 50 metres which we have already made for its self-induction (§ 19), and see whether anything can be done. This is the more desirable because the experiments comparing a round wire, a ribbon of foil, and a strip of gauze come in this portion of the record (see §§ 22 and 23). It is probable that for most of these cases the first approximation alone will serve fairly well, since the resistance of the whole circuit was but small.

Wire v. *Ribbon.*

Referring back to p. 292, therefore, and using the dimension of wire and foil recorded in § 22, I calculate the self-induction of the piece of No. 12 wire by equation (7) as being 26 metres; which becomes 25 metres if the current keeps to the surface.

For the foil, if the current were distributed uniformly all through it, the self-induction would be

$$L = 4\pi r\left(\log\frac{8r}{\cdot 2656\, b} - 2\right) = 436 \times 3\cdot10 = 13\cdot5 \text{ metres};$$

where b is its breadth (6·4 centimetres), its thickness being very small. If the current keeps to the surface, L is a trifle less than this, and if it distributes itself unequally, so as to make L as small as possible, as in practice it will, there will be a further reduction; but its amount I do not quite know how to reckon. So I shall take the self-induction as 25 metres for the wire, and 13 metres for the foil.

Hence, using them as alternative paths, and applying the first approximation, with L assumed 50 metres, we should have:

$$\frac{V}{V_0} = \frac{L_0}{L + L_0} = \frac{25}{75} \text{ for the wire.}$$

and

$$\frac{13}{63} \text{ for the foil.}$$

Comparing these and similarly made calculations with the observations recorded in sections 23 and 24, as made with two-gallon jars and various alternative paths, we have the following table, comparing theory and experiment:

Alternative path.	$\frac{L_0}{L \times L^0}$.	Theoretical *B* spark between flat plates.	Calculated length of *B* spark between knobs used.	Observed length of *B* spark.
Circle of No. 12 wire .	·333	·500	·53	·55
Ribbon of copper foil .	·206	·310	·33	·27
Ribbon of copper gauze	—	—	—	·27
Long No. 19 copper wire	·920	1·38	1·62	1·65
No. 27 iron wire . . .	·925	1·39	1·65	1·68
No. 1 copper wire . .	·886	1·33	1·54	1·63
No. 27 iron wire . . .	·925	1·39	1·65	1·63
No. 18 iron wire . . .	·921	1·38	1·62	1·68
No. 19 copper wire . .	·920	1·38	1·62	1·68

This is not a complete account of the matter, because the experiments recorded in § 24 as having been made with pint jars are omitted. The first approximation by itself shows no reason why the capacity of the jars used should matter; but, experimentally, it did matter somewhat.

The calculated spark length for the copper foil is too great;

probably I have not allowed enough for the freedom of distribution of current inside a ribbon. The real self-induction would be less than what I have reckoned it. The slight discrepancy of the No. 1 copper I see no reason for. The self-induction of the strip of gauze I have not yet reckoned.

Earlier Experiments reduced by help of § 21. — Again, going further back still in the record of experiments to §§ 16 and 17, and reducing the spark lengths there given to centimetres between flat plates, we have $A = 1{\cdot}87$ centimetre, and for B the following numbers:

Alternative path.	Observed B spark reduced to flat plates.
No. 1 copper	1·66 cm.
„ 27 iron	1·51 „
„ 1 „	1·53 „
„ 19 copper	1·62 „
„ 18 iron	1·53 „
Two gutta-percha leads in concurrent series	1·80 „
„ „ „ in opposing series	1·48 „
„ „ „ in parallel	1·48 „
One lead only	1·48 „
One lead, with the other one closed	1·41 „

Applying the first approximation theory to the No. 1 copper lead, and taking L as 50 metres, we get as the theoretical B spark, for this case, 1·65 centimetre; which agrees with observation. The other leads seem to require the second approximation to bring out their slight differences.

We will now proceed with the record of further experiment.

Record of Further Experiments, August, 1888.

41. A circuit was now prepared, of which every detail could be easily determined, and whose self-induction could be varied within certain limits at pleasure. The main part of it consisted of two wires stretched parallel to one another, about four inches apart, being supported at their ends by silk threads and glass pillars. To the ends of these parallel wires, and as prolongations of them, were arranged two very small Leyden jars, such as are supplied with an ordinary Voss machine, clipped in a couple of retort stands standing on a wooden

table. From the outer coats of these jars, thus woodenly connected, two short wires led straight to the spark micrometer.

Connection with the machine was effected through wooden sticks, so that the machine and leading wires are quite out of the circuit as far as discharges and oscillations are concerned (see Fig. 42).

It was proved experimentally that none of the circuit beyond the wooden semi-interruption made the least difference. Nor did it matter whereabouts the wood made connection with the jar circuit. An *A* spark was provided by the universal discharger, which was placed under the two long wires so as nearly to bridge across the gap. This dis-

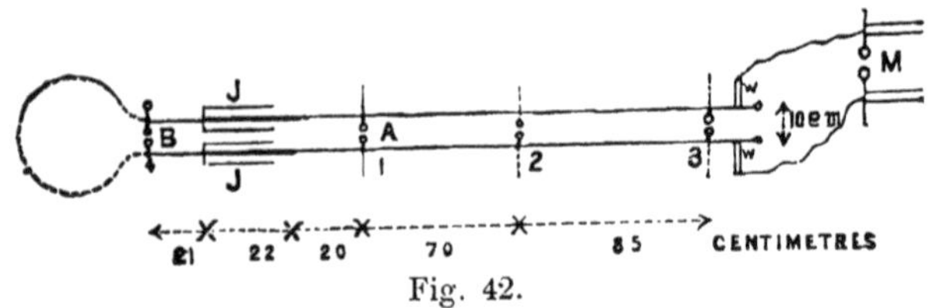

Fig. 42.

J are the two Voss jars arranged as shown in figure, with long wire continuations, two No. 17 copper wires a decimetre apart. *M* is the machine and *W* are wooden connectors. *A* is the universal discharger, which can be placed in either one of three positions, 1, 2, 3. *B* is the spark micrometer, with an alternative path completing the circuit shown by a dotted line. Linear dimensions are expressed in centimetres.

charger could be shifted along the whole length of the wires, and thus the self-induction of the jar circuit varied. Ordinarily three definite positions were employed, one near each end and one near the middle, labelled respectively 1, 2, and 3, as shown in the diagram. The knobs of the discharger and of the spark micrometer are the same as have already been described and experimented on, as recorded in § 21. The dimensions of the whole arrangement are best represented in the diagram.

It is necessary to specify the thickness of the different parts of the circuit:

Thickness of No. 17 wires	·15 centim.
Thickness of rods inside jars	·98 ,,
Thickness of *A* discharger rods . . .	·60 ,,
Thickness of *B* discharger rods . . .	·58 ,,

Capacity and Self-Induction.

The capacity of the two Voss jars in series, and of the two No. 17 wires, amounted altogether to 143 centimetres electrostatic units. This, therefore, is the value of S when the Voss jars are used.

Applying calculation to the rectangular circuit included between the A and the B knobs I make out that its self-induction in electro-magnetic measure is as follows:

For the A knobs in No. 1 position . $L_1 = 1290$ centimetres.
" " No. 2 " . $L_2 = 2677$ "
" " No. 3 " . $L_3 = 4397$ "

Alternative Paths.

42. A number of alternative paths were now prepared, all consisting of the same length of wire, namely 235 centimetres, carefully bent into a circle about 2ft. 6in. in diameter, with ends arranged so as to be easily connected direct to the rods of the spark micrometer. They were all the same *length*, but of very different thickness and material, the finer wires being supported on a hoop of cardboard.

Table A shows their thickness, their ordinarily measured

TABLE A.

Description of wire.	Thickness.	Resistance.	Self-induction L_0.
	Centimetre.	Ohm.	Centimetres.
No. 5 copper	·535	·0048	2,340
No. 2 iron	·71	·0166	2,230
No. 2½ brass	·68	·0051	2,240
No. 18 iron	·11	·254	3,100
No. 18 copper	·12	·092	3,060
No. 24 brass	·05	·89	3,470
No. 25 iron	·045	2·03	3,520
No. 23 copper	·065	·29	3,340
No. 40 copper	·01	2·60	4,230
"Longer leads" . . .	·15	1·95	100,000

resistance, and their calculated self-induction, for cases where the current keeps to their outer surface.

Sometimes much longer leads were used, consisting of a

pair of No. 17 copper wires, each 38·2 metres long, arranged along and across the room, parallel to one another, so as to enclose a distorted rectangle half a metre broad. Particulars of these are entered in the last line of the above table. The electrostatic capacity between one wire and the other was 147 centimetres.

Effect of Light on the Sparks.

43. A few rough preliminary experiments with these alternative paths were made to begin with, when it was noticed that the *B* spark was affected by the presence or absence of an opaque partition between it and the *A* spark, and not only an opaque partition—even a glass plate made a decided difference. The *B* spark was longer when the light from the *A* spark was suffered to fall upon it, and shorter when it was screened. This is manifestly the same effect as had been previously discovered by Dr. Hertz in the course of his experiments with an induction coil. He showed that it was ultra-violet light that was most effective in helping the passage of a spark, and hence the action of a glass screen. Some readings are given in § 44 and also in § 67.

The experiment is interesting now as proving that the *A*

TABLE B.

Alternative path.	Length of *B* spark for different positions of *A*.		
	Position 1.	Position 2.	Position 3.
None	·638	·638	·638
No. 5 copper	·288	·210	·148
No. 2 iron	·271	·163	·090
No. 2½ brass	·269	·170	·105
No. 18 iron	·290	·120	·056
No. 18 copper	·310	·148	·148
No. 40 copper	·300	·275	·173
No. 24 brass	·260	·200	·128
No. 25 iron	·275	·068	·078
No. 23 copper	·291	·245	·175
"Longer leads"	·338	·183	·183
None	·525	—	·523

spark occurs infinitesimally before the *B* spark; sufficiently before it for light to have time to travel the interval between the two, though not necessarily any more than this. A screen of black card was now arranged so as to screen off the *B* knobs from the light from *A*. The varying illumination of the day was not found to make any appreciable difference, so the window light was ignored.

Results.

44. A systematic set of experiments was now made. The *A* knobs were set permanently 1·35 centimetre apart, and the *B* knobs adjusted by their micrometer-screw until about half the sparks failed and half passed. Their distance apart was then read, and the discharger *A* shifted to a fresh position (1, 2, or 3 in the above diagram), always keeping the *A* spark-length 1·35. Then *B* was adjusted again, and so on. Table B is a record of the readings.

TABLE C.

Alternative path.	Critical length of *B* spark for different positions of *A*.	
	Position 1.	Position 3.
None	·800	·745
No. 25 iron	·330	·100
No. 5 copper	·345	·100
No. 2 iron	·323	·118
No. 25 iron	—	·173 Unilluminated.
,, ,, ...	—	·293 With magnesium burning close to.
,, ,, ...	—	·275 Not screened from view of *A* spark.
,, ,, ...	—	·243 Screened off again.
Longer leads	·550	·323
Longer leads with far ends disconnected	·400	·300
Longer leads with ends joined again	·563	—
Ditto	·580	Sparks just ceasing.
Ditto	·450	Sparks about half-and-half.
Ditto	·240	Sparks never failing.
None	·800	Sparks just ceasing.

The iron conductors gave the sharpest and most definite value for the *B* spark length, and the coppers gave the least definite value of all the alternative paths. When no alternative path was used the critical spark length was much more vague, considerable variation being possible before the *B* sparks either wholly failed or always passed.

45. Next day the retort holders were removed from the jars so as to get rid of a possible effect of their capacity, and the jars were mounted on paraffin blocks instead; but when no alternative path was used their outer coatings were connected imperfectly by a wooden penholder or something of the sort laid across them, in order that they might have no difficulty in charging.

The *B* interval was now usually adjusted until the sparks there were just ceasing, instead of being half-and-half. The same Voss jars were used, and *A* was still 1·35 centimetre. Table C shows the result.

The discharger *A* was then removed from any of its three regular positions, and put right away at the end of the "longer leads" described in § 42. In this case an *A* spark 1·35 centimetre long could not be obtained, by reason of leakage; so it was shortened to ·97, and the following readings taken. The *B* spark was now relatively enormous when there was no alternative path, although *A* had been so much shortened.

Alternative path	Critical *B* spark, with *A* at far end of leads.
No. 25 iron	·040
No. 5 copper	·031
None	1·620
No. 2 iron	·028

As a variation the *A* discharger was restored to its old position, A_1, and the *B* knobs put right away at the end of the longer lead. The *A* spark being still ·97, the following readings were taken:

Alternative path.	Length of *B* spark at far end of leads.	
	A in Position 1.	*A* in Position 3.
None	·271	·430
No. 25 iron . . .	·069	·054

It must be noted that the here used *A* spark lengths, 1·35 and ·97 centimetres, become, when reduced to flat plates, 1·1 and ·85 respectively.

Record of Further Experiments in September, 1888.

46. The arrangement depicted in Fig. 42 was still employed; but a couple of pint jars were substituted for the two small Voss jars. The static capacity was by this change raised from 143 centimetres to 670 centimetres, or nearly fivefold. The *A* spark was 1·35 centimetres long between the knobs of the universal discharger (or reduced to flat plates, 1·1 centimetre). It could be placed in either of the three positions shown in the figure, and its light was always screened off the *B* knobs (see § 43). These were adjusted till the *B* spark between them was just failing, passing occasionally, but missing much more often. The alternative paths used were wire circles, about 2ft. in diameter, whose particulars are already recorded in § 42. The following are the results:

Pint Jars.

Alternative path.	Length of *B* spark for different positions of *A*, in millimetres.		
	Position 1.	Position 2.	Position 3.
None	6·50	7·35	8·50
No. 25 iron . . .	5·8	4·7	3·65
No. 2 iron . . .	5·0	—	2·0
No. 5 copper . .	5·0	—	2·0
No. 23 copper . .	5·85	—	2·5
No. 40 copper . .	6·3	—	4·5
No. $2\frac{1}{2}$ brass . .	5·3	—	2·15
No. 24 brass . . .	5·9	—	3·55

The pair of wires called "longer leads" (§ 42) were now inserted between the discharger and the jars, and the *A* spark had to be shortened to 7·5 millimetres. The following were now the readings of the *B* spark:

Alternative path.	*B* spark in millimetres.	Alternative path.	*B* spark in millimetres.
None. . . .	10·55	No. 23 copper.	·44
No. 5 copper .	·22	No. 25 iron. .	·45
No. 2 iron . .	·25	No. 40 copper.	·86

Application of Theory.

47. Applying theory to these experimental results, we deduce the following values for the rate of alternation per second, and for the maximum strength of current passing through the conductors.

The rate of alternation is slightly different for the different alternative paths. The extreme values are those for the thickest iron and the thinnest copper; so, taking them as examples, we have—

Pint Jars.

When the thickest or No. 2 wire is used for alternative path.

Position of A knobs.	Rate of alternation.	Strength of max. current.	Impedance of alternative path.
No. 1.	3·1 millions per sec.	3,000 amperes	7 ohms
No. 3.	2·2 millions per sec.	2,500 amperes	5 ohms
End of longer leads	0·52 million per sec.	[670 amperes]	1·2 ohms
When the thinnest or No. 40 wire is used as alt. path.			
No. 1.	2.4 millions per sec.	2,700 amperes	10 ohms
No. 3.	2·0 millions per sec.	2,000 amperes	8·5 ohms
End of longer leads	0·52 million per sec.	[660 amperes]	2·2 ohms

The strength of current put in square brackets in the third column is what corresponds to the same A spark length as in the other two columns. In actual fact this spark length had to be reduced, because of leakage, and accordingly the current was reduced too in about the same proportion.

With the Voss jars the rate of alternation will be rather more than double, and the current amplitude about half, what is recorded for the two pint jars.

Comparison between Theory and Experiment.

48. Calculating absolutely to the length of B spark theoretically to be expected, by means of the first approximation (§ 3), which ought to serve for this case, I do not find a good agreement when small jars are used, the Voss jars giving

greater disagreement than the pint jars. The sparks observed are shorter than $\frac{L_0}{L+L_0}$ times the A spark. I was at one time almost prepared to say that this was probably because loss by radiation damps out the vibrations with these small capacities more quickly than has hitherto been supposed, and hence that an extra damping term of some vigour comes into these small jar operations. I now think it more likely that it is a mere question of static capacity; and the potential falls when a small jar has to empty itself over wires of even distantly comparable capacity.

But it may be worth noticing that, taking merely relative numbers, the agreement is very fair. Thus, if we multiply the value $\frac{L_0}{L+L_0}$ by the factor 7·6 (instead of by 11, the length of the A spark between flat plates in millimetres), we get the column called "calculated (relative)" in the annexed table.

Pint Jars.

Alternative path, and position of A.		Length of B spark when half fail.	
		Calculated (relative).	Observed.
No. 25 iron	Position 1	5·9	5·8
	Position 2	4·6	4·7
	Position 3	3·6	3·7
No. 2 iron	Position 1	5·1	5·0
	Position 3	2·7	2·0
No. 5 copper	Position 1	5·2	5·0
	Position 3	2·8	2·0
No. 23 copper	Position 1	5·8	5·9
	Position 3	3·5	2·5
No. 40 copper	Position 1	6·2	6·3
	Position 3	3·9	4·5
No. $2\frac{1}{2}$ brass	Position 1	5·2	5·3
	Position 3	2·8	2·2
No. 24 brass	Position 1	5·9	5·9
	Position 3	3·5	3·6

The agreement here is not perfect, but it is not very bad.

It is to be noticed that the distance at which half the *B* sparks fail is not the best thing to observe. What theory indicates is the distance at which they all fail. This would not account for much of the defect in agreement between observed and calculated absolute values.

Whether we can improve it, and obtain fair absolute, instead of only relative concordance, by taking into account the radiational dissipation of energy, we will consider later.

49. When the longer leads (§ 42) are inserted, with a self-induction of about 1,000 metres, the same mode of calculation gives us the theoretical *B* spark lengths as closely proportional to the self-induction of the alternative paths, but it is improbable that the first approximation is now sufficient, especially with the No. 40 wire. It is scarcely worth troubling about the second approximation in such a case as this, so I will proceed with the record of experiment, using the gallon jars instead of the pint. The *A* spark is still 1·35, and everything remains the same as before, the light being screened off.

The capacity charged is now 2,890 centimetres, except when the long leads are used, and then it is 3,040.

Gallon Jars.

Alternative path.	*B* spark length when half sparks fail, for different position of *A*.		
	A_1.	A_2.	A_3.
None	8·2	—	9·1
No. 5 copper . . .	6·35	4·10	2·57
No. 2 iron . . .	5·72	3·97	2·77
No. 25 iron . . .	7·05	5·45	3·70
No. 23 copper . .	6·90	5·03	3·95
No. 40 copper . .	7·35	5·84	4·60
No. 24 brass . . .	6·95	5·26	3·80
No. 2½ brass . . .	6·25	4·10	2·45
No. 18 iron . . .	6·40	4·60	3·14
No. 18 copper . .	6·80	4·95	3·12

50. Repeat with "longer leads" joined up to jar, with their far ends disconnected. *A* spark shortened to ·96, because it could not be got to form.

Alternative path.	*B* spark length for various positions of the *A* knobs.		
	Position 1.	Position 3.	Position 4. (Far end of long lead.)
None	6·8	7·0	19·25
No 2 iron . . .	4·55	2·10	·12
No. 5 copper . . .	4·80	1·80	·20
No. 40 copper . .	5·30	3·18	·48
No. 25 iron . . .	4·95	2·85	·31
No. 23 copper . .	5·50	3·00	·25
No. 18 copper . .	5·28	2·70	·19
No. 18 iron . . .	4·80	1·80	·16
No. 2½ brass . . .	4·84	1·71	—
No. 24 brass . . .	5·28	2·92	—

Comparison with Theory.

51. Taking the ratio $\frac{L_0}{L+L_0}$ and multiplying it by the length of the *A* spark, we get very fair accord between theory and experiment. But the agreement is still closer if instead of multiplying by 11 we multiply by an arbitrary factor 10 for the longer *A* spark and by 7·1 instead of 8·5 for the shortened one (see end of § 45). The table on the next page shows the comparison between calculation and observation for the longer *A* spark :

52. Similarly, comparing theory and experiment with the shortened *A* spark, and with the long leads as part of the charged conductors, we get the table printed on the next page for each of the three positions tried :

Historical Note.

53. It has been everybody's opinion, I believe, that the original alternative-path experiment was performed by Faraday ; but Mr. Oliver Heaviside has been good enough to send me the following extract, showing that it is much older and is due to Priestley :

Extract from John Cuthbertson's "Practical Electricity and Galvanism," dated 1807. *Printed for J. Callow, London.*

"An electric discharge prefers a short passage through the air to a long one through good conductors.

Gallon Jars.

	$\left\{\frac{10\ L_0}{L+L_0}\right\}$ Calculated (relative).	Observed B spark.
No. 25 iron . . . 1	7·2	7·1
2	5·7	5·5
3	4·5	3·7
No. 2 iron . . . 1	6·3	5·7
2	4·5	4·0
3	3·3	2·8
No. 5 copper . . 1	6·4	6·4
2	4·7	4·1
3	3·4	2·6
No. 23 copper . . 1	7·2	6·9
2	5·5	5·0
3	4·3	4·0
No. 40 copper . . 1	7·64	7·4
2	6·1	5·8
3	4·8	4·6
No. 2½ brass . . . 1	6·4	6·3
2	4·6	4·1
3	3·5	2·5
No. 24 brass . . . 1	7·2	7·0
2	5·6	5·3
3	4·3	3·8

Gallon Jars and Long Leads.

	Calculated (relative).	Observed.
No. 2 iron . . I.	4·4	4·6
III.	2·3	2·1
IV.	0·16	0·12
No. 5 copper . I.	4·5	4·8
III.	2·4	1·8
IV.	0·17	0·20
No. 40 copper . I.	5·4	5·3
III.	3·4	3·2
IV.	0·30	0·48
No. 25 iron . . I.	5·1	5·0
III.	3·2	2·9
IV.	0·25	0·31
No. 23 copper . I.	5·1	5·5
III.	3·0	3·0
IV.	0·24	0·25
No. 18 copper . I.	5·0	5·3
III.	2·9	2·7
IV.	0·21	0·19
No. 18 iron . . I.	5·0	4·8
III.	2·9	1·8
IV.	0·21	0·16
No. 2½ brass . . .	4·5	4·8
	2·5	1·7
	0·16	—
No. 24 brass . . .	5·1	5·3
	3·0	2·9
	0·25	—

"Exp. 106.—Bend a wire, about 5ft. long, in the form of an $\underset{a\ b}{\Omega}$, that the parts *a b* may stand about a quarter of an inch from each other, then connect the end *a* to the outside of a battery or coated surface sufficient to melt wire. When it is charged set one leg of the insulated discharge upon *a* and make the other end to touch one of the knobs of the battery: it will be discharged. At the time of the explosion a spark will be seen between *a* and *b*, which shows that the electric fluid prefers a short passage through the air to a long one through the wire. This spark at first was supposed to contain the whole discharge, but the contrary is proved by the following experiment:

"Exp. 107.—Lay a small wire from *a* to *b* of such a thickness as the battery is capable of melting when charged to the same height as before; discharge it, and the small wire will not be melted; cut the large wire in two at *c* so as to disconnect the circuit, make a discharge as before, and the small wire will be melted by the same explosion which before scarcely made it red hot. In this manner Dr. Priestley, who is the inventor of this experiment, states the conducting power of different metals may be tried, using metallic circuits of the same length and thickness, and noting passage through the air in each case."

Alternative-Path Experiments in September, 1888—(*Continued*).

Comparison of Various Metals.

54. *September* 26, 1888.—I now arranged a new set of wires round the room, making them all the same length, and each enclosing a rectangle 475 × 875 square centimetres. The wires were of iron, copper, and lead respectively, a thick and a thin wire of each. The length of each wire was 2,686 centimetres. The following are the particulars:

Description of wire.	Thickness.	Common resistance.	Self-induction.
	Millims.	Ohms.	Centimetres.
No. 12 iron . . .	2·40	·73	44,820
No. 27 ,, . . .	·40	22·6	54,540
No. 12 copper . .	2·35	·31	44,820
No. 27 ,, . .	·40	4·1	54,540
No. 11 lead . . .	2·90	·93	44,820
No. 23 ,, . . .	·70	13·7	52,000

These wires were to be used as alternative paths, and the rest of the circuit was to be varied as seemed good. The object of this series of experiments was to examine more closely into the question of iron *versus* copper, to see whether the apparently anomalous result of iron wire affording a better alternative path than thick copper could be definitely repeated, and to determine the conditions under which the result occurred. At Bath I had promised to go into this matter more carefully, and at that time I had not worked out at all fully the theory of the alternative path experiments, nor had I taken into account the damping in the way we have now done, so that I was in the dark as to the conditions under which the apparently anomalous result was to be expected.

The gallon jars already specified (§ 12) were employed. The *A* knobs were those of the universal discharger already described (§ 21), and the machine was joined up to them with the intervention of pieces of wood pretty much as in Fig. 42. The *B* knobs belonged to the spark micrometer (§ 20).

The *A* spark length was 2·44 centimetres; its light not screened off the *B* knobs. The *B* knobs were adjusted till about half the sparks passed and half failed. The alternative paths were joined up to the *B* knobs by pieces of wire, each 122 centimetres long; at first of No. 12 copper. (Table printed opposite.)

Thus, then, under these circumstances, iron showed often some slight disadvantage as compared with copper, only in one case showing an advantage, and that not great. But it was noticed that the noise of the *A* spark was always distinctly enfeebled when the thin iron wire was in circuit.

The object of interposing different conductors between the *B* knobs and the main length of alternative path was to see if the first portion of this path had an exaggerated influence on the result. Hence the use first of No. 12 copper leads and then of No. 27 iron. No such exaggerated influence of the first portion was detected, however; nor is it now pointed to by theory.

I see no clear reason why introducing longish wires into the *A* part of the circuit should in one case lengthen the *B* spark and in the other case shorten it. The only obvious difference between the two cases is that in one case the wires were charged, forming part of the inner coats of the jar, while in the other case they formed part of the outer coats and were uncharged. The *de*crease they cause is natural (extra self-

Details of circuit.	Length of *B* spark with different alternative paths.			
	No. 12 copper.	No. 12 iron.	No. 27 copper.	No. 27 iron.
A knobs connected direct to jars .	1·60	1.65	1·64	1·67
A knobs connected to jars by means of a couple of lengths of No. 17 copper wire, each 230 centimetres long and 0·15 centimetres thick, opened out to form a rough circle	1·88 1·83	1.87 1.83	1·77	1·745
Same, but with a couple of No. 27 iron wires used to join up the alternative paths to the *B* knobs (each wire 122 centimetres long) instead of the No. 12 copper pieces	1·76	1·74	1·76	1·79
Same, but with relative position of jars and connecting wires in the *A* part of the circuit interchanged	1·575	1·575	1·61	1·62
Long wires between jars and *A* knobs removed	1·535	1·535	1·57	1·62

induction), but the only obvious reason for the *in*crease is their static capacity, which one would not suppose sufficient to make so much difference.

55. *September* 27.—Next day the gallon jars, standing ordinarily on a table, were connected direct to the knobs of the machine, these knobs being used as *A* spark (as in Fig. 38, p. 274), and the outer coats of the jars were connected to the alternative paths round the room by a yard or two of No. 18 copper wire. Length of *A* spark being about 3·52 centimetres, the following are the *B* spark readings when balanced against different alternative paths:

Alt. path.	*B* spark.
No. 12 copper . .	The greatest length of *B* spark attainable was 4·15; it then nearly always failed. At 3·1 it nearly always passed. Sometimes it could be screwed up to the maximum without failing, and then, once it failed, it had to be screwed down to 3·6 or so before it could go again.
No. 12 iron . . .	3·1 or thereabouts. It fails mostly even at 2·9.
Thick copper again	Went up to 3·4 or so.

Alt. Path.	*B* Spark.
No. 27 iron . . .	At 2·65 it was half and half. At 2·8 it generally failed.
Thick copper again	Generally passed at 2·85, nearly always at 2·8.
Thin iron again .	Always failed at 2·8.

To compare these two leads more particularly, a couple of switches or bridges were employed, one of which switched in the thick copper lead, the other the thin iron; and it was clearly perceived many times that the *B* spark was both longer and stronger with the thick copper alternative path than with the thin iron. With the thin iron path it failed more frequently at a set distance between the *B* knobs, and when it occurred it sounded weaker, "showing distinctly that iron *does* cause a shorter and weaker spark than copper; but why? It must be the surgings in the conductor one is observing, and they are weaker in the iron." This is a quotation from my notebook at the time, and shows that the true explanation, subsequently confirmed by theory, was first suggested by experiment.

Experiments in Answer to Criticism.

56. *September* 28.—At Bath, Prof. Forbes questioned my mode of connecting up the alternative path to the outside of the jars, and though the objection does not bear critical examination, it was worth trying what difference there would be if the long wire were charged instead of remaining at zero potential. Accordingly the connections of Fig. 37 was adopted. The *A* spark being set at 1·89, the following were the *B* spark lengths when balanced against the stated alternative paths, and just failing:

No. 12 copper round room	1·60
No. 27 iron ,, ,, ...	1·63
No. 27 copper ,, ,, ...	1·61

Practically all the same, but a trifle longest with the iron. To be sure that the wire was acting conductively, and not merely by static capacity, it was cut in the middle of its length, *i.e.*, at the far end of the room, and a six-inch length of silk thread interposed. The *B* spark length was now 3·5, everything else remaining the same. At the break in the wire long sparks could be got to discharging tongs, etc., and there was a slight noise or click at the cut gap at every failure of a *B* spark, a noise noticed by an observer standing near the gap before the sound of the *A* spark was audible across the room.

Experiments on Effect of Surrounding Medium.

57. At Bath, Mr. Preece suggested that the nature of the medium surrounding the wire would probably have an important influence in these alternative path experiments; and although I saw no reason for the supposition it seemed quite possible, so to test it the following experiments were made at beginning of October, 1888:

A couple of pieces of wood, $17\frac{1}{2}$in. by 16in. by $\frac{7}{8}$in. thick,

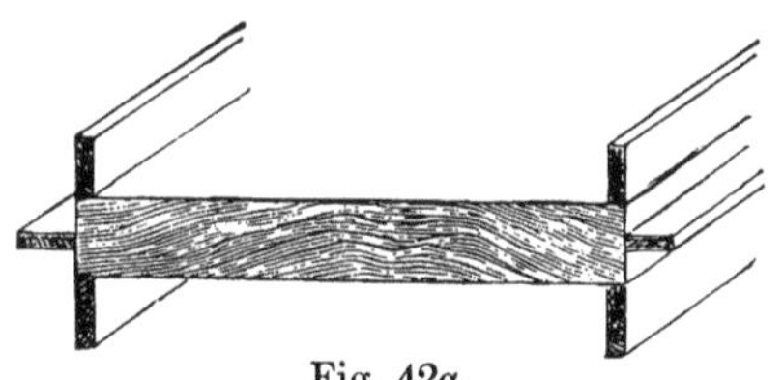

Fig. 42*a*.

were edged with thin wooden flanges or projections, so that when wound with wire the wire would have a half-inch clear space between it and the wood. A length of 302in. of No. 12 copper wire was then wound on each frame in 8 coils, each coil having 2in. between it and the next, the pair of frames being wound exactly alike.

Their equality was then tested by using them as alternative paths. The following are the *B* spark lengths (length of *A* spark, 1·51 centimetres):

	No. 1 coil.	No. 2 coil.	No. alternative path.
Length of *B* spark, using pint jars . .	·928 ·928	·928 ·928	1·346
Using gallon jars. .	1·040 1·034	1·045 1·043	1·505

They are practically exactly equal.

No. 2 was now immersed in melted paraffin, in such a way

that when solid it should be completely embedded in a cake of solid paraffin, 1in. thick above the wire everywhere.

They were then tested again.

		No. 1 coil.	No. 2 (paraffined) coil.
Pint jars . . .	*A* spark 2·0	1·054	1·074
	,, 1·5	·917	·906
	,, 2·5	1·25	1·25
Gallon jars . .	,, 1·5	1·03	1·03
	,, 2·0	1·26	1·26

Their equality is, as far as one can tell, absolutely undisturbed. The change of surrounding medium from air to paraffin makes no appreciable difference.

To see if a *conducting* covering would have much effect, the paraffined coil (which looked like a solid block of paraffin) was now coated all over with tinfoil, except just at the places where the terminals protruded. The following were now the *B* spark lengths:

	No. 1 or uncovered coil.	No. 2 coil with tinfoil insulated.	No. 2 coil with tinfoil earthed.	No. 2 with tinfoil to one of the jars.
A spark 1·5 .	·906	·841	·828	...
A spark 2·0 .	1·048	·886	1·026	·884
A spark 2·5 .	1 258	1·104	1·098	...
Covered No. 2 up now in soldered tinplate, in addition to the tinfoil, except for the terminals.				
A spark 1·5 .	·860	·775	·775	...
A spark 2·0 .	1·068	·933	.935	...

Thus in this case there is a distinct though not very great diminution of the *B* spark caused by the surrounding metal, whether it be earthed or not. The armour-plated coil makes a *better* alternative path than the bare wire, a result which is natural enough, since the space round it, and therefore its inductance, is somewhat less.

Effect of Iron Introduced into Alternative Path Spirals.

58. Referring back to § 18, where iron inside a circuit is reported as having no effect on the length of the *B* spark, I proceeded to test the matter further.

September 28, 1888.—Wound a spiral of uncovered No. 12 copper wire on a long glass tube, with interspaces of between

$\frac{1}{8}$in. and $\frac{1}{4}$in., and made a long paraffined bundle of fine iron wires, say 2ft. long and $\frac{3}{4}$in. diameter, to fit the tube. Used this spiral as alternative path both with and without the iron core. A spark between machine knobs 4·1 centimetres long.

	B spark with spiral as alternative path.	
Without iron core .	2·27	Centim. length at which *B* sparks just fail.
With iron core inserted	2·27	They still just fail.
Ditto	2·12	They occur and fail about equally often.
Remove iron core .	2·12	They do just the same.

But this is noticeable, that whenever the *B* spark fails, so that the discharge has to go round the spiral, *the noise of the A spark is distinctly less when the iron core is in the spiral than when it is withdrawn.* Repeated this many times, and made quite sure of the observation. The difference is very marked: without the iron core the discharge is much noiser. Compare this with the effect of an iron wire as alternative path as reported in § 55, and for theory see § 62.

Magnetization Observations.

59. Tried whether the iron bundle was magnetized after this treatment. It was. It lost some on tapping, but not much. Inserted it the other way into helix, and passed a spark: no change in the magnetism. Passed three sparks: magnetism destroyed. Passed more sparks: magnetism put back in old direction. Passed still more: magnetism reversed.

Put the core in now the original way, and sent sparks: magnetism not destroyed till after several sparks, then it became reversed again. On the whole it appears that the magnetization caused by discharge current is somewhat accidental, but that there is a tendency in one direction.

October 1.—Tried this further with a set of 4 steel knitting needles, kept well separated by corks and inserted inside spiral.

Whenever the *B* spark failed, so that the whole discharge had to pass round the spiral, the magnetization produced was feeble but in the correct direction. It took several sparks to produce much effect.

Wherever the *B* spark occurred, so that probably all the oscillations after the first were short-circuited through the air gap, the magnetization excited was much more vigorous, but it was almost a toss-up which polarity it had; a preponderance was found for the same direction as before. Often, however, it was suddenly and violently reversed.

Lengthening the *B* gap so that all the flash had to go by the helix again, the magnetism was gradually set right even if it had been started wrong, but the residual magnetism was now only feeble.

To sum up:—The entire discharge gave feeble magnetism definite in direction; while, when partially shunted by a *B* spark, it gave strong magnetization uncertain in direction.

Further Alternative Path Experiments comparing Iron, Copper, and Lead.

60. *October* 13, 1888.—Having found that the effect of iron wire used as alternative path was to shorten the *B* spark, and to diminish the noise of the discharge altogether, I proceeded to examine and make sure of this; and for this purpose had had the new wires rigged up round the room as reported in § 54. Six wires altogether, of iron, of copper, and of lead, a pair of each metal, one thick and one thin, their particulars being given above, p. 324.

Gallon jars standing on table connected up with short wires to machine, their outer coats connected to knobs of spark micrometer (where the *B* spark occurs).

Distance between the *A* knobs (those of the machine) 4·0 centimetres. (Table printed on next page.)

Summary of Table.

61. It is here to be noticed that the *B* spark with iron as its alternative path is shorter than with either of the other metals, as if it formed an easier path for the discharge; the real reason being its enormous resistance or throttling to an alternating current. Thereby the vibrations are rapidly damped out, and the difference of potential considerably reduced even in the time of a quarter-period, *i.e.*, the time which elapses after an *A* spark before the *B* spark occurs. This same resistance it is which so markedly reduces the noise and violence of the *A* spark, *i.e.*, of the discharge altogether, when iron forms part of the circuit. A thin lead wire has something of the same effect, but a thick lead wire permits nearly as noisy a discharge as does copper. The throttling effect of even a thick iron wire is very marked.

Alternative path.	Length of *B* spark in cms.	Remarks.
None	5·354	
No. 12 copper wire	5·20	*B* spark occurred sometimes. } Copper is always less definite than iron in fixing a *B* spark length.
Ditto	4·425	Spark still often fails.
No. 12 iron . .	3·074	Pretty definite. Spark on the verge of failing altogether.
No. 12 lead . .	3·925	Spark fails to pass 4 times out of 5.
No. 12 copper again	4·075	Fails 3 times out of 4.
No. 12 lead . .	No perceptible	difference in noise, or anything between lead and copper.
No. 12 iron . . .	2·975	Fails 9 times out of 10. Noise of spark much less.
No. 12 lead . .	4·074	Fails 5 times out of 6.
Thin lead . . .	3·250	Fails 3 times out of 4. Nearly as quiet as iron.
Thin iron . . .	3·250	Fails 4 times out of 5.
Thin copper . .	3·600	Fails 1 out of 2, but pretty definite. Noisier than either iron or thin lead.
To compare with	some old alte	rnative paths.
Old circle of No. 5 copper (see § 42)	0·952	Fails 3 times out of 4.
Ditto No. 2 iron .	·915	Fails 7 times out of 9.
Same length of No. 12 lead . . .	1·00	Fails 3 times out of 4.

It is noteworthy also that the adjustment for the critical *B* spark length is decidedly sharper and more definite for iron or thin lead than it is for copper, especially for thick copper.

Singular Effect of Iron Cores.

62. It is further to be noticed that the introduction of iron into a helix, through which the discharge has to pass, although it does not affect the difference of potential required to propel the discharge through that helix, yet exerts a very marked effect in quieting the discharge. It may be understood as increasing the radiating power of the circuit, and thus aiding the quiet dissipation of energy. But if it did this alone it ought to shorten the *B* spark, as iron does when forming part of the circuit. It does not affect the *B* spark at all, so far as I have been able to ascertain, when inserted into helices; hence it would appear that it must increase

the impedance of the alternate path, as well as the damping factor, and that the two effects, the one tending to lengthen the B spark, the other to shorten it, just neutralize each other.

I see no theoretical reason why these two different effects should just balance each other, and hence was at first disposed to explain the constancy of the B spark, when iron cores were introduced or removed, by supposing that they produced no effect whatever. But they do produce some effect, as is evidenced by their quieting of the noise of the discharge, § 58.

It is therefore singular that the E.M.F. needed to drive a discharge through a coil should be independent of the presence of iron in its core.

Referring back to § 34, theory gives for the ratio of the B spark potential to the A spark potential

$$\frac{B}{A} = \frac{\sqrt{(p_0^2 + R_0^2)}}{p + p_0} e^{-\frac{\pi(R - R_0)}{4(p + p_0)}}$$

where p_0 is the inertia part of the impedance of the alternative path, and p the same thing for the rest of the circuit; and where R_0 is the resistance or dissipation of energy constant for the alternative path (inclusive of its radiating power), and R the same thing for the rest of the circuit. Now an increase of p_0 alone would almost always increase the ratio B/A, but an increase of R_0 alone need not. The effect of an iron core may naturally be supposed to consist in an increase of both p_0 and R_0; and experiment suggests that while R_0 is largely increased by the insertion of iron cores, there is also a corresponding increase in p_0 of such magnitude that the above ratio $\frac{B}{A}$ remains appreciably constant.

Calculation of the Effect of Radiational Dissipation of Energy on the Length of the B Spark.

63. In order to see whether the suggestion made in the last paragraph, that the effect of iron cores may be to increase the radiational dissipation of energy by the circuit, let us see what the effect of this term is in general. The rate at which a discharging condenser loses its energy by radiation may be expressed in terms of a dissipation resistance coefficient, extra and added to any resistance which the metals of the circuit may have. Call this resistance ρ, then the rate of loss of energy at any instant is

$$\frac{d\left(\frac{1}{2} S V^2\right)}{d t} = \frac{V^2}{\rho} = \frac{V^2}{3 \mu v \left(\frac{\mathrm{L}}{\mu l}\right)^2},$$

the last term being obtained in the "Phil. Mag.," July, 1889, p. 56; l being the length of the circuit, L its inductance, and μv being 30 ohms.

This conclusion is independent of the capacity of the jars; for, while the rate at which a discharging circuit loses energy depends only subordinately on the circumstances of the conducting portion, so long as it has a simple open contour, it depends not at all on the capacity of the condensing portion.

But though the loss of energy per second is independent of the capacity of the jars used, the *proportionate* loss of energy, on which damping depends, is very much otherwise. For the potential amplitude at any instant falls exponentially thus

$$V = V_0 e^{-\frac{t}{S \rho}},$$

and the damping coefficient effective in these alternative path experiments is just this same exponent factor, with t equal to a quarter period or $\frac{1}{2} \pi \sqrt{(L S)}$.

Hence, the extra or radiational damping term is

$$e^{-\frac{\pi}{2 \rho}\sqrt{\frac{L}{S}}} = e^{-\frac{\pi \mu v}{2 \rho}\sqrt{\frac{L/\mu}{S/K}}} = e^{-\frac{\pi \mu^2 l^2}{6 L^2}\sqrt{\frac{L/\mu}{S/K}}}.$$

64. To find what sort of value this term has in ordinary open circuits without iron, we can refer back to the experiments, recorded § 44, page 315. The values of S for the cases there treated were—

For the gallon jars (§ 49) . 2,890 K centimetres.
For the pint jars (§ 46) . . 670 K centimetres.
For the voss jars (§ 41) . . 143 K centimetres.

The self-induction varied with the alternate path and with the three positions of the discharger (Fig. 42). Its smallest value for the No. 2 wire and No. 1 position is 2,230 + 1,290 = 3,500 μ centimetres, or, say, as follows:

Values of L/μ in Centimetres.

	Position 1.	Position 2.	Position 3.
Thick wire paths . . .	3,600	5,000	6,700
Medium paths	4,600	6,000	7,700
Thinnest (No. 40) path .	5,500	6,900	8,600

So the values of $\sqrt{\left(\frac{L/\mu}{S/k}\right)}$ are about

8 for the thinnest wire, smallest jars, and No. 3 position.

5 for the thick wire, smallest jars, and No. 1 position.

$3\frac{1}{2}$ for the thick wire, pint jars, and No. 3 position.

$1\frac{1}{2}$ for the medium wire, gallon jars, and No. 2 position.

In none of the cases is this term bigger than 8. As to the value of l, we may suppose that to be about 10 centims. only, if we measure it from jar to jar, as the effective length of a horseshoe magnet is measured; and in that case $L = 500\,\mu\,l$. But even if we took it as the whole length of the circuit, so that $L = 10\,\mu\,l$, the radiational resistance would still come out 9,000 ohms, or much too big for the radiational damping term to have any appreciable effect in these particular cases.

This teaches us that while radiation is the principal way in which small linear oscillators like Hertz vibrators get rid of their energy, it is otherwise with our closed circuits with electrostatic ends close together, and therefore of great capacity. These dissipate the greater part of their energy by heating the wire; the waves they emit are very long, and the portion of wave broken off and abandoned to space contains but a small fraction of the total energy.

65. Hence, for an explanation of the fact noticed in § 48, that the B sparks from small jars fall below their calculated value, we must look, not to their larger radiation damping term as there suggested (because, though it is larger for small jars than big ones, it is not large enough to be effective), but to the fact that the electrostatic capacity of the B portions of wire has in the calculation been ignored. The capacity of these wires is not usually enough to have any appreciable influence on the discharge of large jars, but with small ones an appreciable fraction of their relieved charge is occupied in charging up the B leads, and hence the energy of its rush is diluted. That seems to me now the most probable explanation of the very satisfactory agreement with theory shown by large jars, and of the distinctly defective vigour of small ones.

Experiments on the Deflagration of Different Wires.

66. To see how the material of a lightning conductor affects its liability to be destroyed by a flash, a set of exceedingly fine wires supplied by Johnson and Matthey, all of the same diameter, of gold, silver, platinum, and iron, were

employed. A scrap of each was laid down on a 4in. glass plate in four parallel lines, between two tinfoil strips, so that the length of each wire was 1·5 centimetres. A number of such plates were prepared. One of the large glass condensers was employed, capacity ·02 microfarads, and the four wires were arranged in its circuit.

Four Wires Parallel.

Spark length of discharge sent through them.	Result.
10 millim.	All four wires deflagrated.
5 ,, ...	ditto
1 ,, ...	No effect on any.
2 ,, ...	Silver deflagrated and smeared on the glass. Copper disappeared bodily, without deposit. Iron and platinum uninjured.
Another plate, 1·5 millim.	Silver melted at one end. Copper oxidized. Other two untouched.
Same plate, repeated, 1·5 millim.	Copper disappeared with very slight trace. Iron and platinum remain.
Same again, 1·5 millim. .	Platinum is deposited in small globules. Iron all right.
Same again, 1·5 ,, .	Iron deflagrated and broadly dispersed.

This is typical of what happened in many trials. When in parallel the best conductor took most of the discharge and was most damaged. The iron having much the highest throttling resistance took very little, until there was no other path open. Even platinum got destroyed before iron.

When the wires are in series, however, so that the same current has to pass through all, the order of their disappearance is inverted, and the iron is distinctly more heated than the others.

Four Wires End to End.

Spark length.	Result.
2 millims.	Platinum melted into globules. Iron deflagrated ; other two uninjured.
Another plate, 1 millim. .	Platinum broken in two near the middle. Iron has evidently been very hot, but is not destroyed. The copper and silver are all right.

Effect of inserting Long Leads in the A portion of the Circuit.

67. So far, the various long wires round the room had been used as alternative path, *i.e.*, had been made to shunt the *B* spark. I now proposed to insert them into the circuit in another place, so that while they exerted their full effect on the frequency and damping of the oscillations, they should not otherwise directly affect the *B* spark, which was to be shunted by a distinct and constant conductor.

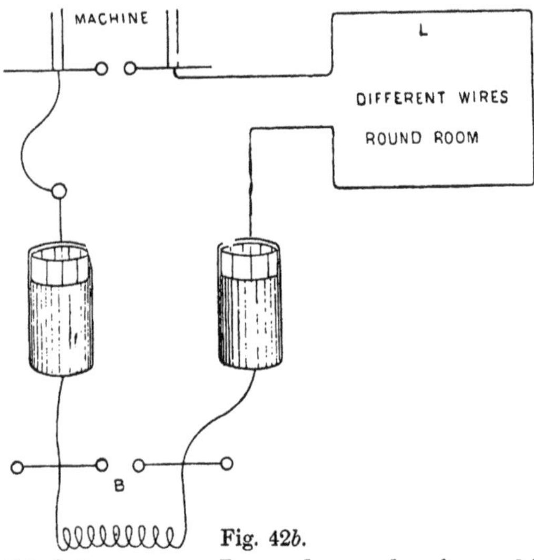

Fig. 42*b*.

For this latter purpose I wound on a glass lamp chimney a spiral of copper wire of the following dimensions:

685 centims. of No. 12 copper wire in 39 turns.

Diameter of each turn 5·5 centims. Thickness of wire ·24 centims. Total length of spiral 28 centims. Free ends, 13 centims. the two, attached to spark micrometer. A pair of No. 16 copper wires, 18 inches each, joined the micrometer to the outside coats of the gallon jars standing on insulating stool.

The spiral was easily detachable, and so the alternate path was either this spiral or nothing.

One or other of the long leads round the room was now used to join the knob of one of the jars to the machine. The knobs of the machine were set 4 centims. apart, and the experiment was ready to begin.

It was found that under these circumstances (as had often been found before), *i.e.*, with a long lead in the main circuit, the jar to which it was connected was very apt to overflow, *i.e.*, to spark round its edge, a distance of 8 inches; this tendency being more marked when the circuit is formed of good conductors than when the resistance and consequent damping of oscillation is considerable.

The leads inserted in the main circuit were the same as have already been described—viz., about 27 metres of No. 12 and of No. 27 of iron, copper, and lead respectively. *October 15th*, 1888. The following was the result :—*A* spark length 4·0.

Lead in main circuit.	Alternative path.	Length of max. *B* spark.	Remarks.
Thin copper.	None	9·178	Jar does not overflow.
,,	Copper spiral	·763	Jar sometimes overflows.
Thick copper	None	9·460	Does not overflow.
,,	Copper spiral	approximately ·9	But jar now overflows every time.
Thin iron .	None	8·0	Decidedly shorter than with copper.
,,	Copper spiral	·745	Jar sometimes overflows.
Thick iron .	None	9·135	
,,	Copper spiral	·80	Overflows rather often
Thin lead .	None	7·525	
,,	Copper spiral	·778	Sometimes overflows.
Thick lead .	None	8·79	
,,	Copper spiral	·850	Overflows nearly every time.

Thus, unless the circuit is closed by the alternative path, there is no overflow of the jar, but the maximum *B* spark is very long, about twice as long as the exciting *A* spark. When the air gap is bridged by the copper spiral, overflow occurs, but much more easily with the thick copper lead than with the others.

The reason why the *B* spark is so short is because the main part of the circuit is so long, *i.e.*, because in the approximate formula

$$\frac{B}{A} = \frac{L_0}{L + L_0},$$

L is much greater than L_0. (L_0 is the inductance of the alternative path ; L is the inductance of the rest of the circuit.)

Long Leads in both A and B Portions of Circuit.

68. The same arrangement of jars, etc., was tried again on October 17th, but some one or other of the long leads was used as the alternative path, while another of them, viz., the thin copper, was included in the main or A portion of the circuit. The connections were modified by removing the spiral marked L_0, and substituting another long wire, like L. The remarks appended in the last column of the following table, showing the number of times the B spark occurs and the number of times it fails under given circumstances, are useful as indicating the sharpness of the adjustment and the range of uncertainty.

Alternative path.	Length of B spark-gap in centimetres.	Remarks.
None . . .	7·90	—
Thin iron .	1·36	2 B sparks occur in 10 attempts.
,, . .	1·35	3 sparks in 6 attempts.
Thick copper	—	Spits over edge of jar every time.
Thin lead .	1·240	2 B sparks pass in 17 attempts.
,, . .	1·225	2 ,, ,, 8 ,,
,, . .	1·213	6 ,, ,, 6 ,,
Thick iron .	1·162	6 ,, ,, 8 ,,
,, . .	1·175	2 ,, ,, 7 ,,
,, . .	1·201	1 ,, ,, 13 ,,
Thick lead .	1·075	3 ,, ,, 3 ,,
,, . .	1·087	3 ,, ,, 6 ,,
,, . .	1·100	1 ,, ,, 10 ,,

With the thick iron the jar spits round its edge pretty often, and with the thick lead it does so nearly every time. With the thick copper, as stated above, it always does so, and hence no satisfactory readings could be taken with it.

Experiments on the Effect of Mutual Induction between A and B Portions of Circuit.

69. But since the two leads used in this experiment both went round the same room, and not more than a foot apart, there is some mutual induction between them; so observations were made, noticing carefully whether the discharge was going the same way round the two wires or the opposite way round. The main circuit conductor was still the thin copper wire, and the alternative path was one of the others.

When the current goes the same way in both I call it the concurrent arrangement; when opposite ways, non-concurrent.

The A spark was shortened to 3·5 centims. in order to avoid so much overflowing of the jars.

Alternative path.	Length of B spark.
Thin iron, concurrent arrangement . . .	1·34
Ditto non-concurrent	1·30
Ditto concurrent again	1·33
Thick copper, concurrent	·95
Ditto non-concurrent	·84
Ditto concurrent again	·95

Thus the effect of mutual induction was slightly to increase the apparent impedance of the alternative path when its current flowed in the same direction as that in the main circuit. But the effect is not great. The fact is that L and L_0 are both affected together, and so the effect on the ratio $\frac{L_0}{L+L_0}$ is unimportant.

Effect of Diminishing Self-induction in B Circuit.

70. A zigzag of wire was now prepared, consisting of the same length of the same wire as those round the room, wound zigzag on pegs on a board. A pair of wires were thus prepared, one of thin iron the other of thin copper. The thin copper wire round the rooms was still kept in the A circuit, and, in fact, everything else remained as it had been.

There was now, of course, no mutual induction between the parts of the circuit.

Alternative path.	Length of B spark.
Thin iron zigzag	·97
Thin copper zigzag	·71

Substitute now in the A circuit the thin copper zigzag instead of the thin copper wire round room; leave all else the same. (A spark still 3·5.)

Alternative path.	Length of B spark.
None	4·95
Thin iron zigzag	4·13

The diminution of L has brought up the length of the B spark tremendously.

Effect of the Light of One Spark in Assisting the Other.

71. A large number of similar experiments were made, combining the zigzags with the wires round the room in various ways, and inserting other things in the A portion of the circuit; but it is unnecessary to reproduce any more of these. As has been previously noticed (see §§ 43 and 44, the adjustment of the maximum B spark was often found to be different when it could see the A spark from what it was when the light of the A spark was screened from it. The effect is not very large, and the conditions have to be such as give fair sensitiveness to the B spark—*i.e.*, those which make it neither very short nor very long; but, under these conditions, the effect of the light of the A spark in assisting the passage of a discharge across the B spark-gap is quite distinct, even when the two sparks are not at all close to each other. The following extract indicates this:

October 24th.—Spiral on lamp glass (§ 67) in A circuit. A spark length 3·5 as before.

Alternative path.	Condition as to light.	Length of B spark.	Remarks concerning B spark.
None.	Light of A screened off	7·0	—
,, ...	Light of A visible at B	7·0	—
Thin copper wire round room	Light visible	2·775	9 sparks pass out of 10
,, ,, ,,	,,	2·800	5 ,, ,, 10
,, ,, ,,	,,	2·825	1 ,, ,, 10
,, ,, ,,	Light screened off	2·675	6 ,, ,, 10
,, ,, ,,	,, ,,	2·700	2 ,, ,, 10
,, ,, ,,	,, ,,	2·725	1 ,, ,, 10
,, ,, ,,	,, ,,	2·750	0 ,, ,, 10
Thin iron round room .	Light visible	2·48	—
Thin lead round room .	,,	2·63	—
Thick lead	,,	2·65	—
Thick copper	,,	2·52	—
Thick iron	,,	2 40	—
Thin iron zigzag. . .	,,	1·88	—
Thin copper zigzag. .	,,	1·58	—
Old No. 5 copper circle, § 42	,,	·20	—
Old No. 2 iron circle .	,,	·19	—
No. 26 iron circle . .	,,	·29	—
No. 26 copper circle .	,,	·24	—
No. 18 iron circle . .	,,	·24	—
No. 18 copper circle. .	,,	·23	—

The effect of the light of one spark in helping the other is plain enough in the above "thin copper wire round room";

but it was repeated and verified with many other arrangements. It had been previously observed in another manner by Hertz.

Experiment to Compare Zigzags with Spirals of the same Length.

72. The same kind of wire as had been used for the zigzags (27 metres) was now wound on an open wooden frame in a spiral, say, a yard long and a foot square or thereabouts, with about an inch between the different turns.

The small spiral on lamp glass of § 67 was put in A circuit, and A spark adjusted to 3·5 centims. [Setting of A spark not precisely accurate from one day to another.] The following is a typical selection from the readings:

Alternative path.	Length of B spark.	
None	7·93	
Thin copper zigzag	1·70	
Thin iron zigzag .	1·88	
Thin iron spiral .	3·36	
Thin copper spiral.	4·03	but overflows jar nearly every time.
Both zigzags in series	2·2	

Thus, the effect of increasing L_0 on the length of B spark, everything else remaining the same, is very marked. The figures show that either spiral offers far more impedance than both the zigzags (of the same wire exactly) in series.

This is the end of the account of my experiments on "The Alternative Path."

CHAPTER XXVII.

OTHER EXPERIMENTS ON THE DISCHARGE OF LEYDEN JARS.

THE following experiments among others were made in the course of 1888, beginning in February of that year. A brief account of the early experiments, with some of the deductions from them, was given in the above lectures to the Society of Arts on Lightning Conductors; and in the "Electrician," vols. xxi., xxii., xxiii., under the same title, a number of others were published at length, viz., the series of experiments relating to the "alternative path." But an account of the rest of the experiments has only now been communicated to the Royal Society. This account follows here.

Description of Jars Used.

1. The pattern of jar ordinarily used was an open cylinder without lid or neck, with the charging rod firmly supported from the interior and quite free from the glass above the tinfoil.

They were of two principal sizes, which I call for short "gallon" and "pint."

Each gallon jar was 40cm. high and 13cm. diameter, coated to within 10·5cm. of the top; and the capacity of the pair chiefly used was 0·0062 microfarad each. Two in series had a capacity of 28 K metres. Each pint jar was 16·5cm high and 8·2cm. diameter, and was coated to within 5cm. of the top. The capacity of the one chiefly used was 0·0016 microfarad. Two pint jars in series had a capacity of 6·6 K metres.

In addition to these ordinary jars, a couple of large condensers were made, each consisting of 16 pairs of 11-inch square tinfoil sheets, separated by double thicknesses of window glass, each pane about $\frac{1}{10}$ inch thick, and with a good margin; tinfoil strip connectors protruding on alternate sides, and copper wire prolongations, with all joints soldered, terminating in a pair of knobbed rods projecting upwards through stout glass tubes more than a foot apart; the whole thoroughly soaked and embedded in a mass of paraffin, poured molten into a strong teak outer case 22 × 20 × 13 inches, the whole when finished weighing about 3 cwt.

The capacity of one of these condensers was 0·028, of the other 0·02, microfarad. Single glass thickness would have given much greater capacity, but preliminary experiments showed that single thicknesses of glass were punctured by very modest sparks.

It is important in these experiments to have joints better made than is usual for high-tension electricity. Fizzing or sparking inside jars is abominable.

Account of the Long Conductors used in the Early Experiments.

2. Round the Physics Lecture Theatre in University College, Liverpool, supported on four vertical posts a good way from every wall, were stretched and supported, either by silk thread or silk ribbon according to the strength demanded, four or five wires, two of them of copper, one thick (No. 1 B.W.G.) and the other thin (No. 19); two of them of iron, one thick (No. 1) and the other thin (No. 18). They are called respectively "long thick copper," "long thick iron," "long thin copper," "long thin iron." Sometimes a "thinnest iron" of No. 27 B W.G. was used too. The thick wires formed a rude rectangle 840 × 515cm.; being joined mechanically not far from their ends by a foot or so of silk ribbon, and sufficient free ends being left to connect directly with jars or machine; connection being generally made by wrapping tinfoil tightly round the joined conductors. The thinner wires formed rather larger rectangles.

Particulars of these conductors here follow:

	Length.	Diameter	Ordinary resistance.	Approximate effective inductance.	Approximate capacity.
	metres.	centim.	ohms.	metres.	metres.
No. 1 copper	27·1	0·74	0·025	390	5
No. 1 iron .	27·1	0·71	0·088	390	5
No. 19 copper	30·3	0·085	2·72	570	3½
No. 18 iron .	30·3	0·12	3·55	550	3½
No. 27 iron .	30·3	0·035	33·3	630	3

The copper is commercial quality and evidently of miserable conductivity. I afterwards got some real copper from Messrs. Thos. Bolton and Sons, and with it the phenomena are still better marked.

Early Experiments.

3. The large glass condenser (0·028mfd.) was charged through one or other of the long wires, and a choice was offered the discharge, so that it might go either round the wire or leap an air gap, as it chose; as shown in Fig. 43.

A are the ordinary terminal knobs of the Voss or Wimshurst machine where the spark occurs; *B* is the discharge interval acting as a shunt to the wire or other resistance. *M Q N* represents diagrammatically one of the wires round the room. The spark-length *B* was adjusted so that it was an off chance whether the discharge chose it or the wire. It was noticed that when the discharge chose *B* the *A* spark was strong, but when the discharge chose the wire the *A* spark was weak. The difference appeared to be only in the noise or suddenness of the spark, for when a Riess's electro-thermometer was inserted in the circuit it indicated about the same in either case.

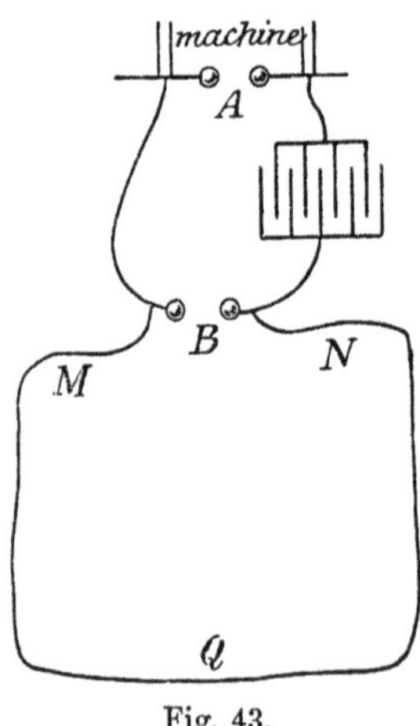

Fig. 43.

A capillary tube was filled with very dilute acid so that its resistance was about $\frac{1}{4}$ megohm, and was connected across the *B* knobs instead of the long wire. When this acid tube was thus made the alternative path, and the *B* knobs placed so far apart that the discharge was obliged to choose it, the *A* spark was very weak, being reduced to a quiet spit, which could be analyzed by a slowly rotating mirror into several detached sparks.

After a number of readings of spark-length, which have been elsewhere published (and which showed among other things that it made very little difference whether the alternative path were copper or iron), a common Leyden jar was substituted for the condenser, and similar results were obtained with it.

But it was now noticed, in addition, that the jar frequently overflowed by sparking over its lip; and that when this happened a spark still occurred at *B* though not at *A*.

A special overflow or short-circuiting path was then provided, equivalent to a pair of discharging tongs; calling this air gap *C*, it was found that, according to the adjustment of the width of spark gaps, flashes at *B* and *C* could be got without *A*; or at *A* and *B* without *C*; or at *C* only. (This was the beginning of experiments on overflow.)

Putting acid resistance into the circuit at *M* or at *N* weakens but does not stop the *B* sparks; and it has the same effect at *M* as at *N*. But inserting resistance at *Q* does not weaken the *B* spark perceptibly; neither does cutting the wire there; only, of course, in order to permit the charging of the jar in this case, the *B* gap has to be bridged by some imperfect conductor; this shunt high resistance, which may be a piece of dry wood or anything just sufficient to convey the *charging* current, having no appreciable effect upon the *B* spark.

But it was noticed that when the wire was cut at *Q* a singularly long spark or strong brush discharge attempted to jump the space there whenever the machine spark occurred. (This was the beginning of experiments on "recoil kick.")

It was also found that connecting the machine side of the jar to earth (the long wire, not interrupted anywhere, being insulated) increased the strength of the *B* sparks very much, and made them easier to get. Evidently the wire was acting as one coat of a condenser, the wall being the other coat. Even when the jar was discarded, no connection being made in its place, and the wire alone used, sparks occurred at *B* perfectly well whenever the machine gave a spark at *A*. (This led to experiments on the "surging circuit.")

Experiments on Overflow (February, 1888).

Small Jar.

4. Tried the arrangement shown in Fig. 44, the jar being pint size, as described above, of plain cylindrical shape, open at top, with its lip projecting 2 inches above the tinfoil so that the overflow distance was 4 inches. The long wire was the 30 yards of No. 1 copper. In addition to the machine spark gap *A*, a couple of other intervals, labelled *D* and *F*, were also provided; the spark gap *D* being led up to through the long thick wire, the spark gap *F* through the capillary water tube of high resistance already mentioned. The *A* knobs were each 2·34cm. diameter. The size of the others does not seem to be recorded.

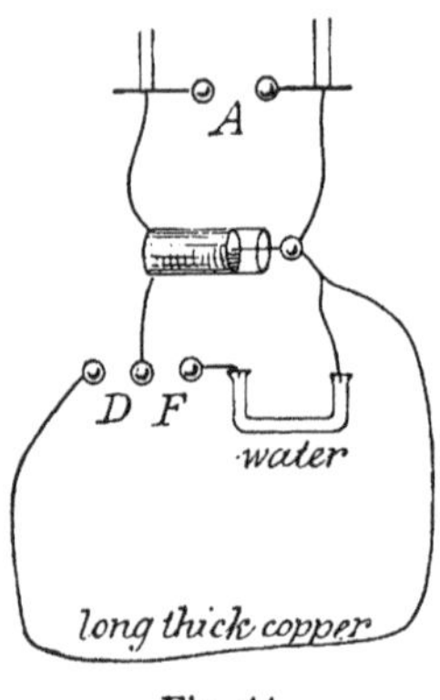

Fig. 44.

Separating the machine knobs too far for a spark there, sparks could be got either at *C* or at *F* or over the lip of the jar, or in two or three places at once. The lengths were $D = 0{\cdot}72$ inch, $F = 0{\cdot}68$ inch. Bringing the *A* knobs nearer together, a distance of 0·57 inch, it went there too. The *A* spark is the noisiest, then *D*, and lastly *F*; *F* is in fact quite weak. When it sparks at *D* it mostly goes at *F* too, and likewise overflows the lip of the jar too, but not always.

Shorten all the air gaps so as to avoid overflow, and they spark simultaneously at the following distances:

A.	*D.*	*F.*
0·435	0·565	0·575

Modified the plan of connections to that shown in Fig. 45; the second water resistance or "leak" being now introduced merely in order to give the jar the possibility of charging.

Whenever an *A* spark occurs, a considerable range is permissible with the others. As to *F*, it does not matter how short that is made; it is affected by the others, but has no effect on them. The overflow of jar specially accompanies a spark at *D*. Frequently sparks occur in all four places at once; and at times the overflows of jars are violent and numerous, so that, when *A* and *D* are both pretty long, flashes fly from cork and wood and almost anything that happens to be in contact with the jar. (The jar stood on a wooden block on an insulating stool: it was principally from this that flashes sprang sometimes.)

The following readings give an idea of the range of adjustment permissible; all the flashes in a horizontal line occurring simultaneously:

Length of Sparks (in inches).

A.	*D*.	*F*.	Jar lip.	Remarks.
0·48	0·53	0·48	Overflowed (4 inches).	
0·48	—	0·48	Quiet.	
0·48	0·42	0·37	Overflowed.	
0·69	0·32	0·45	Overflowed.	Here *F* began to fail.
0·69	1·03	0·0	Overflowed violently.	Here *D* began to fail.
0·69	1·03	0·9	Flashing from wood or anything.	Here *F* began to fail again, or to be replaced by other flashes.

Thus, with a long *D* spark, *F* could be anything up to nine-tenths of an inch; whereas, with a short *D* spark, it failed at half that distance. The jar-overflow is precipitated by a moderate *A* spark if *D* occur too. *D* can be much longer than *A*. If both *A* and *D* are long, the overflow is violent.

Larger Jars.

Now replace the first pint jar by one of the large "gallon" jars of similar open shape, but with the glass protruding 4 inches above the coatings, so that its overflow flash was 8 inches long.

(The capacity of the jar was 0·0062 microfarad.)

With *A* spark 0·62 inch long, the *D* and *F* gaps might be anything, but so long as the *D* spark was allowed to pass the jar overflowed every time the machine gave a spark at *A*.

On putting one terminal of the machine to earth (the one not attached to the jar), the *D* spark is considerably lengthened; and, even when the knobs are widely separated, brushes leap from each into the air whenever an *A* flash occurs.

Simplified Connections.

5. Tried now this same gallon jar connected up to the machine in the simplest possible manner, either direct by a foot or so of ordinary wire, or else by the long thick copper round room or some other long wire, or sometimes by both,

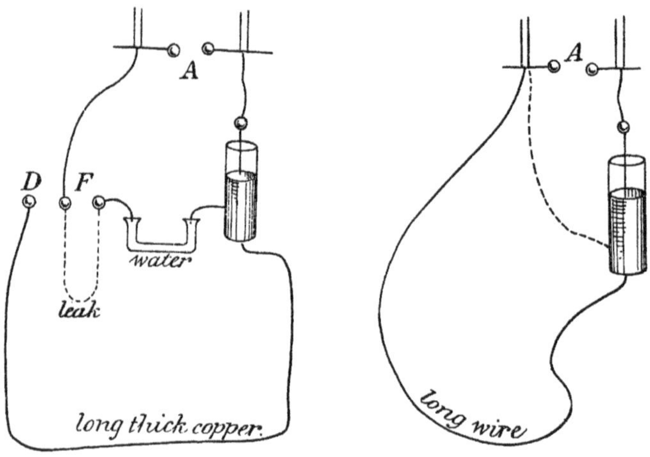

Fig. 45. Fig. 46.

as shown in Fig. 46, so as to see what difference the length of connecting circuit made to ease of overflow.

The machine's knobs were gradually separated until the jar flashed over its lip, and then their distance apart was read. It was found that with the long connector a very much shorter *A* spark was sufficient to cause overflow than with the short-circuiting wire. And not only was it shorter, it was incomparably quieter; the jar seemed to overflow without any trouble or violence when attached to the long circuit, whereas, when this was short-circuited out, the *A* spark had to be long to cause an overflow, and when it occurred its violence was great, as if threatening to smash the jar. If, under these circumstances, the short circuit was removed and the long wire replaced, the jar overflowed, not in one streak, but in a torrent or cascade of sparks; the number of these splashes gradually decreasing down to one again as the spark *A* was shortened.

It was also found that after an overflow another was more likely, whereas after a failure another failure was probable: that there was, in fact, a kind of hysteresis, the conditions of overflow being easier for a decreasing *A* spark than for an increasing one of the same length. This seemed especially noticeable when the long connector was thin copper, instead of being so thick and massive as the No. 1 copper on the one hand, or so highly resisting as thin iron on the other.

The following table summarizes the readings. The full contrast does not come out strong in the early numbers: there is some caprice about whether the jar overflows or not, probably having something to do with the state of the glass surface.

The contrast comes out best towards the middle of the table. The "thick copper" and other long leads referred to are those specified in § 2.

Conductor used between machine and outer coat of jar.	Length of *A* spark able to make jar overflow (in tenths of inch).	Remarks.
Short wire	7·0	
Long thick copper wire	5·5	
Long thin copper wire .	from 6·65 to 7·4	According to which it did last.
Long thin iron wire . .	7·8	
Short wire again . . .	9·5	
Long iron shunted by short wire	11·5	No overflow.
Long iron alone . . .	11·5	Still no overflow.
Thick copper again . .	6·4	Overflow every time until gap is shortened to this.
Thick copper shunted by short wire	17·0	Does not overflow till this long and noisy spark is reached.
Long thick copper alone	6·2	Still overflows even at this, the spark being gentle.
Retain thick copper. Earth one knob of machine	5·25	Jar still overflows.

Table continued on next page.

Conductor used between machine and outer coat of jar.	Length of *A* spark able to make jar overflow (in tenths of inch).	Remarks.
Retain thick wire, but earth jar end of it	5·9	
Now earth machine end of it	6·25	
Short-circuit it once more	17·0	Still does not overflow.
Simple thick wire alone once more	5·6	Overflows.
Thin copper wire . . .	min. 6·4, max. 7·1	A little indeterminate, according to whether overflow or failure happened last, that which happened last being easiest to get again.
Short-circuit again . .	—	*A* has to be enormous before it overflows.
Thin iron wire . . .	9·2	With this thin iron wire the overflow point seems definite, whereas with the thin copper it was not.
All three long wires in parallel	6·4	
Thick wire again, but with a bridge across trying to shunt out all but about 3 yards of it	6·5	
Short-circuit again added	10·3	
Remove the bridge but leave the short-circuit	from 8·7 to 10·2	No apparent reason for this shortness.
Disconnect one end of thick wire, but leave short-circuit	9·4	
Disconnect both ends, having only short-circuit	9·4	So now evidently the jar is easier to spark over, as it was at the beginning.
Restore thick wire simply	5·5	

Spiral Conductor.

6. Another connecting path was now made, consisting of 8 yards of the No. 1 copper wound into an open spiral about a foot in diameter, and suspended in air by ribbon, as indicated by the dotted line in Fig. 47; when in use, its two ends were led, one to a machine terminal, the other to outer coat of gallon jar, whose inner coil was connected to the other machine terminal.

This being so, the lengths of machine spark needed to make the jar overflow (round its lip always) under different circumstances were again read as follows:

	Kind of connector used.	Length of a spark needed for overflow.
Gallon jar.	Thick copper spiral	0·61 inch.
	Short circuit	1·50 ,,
	Spiral again	0·63 ,,
	Long thick wire round room	0·57 ,,
	Both this and spiral in series	0·56 ,,
	The two in parallel	0·62 ,,
	The spiral alone again	0·61 ,,
Pint jar.	Thick copper spiral	0·58 to 0·52 inch.
	Thick wire round room	0·51 inch.
	Spiral	0·53 ,,
	Short circuit	1·1 ,,
	Spiral	0·54 ,,
	Thick iron wire round room	0·66 ,,
	Iron and copper round room in parallel	0·62 ,,
	Iron alone	0·67 ,,
	Copper alone	0·52 ,,
	Short circuit	1·4 ,,
	Copper again	0·52 ,,

Effect of High Resistance.

7. Interpose the capillary liquid tube ($\frac{1}{4}$ megohm) in the circuit of the thick copper wire, putting it at one or the other end of it, and the jar refuses to overflow, although the spark-length A is increased to $2\frac{1}{2}$ inches.

The spark is quiet, long, and zigzaggy. The resistance has the same effect at either end, but the spark seemed straighter when the resistance was at jar end of long wire.

To test effect of putting resistance into the *middle* of a long connector, both the thick wires round room (one copper, the other iron) were joined in series and used as connector. Overflow began when $A = 0{\cdot}6$ inch. The wires were now disconnected at their far ends, and the capillary tube made to bridge the gap. The jar now refused to overflow, though A was more than trebled in length. (Fizzing stopped it at that point.)

Contrast between C Path and Overflow.

8. But when an artificial overflow path is supplied to the coatings (as indicated by the strong line to C knob in Fig. 47) the matter is different. It does not now feel the effect of a long circuit as different from that of a short one. The space at C being 0·94 inch, a spark jumped there sometimes and sometimes at $A = 0{\cdot}75$, with the high resistance interposed in the two long leads; and just the same happened when the resistance was removed and the long wires directly connected.

When A was shortened to 0·64, the overflow was unable to select C, but it jumped the lip of the jar instead. It preferred 8 inches of jar-lip to 1 inch between the C knobs. When strong enough it would seem to go at C; when too weak for that it jumps the edge; but this is not a clear account of the matter. A better statement is the following:

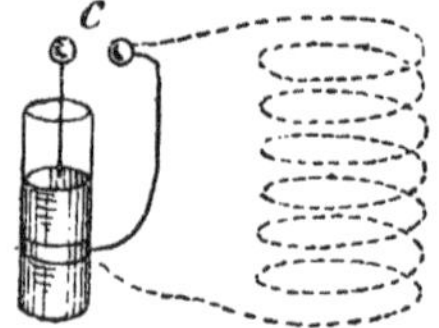

Fig. 47.

An A spark precipitates an overflow (*i.e.*, over the lip of the jar), but it does not precipitate a C spark. When a spark occurs at C there is quiet at A. The A and C sparks are alternative, not simultaneous. Moreover a C spark does not cause overflow. An A spark can easily occur without the edge of the jar being jumped, but the edge is never jumped without an A spark. (Connections being as in Fig. 46, with the addition of a short C or artificial overflow path, as shown by the thick line in Fig. 47.)

Long Connector in C Circuit.

9. But now the thick copper spiral above mentioned (§ 6) was arranged to connect one of the C knobs with the outer

coat of its jar (as indicated by the dotted line in Fig. 47, the strong-line shunt being removed), one of the two long thick wires round the room being used to connect up the machine to the same outer coat, as in Fig. 46. Under these circumstances, simultaneous sparks *could* be got at A and at C, and both about the same length, but not when they were too long, say, $A = 0{\cdot}52$, $C = 0{\cdot}57$ inch. But now the jar could be made to overflow by either spark if of sufficient length. Thus if $A = 0{\cdot}61$ or if $C = 0{\cdot}74$, the jar lip gets jumped, and sometimes the A spark occurs, sometimes the C, but not both. Another reading: $A = 0{\cdot}69$ or $C = 0{\cdot}94$; jar overflows in either case.

Restore now the usual short wire to the C knobs, and the C spark still often goes, but it has no effect on the jar. The A spark makes the jar overflow as before.

But if the long lead between machine and jar be short-circuited-out (as by the dotted line of Fig. 46), while the thick copper spiral still joins up to the C knobs (as indicated by the dotted line in Fig. 47), then A cannot make the jar jump, while C can easily.

Thus overflow is always easily produced by the action of the spark occurring in a long good-conducting lead, not in a short or bad-conducting one.

Effect of Iron Core.

10. Using the thick copper spiral as before (§ 6) to make the pint jar overflow, I tried whether inserting large massive iron bars in it as a magnetic core would have any effect. There happened to be three large bars, each about 3 inches in diameter, which were used. They were of soft iron, and intended for the legs of an electro-magnet.

No effect was found. The length of the A spark needed to make the jar overflow was, as near as one could tell, the same whether the iron was in the spiral or not. Thus:

Without iron	$A = 0{\cdot}53$
With one bar in spiral . . .	$0{\cdot}51$
With three bars	$0{\cdot}515$

No difference that one could be sure of. For further experiments see p. 331.

Effect of Capacity.

11. The spiral was now shunted out by a couple of Leyden jars in series, *i.e.*, with their knobs touching either end of it and with their outer coats connected. If the jars only touched one end of the wire, they had no effect; but when they touched both ends, a larger A spark was needed to cause overflow.

With the spiral alone . . . . $A = 0{\cdot}53$
With the capacity shunt . . . $A = 0{\cdot}76$

Experiments on Large Condenser.

12. It was not desirable to expose the large condenser (§ 1) to such conditions as would make it want to overflow, because overflow with it would mean bursting; but one of the pint jars was arranged on it as a safety-valve, and it was then connected up to the machine. On now taking machine spark at A, the pint jar might or might not overflow its 4 inches.

With very short connections . . $A = 0{\cdot}5$ inch did not overflow it.
With wires each a yard or so long . $A = 0{\cdot}4$ inch was sufficient.
And with spiral of thick copper . $A = 0{\cdot}3$ inch was enough.

Iron Core Again.

13. Tried a stout spiral of brass wire (a spiral spring about a foot long and an inch diameter); it made the jar overflow fairly easily. Then inserted in the spiral a bundle of fine iron wires wrapped in paraffin paper, but could detect no difference whatever (cf. § 18, p. 287 and §§ 58 and 62, pp. 328 and 331).

Summary.

14. The noteworthy circumstance in all these experiments is the remarkable action of a long thick and good conductor in causing the jar to overflow, especially if it be insulated, the most powerful conductor for this purpose being one with considerable self-induction and capacity but very little resistance. Evidently such a conductor assists the formation of an electric surging, whose accumulated momentum charges the jar momentarily up to bursting point. Resistance damps the vibrations down, and short wires have insufficient electric

inertia and capacity to get them up. Iron, whether massive or subdivided, shows no effect whatever on the effective inductance of a circuit surrounding it.

It is also noteworthy how far more readily a jar overflows directly between its coatings over the lip than it does through a pair of discharging tongs held round the lip. Probably the sharp edges of the tinfoil contribute to this effect, possibly also dust or other specks on the surface of the glass, or it may be the action of the air film itself, but it seems as if the extremely small inductance of such a path likewise aids what, if it is to occur at all, must take advantage of a flood tide, a millionth of a second's duration.

Confirmatory Experiments (*6th March*, 1888).

15. Two similar jars, each with dischargers, were connected as shown in Fig. 48.

A spark at *A* now caused the distant jar to overflow easily, but had no effect on the near one. Similarly, a spark at *C*

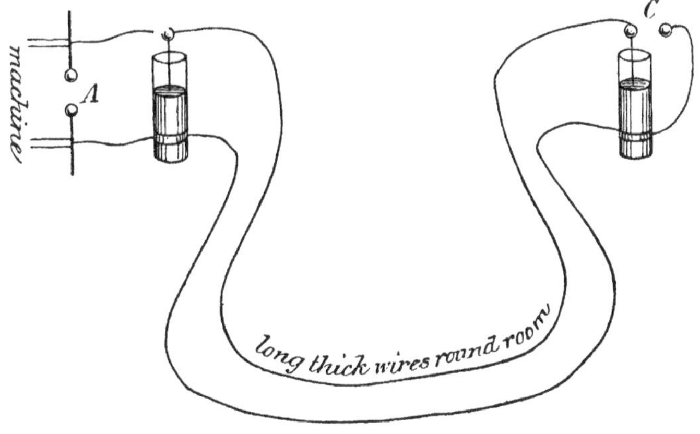

Fig. 48.

caused the jar distant from *C* to overflow easily, but had no effect on its own jar.

An *A* spark never caused a spark at *C*. Sparks occurred either at *A* or at *C* according to which happened to be the narrowest gap, but not at both; and it was always the jar most distant from the spark that overflowed its lip.

16. The explanation probably depends upon the fact that when a spark discharges its near jar the charge from the distant one rushes forward, but, not being able to arrive in time, surges back violently and overflows. The effect can probably be imitated with a long water trough by momentarily opening and suddenly closing a trap-door at one end. It can certainly be observed in a lavatory where there is a constantly dribbling cistern for flushing purposes. By opening and suddenly closing one of the wash-basin taps a surging is set up in the connecting pipe, and the dribble becomes periodic for a second or two, in synchronism with the period of longitudinal vibration of the water in the pipe.

Something apparently of the same sort has been quite recently observed with sinuously alternating currents by Mr. Ferranti in the Deptford mains. But whereas that case can be described as a long stretch of capacity with locally concentrated inductance, mine is a long stretch of inductance with locally concentrated capacity. Accordingly, while he observes an extra current-amplitude, I observe an extra potential.

The phenomenon in another form seems to have been first observed by Sir W. R. Grove, and fully explained by Clerk-Maxwell (see "Phil. Mag.," for March and May, 1868). It was consequently rediscovered by Dr. Muirhead, and explained by Dr. Hopkinson ("Journ. S.T.E.," 1884). A note sent by me to the "Electrician" for 24th April, 1891, contains a summary of the history and explanation.

Discussion of Overflow and Surging Experiments.

17. For the complete explanation of the overflow experiments, the static capacity of the long wire, and the momentum of the pulses rushing along it, must be taken into account, and a wire is more effective when insulated and charged than when lying on the ground.

It does act, however, even when lying on the ground, *i.e.*, when its magnetic momentum is all that can be supposed effective. But the ordinary theory of discharge oscillation will not account for the jar being thereby raised to a higher potential than it was at the beginning of the series; the amplitude of the vibration necessarily decreases. Hence it is probable that the fact of overflow does not prove that the entire potential of the jar is raised; only that the potential of the tinfoil edges is excessive. The charge is probably not uniformly distributed at the extremity of each swing. The fringe of sparkings above the edge of the tinfoil are well known whenever a jar is discharged; and overflow is merely an exaggeration of these sparkings, which usually leap up and subside. In fact they can be seen to jump higher and higher, as the spark is gradually increased, until the lip is leaped.

The idea of the pulses rushing along the connecting wires, and adding their momentum to the oscillation of the jar-discharge, suggests that there must be a best length for the connectors, viz., when the period of their pulses agrees with the period of oscillation of the discharge; and the fact that there is a best length is found experimentally.

The same length of connector is not equally effective with pint and gallon jars. A longer one is best for the larger jar; and if a connector be too long it does not promote overflow any more vigorously than if it were somewhat too short.

The damping effect of resistance no doubt partly comes in here as helping to account for the evil of unnecessarily long connecting wires; and no fine adjustment of length has been found necessary to bring out in a marked manner the surging effects.

If any experimenter should fail to obtain these conspicuously, he probably has his connectors too short or too long. It is advantageous, though not essential, to have the long wire insulated. It is essential to have it highly conducting. Iron is for these purposes by far the worst conducting metal, because it is magnetically throttled.

Another small point is that good contacts aid in causing overflow, especially when the connecting wires are not long enough. Insignificant air spaces suffice to damp out some of the vigour of the subsidiary oscillation to which these effects seem due. With long massive leads, however, good joints are not of so much consequence.

(Parenthetically it may be remarked how well adapted the usual orthodox lightning conductor is to develop violent surging and splashing effects.)

Further Overflow and Surging Circuit Experiments.

18. Two jars standing side by side, and connected in parallel by long wires to the machine, sometimes both overflowed. Sparks taken at the jar knobs with ordinary discharging tongs had no such effect.

The tongs were sometimes arranged over the lip of a jar, so as to help its overflow if possible; but it was not easy to do this. Near the edge of each coating they had the best chance, but the splash usually preferred an immense jump through

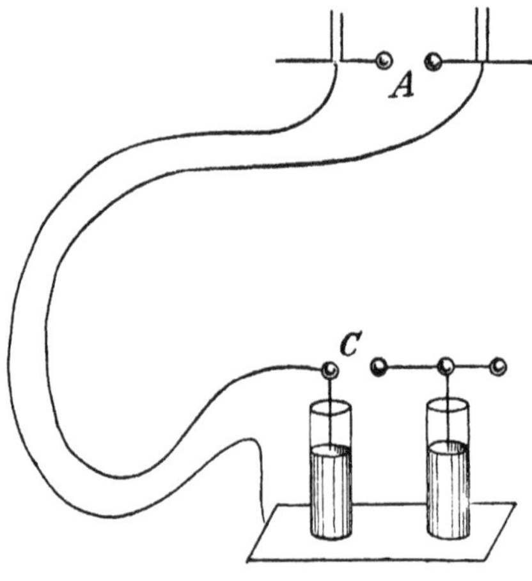

Fig. 49.

air over a glass surface to a much smaller jump through the discharging tongs. Overflow is evidently a very quick effect, and must occur in a hurry or not at all.

A couple of jars standing side by side on the same metal plate had a gap between their knobs as shown in Fig. 49, and one of them was connected by long leads to the machine. There was now often a spark across *C* into the second jar when an *A* spark occurred. But the second jar was not thereby charged. The charge just sprang into it and out again.

Connector without Self-induction.

19. Connected up a jar to the machine with a special anti-induction zigzag of tinfoil, folded to and fro in twenty long layers with several thicknesses of paraffin paper between. Could detect no effect on the jar overflow. It acted like a simple short circuit.

Tried, on the other hand, a high inductance coil, viz., the gutta-percha-covered bobbin of a Wiedemann galvanometer, with an iron-wire core inside: but its resistance was too high: it damped the oscillations.

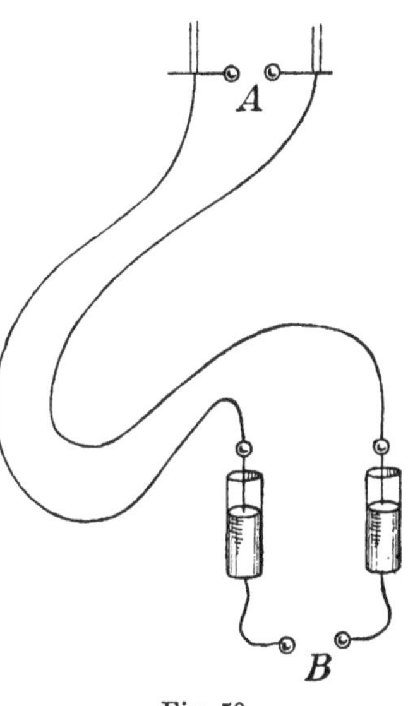

Fig. 50.

Connector with Self-induction.

Interposed between machine and jars two thin wires round the room, and led the outer coats of the jars direct to a discharger, as in Fig. 50; the jars being gallon jars, standing on wooden table. Compared *A* and *B* sparks; *B* was very long. Then substituted short wires for the long ones, and compared again. *B* was nearly as short as *A*. Readings follow:

	Length of *A* spark.	Length of *B* spark.
Jars joined to machine by long wires	0·4 inch	2·2 inches
Short wires substituted	0·4 ,,	0·5 ,,

Overflow of Plate Condenser.

20. Connected a pair of tea-trays to the machine by long thick wires, and fixed them parallel to one another, keeping them asunder by glass or paraffin pillars; the jars standing

on a wooden table, or being otherwise leakily connected so that they might charge. Every machine spark at *A* (Fig. 51) caused long brushes, or sometimes remarkably long flashes between the plates.

A jar standing on bottom plate will receive a flash, but it will not necessarily be thereby charged; a slight residual charge may be found in it, but no more.

Points also get struck, just as noisily as knobs, and no more readily. Crowds of points, and knobs of all sizes, get struck equally well, if of the same height and all equally well connected to the bottom plate. The highest get struck at the expense of the others. Often, however, several get struck at once. A gas-flame burning on the bottom plate gets struck at a much greater distance than does any metallic conductor. The weak hot-air column is precisely what this overflow discharge prefers. It takes it in preference to a metal rod of twice the apparent elevation, and strikes down right through the flame.

But though it thus readily smashes a weak dielectric, it will not take a bad conductor. A wet string or water tube may, in fact, reach right up till it touches the top plate, and yet receive no flash, while the other things shall be getting struck all the time.

When the striking distance is too great for a noisy flash, a crowd of violet brushes spit between the top plate and protuberances on the lower plate; reminding one of some lightning photographs. The effect is still more marked if the top plate is a reservoir of water with a perforated bottom. The rain shower increases the length of these multiple gentle high-resistance purple discharges. Adding salt to the water tends to bring about the ordinary noisy white flash of great length.

Contrast between Path of Discharge under circumstances of Hurry and Leisure.

21. When the plates are arranged as in Fig. 51, so that until an *A* spark occurs they are at the same potential and are then filled by a sudden and overflowing rush of electricity, all good-conducting things of the same height are struck equally well, independently of their shape.

But when, on the other hand, the difference of potential between the plates was established gradually, as in Fig. 52, so that the strain in the dielectric had time to pre-arrange a path of least resistance, then small knobs got struck in great preference to big ones, and points could not be struck at all, because they take the discharge quietly.

An intermediate case is when the charge and discharge of the top plate is brought about by pulling a lever over with string, so as to connect it with the jar, as in Fig. 53.

Sparking Distance between Plates in the Different Cases.

Unless the jars are large, compared with the capacity of the plates, even the conditions of Fig. 51 will not make the rush

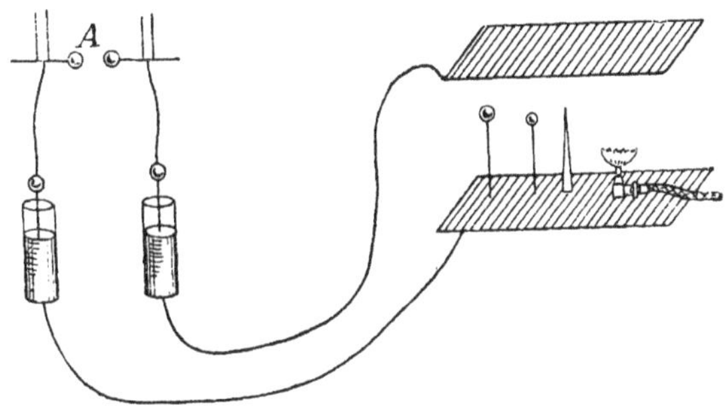

Fig. 51.

quite sudden; and in that case points and small knobs do get struck more easily than large knobs and domes, especially when the top plate is negative.[1] But when the rush is really

Terminal of rod standing on bottom plate.	Sudden rush caused by *A* spark, Fig. 51.	Steady strain, Fig. 52.	Intermediate case, Fig. 53.
Brass knob 1·27 inch diameter	0·93 inch	0·90	0·67
Brass knob 0·56 inch diameter	0·93 ,,	2·95	1·4
Brass point . .	1·03 ,,	At 6 inches it prevented discharge until covered up with a thimble.	—

[1] This fact being emphasized by Mr. Wimshurst.

sudden, no difference as to sign manages to show itself; and even such insignificant advantage as the point happens to show in the first column of the above table disappears.

High resistance, interposed between knob and bottom plate in Fig. 52, alters the character of the spark entirely, making it soft and velvety, but has no effect upon its length nor upon the ease with which its knob gets struck as compared with others connected direct. But the same resistance, interposed in Fig. 51, prevents its being struck altogether.

In other words, sudden rushes strike good conductors, independent of terminal: steady strain selects sharp or small terminals, almost independent of conductivity; the violence of the flash being, however, by high resistance very much altered. The total energy is, doubtless, the same, or even

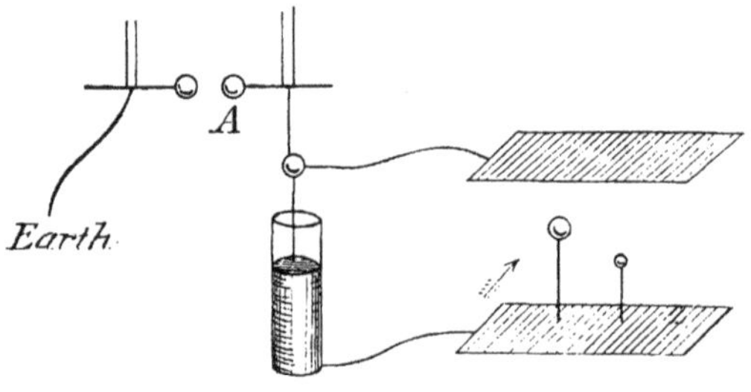

Fig. 52.

greater with the quiet heating spark, because of concentration and no loss by radiation; but the duration of the discharge is what makes the difference. The spark through high resistance, instead of being alternating, can be seen to be intermittent (*i.e.*, multiple), when analyzed in a revolving mirror.

There is no need in these sudden rush experiments for the long leads of Fig. 51, though perhaps they add to the length of the sparks.

22. Sparks thus obtained from the outer coats of jars are convenient for taking under water, or to water; and the phenomena thus seen are singular, and sometimes violent.

Water acts mainly as a dielectric under these circumstances, and, with small electrodes, such as the bared end of a gutta-percha wire, the water between gets burst with extraordinary violence; often breaking the containing glass vessel.

This arrangement of Leyden jars should be handy for blasting operations, because no specially good insulation of the leads is necessary.

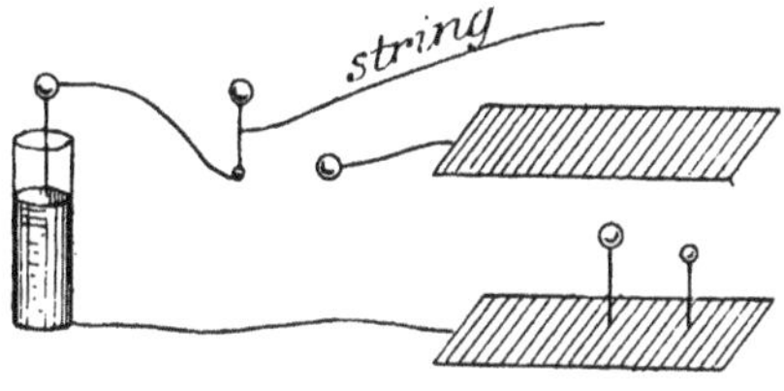

Fig. 53.

Experiments on Surging Circuit Proper.

23. Although all the overflow experiments are controlled by electrical surgings, I have been accustomed specially to apply the name "surging circuit" to the case where sparks are obtained not between two distinct parts of a circuit, but between two points on one and the same good conductor, under circumstances when it does not form the alternative path to anywhere, and when it would ordinarily be supposed there was no possible reason for a spark at all. For instance, in Fig. 54 the loop of wire round the room is a mere off-shoot or appendage of an otherwise complete and very ordinary arrangement, and yet a spark can occur at E whenever the ordinary discharge occurs at A; a spark, too, often quite as long, though not so strong, as the main spark at A.

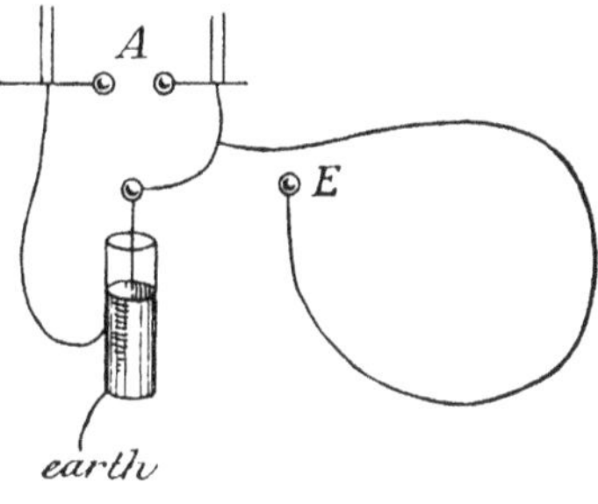

Fig. 54.

The jar is not essential to this experiment; and, in order to analyze it by inserting resistance at various places, it was modified to Fig. 55, and the following readings taken: first, with a thin copper wire, and then with a thick copper wire, round room. The $\frac{1}{4}$ megohm liquid resistance could be inserted at either M, N, O, P, or Q.

The *A* knobs used were the small ones of the universal discharger, 1·4cm. diameter, and 2·4cm. apart all the time

		Length of *E* spark.	Character of *E* spark.
No. 19 copper round theatre.	Resistance inserted at *P*	0·819 cm.	Weak.
	No resistance inserted anywhere	0·597 ,,	Strong.
	Resistance at *P* again .	0·822 ,,	Weak.
	No resistance	0·555 ,,	Strong.
	Resistance at *M* . . .	0·571 ,,	Strong.
	No resistance	0·571 ,,	Strong.
	Resistance at *M* again	0·571 ,,	Strong.
	Resistance at *N* . . .	0·423 ,,	Very weak.
	Resistance at *O* . . .	0·621 ,,	Strong.
	No resistance	0·536 ,,	Strong.
	Resistance at *Q* . . .	No *E* spark at all, and *A* very weak	
No. 1 copper round theatre.	No resistance	0·524 cm.	Strong.
	Resistance at *N* . . .	0·379 ,,	Very weak.
	Resistance at *M* . .	0·638 ,,	Strong.
	Resistance at *Q* . . .	No spark at all, and *A* weak.	
	Resistance at *P* . . .	0·793 cm.	*E* weak but *A* strong.

(equivalent to 1·5cm. spark-length between flat plates). The *E* knobs were those of a spark-micrometer, and were 1·96cm. diameter.

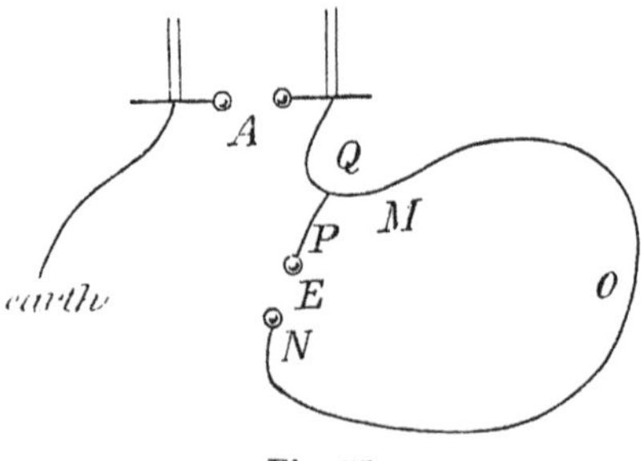

Fig. 55.

This table evidently shows that the main part of the *E* spark is the rushing of the charge in the *N* part of the wire back to the discharged *A* knob. It has two paths, through the wire *viâ* O, and direct across the spark gap E. Most of it chooses E, except when there is high resistance at N or P. Resistance at O interferes but little, and in fact it may help more across E; and resistance at M must certainly have this effect. Resistance at Q prevents any sudden effect of the A spark on the long circuit, and therefore never calls out a spark at E at all: the charged wire discharges leisurely through resistance at Q, and accordingly (there being no jar) the spark at A is quiet.

The fact in the table not immediately intelligible is the extra length of E spark caused by insertion of resistance at P, or, to a less extent, at O. It would appear to indicate the effect of surgings in the conductor, which accumulate a momentary opposite charge on one of the knobs before the one partitioned off by high resistance has had time appreciably to discharge.

Experiments on Recoil Kick.

A number of other experiments, in which the wave-length along wires is measured, are of more purely scientific interest, and are, therefore, not related in detail here. They appear in the Proceedings of the Royal Society for 1891.

CHAPTER XXVIII.

LIGHTNING CONDUCTORS FROM A MODERN POINT OF VIEW.[1]

A LIGHTNING conductor used to be regarded as a conduit or pipe for conveying electricity from a cloud to the ground. The idea was that a certain quantity of electricity had to get to the ground somehow; that if an easy channel were opened for it the journey could be taken quietly and safely, but that if obstruction were opposed to it violence and damage would result. This being the notion of what was required, a stout copper rod, a wide-branching and deep-reaching system of roots to disperse the charge as fast as the rod conveyed it down, and a supplement of sharp points at a good elevation to tempt the discharge into this attractive thoroughfare, were the natural guarantees of complete security for everything overshadowed by it. Carrying out the rainwater-pipe analogue, it was natural also to urge that all masses of metal about the building should be connected to the conductor, so as to be electrically drained to earth by it; and it was also natural to insist on very carefully executed joints, and on a system of testing resistance of conductor and "earth" so as to keep it as low as possible. If ever the resistance rose to 100 ohms it was to be considered dangerous.

The problem thus seemed an easy one, needing nothing but good workmanship and common sense to make accidents impossible. Accordingly, when, in spite of all precautions, accidents still occurred, when it was found that from the best

[1] "Industries," 13th June, 1890.

constructed conductors flashes were apt to spit off in a senseless manner to gun-barrels and bell-ropes and wire fences and water butts, it was the custom to more or less ridicule and condemn either the proprietor of the conductor, or its erector, or both; and to hint that if only something different had been done—say, for instance, if glass insulators had not been used, or if the rod had not been stapled too tightly into the wall, or if the rope had not been made of stranded wires, or if copper had been used instead of iron, or if the finials had been more sharply pointed, or if the earth-plate had been more deeply buried, or if the rainfall had not been so small, or if the testing of the conductor for resistance had been more recent, or if the wall to which the rod was fixed had been kept wet, or etc., etc.—then the damage would not have happened. Everyone of these excuses has been appealed to as an explanation of a failure; but because the easiest thing to abuse has always been the buried earth connection, that has come in for the most frequent blame, and has been held responsible for every accident not otherwise explicable.

All this is now changing or changed. Attention is now directed, not so much to the opposing charges in cloud and earth, but to the great store of energy in the strained dielectric between. It is recognized that all this volume of energy has somehow to be dissipated, and that to do it suddenly may be by no means the safest way. Given a store of chemical energy in an illicit nitro-glycerine factory, it could be dissipated in an instant by the blow of a hammer, but a sane person would prefer to cart it away piecemeal and set it on fire in a more leisurely and less impulsive manner. So also with the electrical energy beneath a thundercloud. A rod of copper an inch or a foot thick may be too heroic a method of dealing with it; for we must remember that an electric discharge, like the recoil of a spring or the swing of a pendulum, is very apt to overshoot itself, and is by no means likely to exhaust itself in a single swing. The hastily discharged cloud, at first suppose positive, over-discharges itself and becomes negative;

this again discharges and over-discharges till it is positive as at first, and so on, with gradually diminishing amplitude of swing, all executed in an extraordinarily minute fraction of a second, but with a vigour and wave-producing energy which are astonishing. For these great electrical surgings, occurring in a medium endowed with the properties of the ether, are not limited to the rod or ostensible conduit; the disturbance spreads in all directions with the speed of light, and every conducting body in the neighbourhood, whether joined to the conductor or not, experiences induced electrical surgings to what may easily be a dangerous extent. Not only is there imminent danger of flashes spitting off from such bodies for no obvious reason—splashes which, on the drainpipe theory, are absolutely incredible—flashes sometimes from a perfectly insulated, sometimes from a perfectly earthed, piece of metal; but, besides this, remember that near any considerable assemblage of modern dwellings there exists an extensive metallic ramification, in the gas-pipes, that these are in places eminently fusible, and that the substance they contain is readily combustible.

On the drain-pipe theory, the gas-pipes, being perfectly earthed, would be regarded as entirely safe, so long as they were able to convey the current flowing along them without melting; but, on the modern theory, gas-pipes constitute a widely spreading system of conductors able to propagate disturbance underground to considerable distances, and very liable to have some weak and inflammable spot at places where they are crossed by bell wires or water-pipes or any other metallic ramification.

Above ground we have electrical waves transmitted by the ether, and exciting surgings throughout a neighbourhood by inductive resonance. Below ground we have electrical pulses conveyed along conductors, leaking to earth as they go, but retaining energy sufficient to ignite gas wherever conditions are favourable, even at considerable distances.

The problem of protection, therefore, ceases to be an easy

one, and violent flashes are to be dreaded, no matter how good the conducting path open to them. In fact, the very ease of the conducting path, by prolonging the period of dissipation of energy, tends to assist the violence of the dangerous oscillations. The drain-pipe theory, and the practical aphorisms to which it has given rise, would serve well enough if lightning were a fairly long-continued current of thousands of amperes urged by a few hundred volts, or if there were no such thing as electro-magnetic inertia; but, seeing that the inverse proportion between amperes and volts better corresponds to fact, and seeing that the existence of electro-magnetic inertia is emphasized by multitudes of familiar experiments, the drain-pipe theory breaks down hopelessly, and only a few of its aphorisms manage to survive it.

What, then, are we to set up in place of this shattered idol? First of all we can recognize, what was virtually suggested by Clerk-Maxwell, that the inside of any given enclosure, such as a powder magazine or dynamite factory, can, if desired, be absolutely protected from internal sparking by enclosing it in a metallic cage or sheath, through which no conductor of any kind is allowed to pass without being thoroughly connected to it. The clear recognition of the exact, and not approximate, truth of this statement is a decided step in advance, and ought to be satisfactory to those who have to superintend the practical protection of places sufficiently dangerous or otherwise important to make the aiming at absolute security worth while. Similarly, for wire-covered ocean cables absolute protection is possible.

But not for ordinary buildings, any more than for ordinary land telegraph offices, is such a plan likely to be adopted in its entirety. Some approximation to the cage system can be applied to ordinary buildings in the form of wires along all its prominent portions; and such a plan I suggested, and I understand it was carried out, for the entrance towers and part of the main body of the recent Edinburgh Electrical Exhibition, Mr. A. R, Bennett having asked me to recommend a plan to the committee as a sort of exhibit. For chimneys a set of four galvanized iron wires, joined by hoops at occasional intervals, and each provided with a fair earth, seems a satisfactory method; but it is to be noted that a column of hot air constitutes a surprisingly easy path, and that it is well to intercept a flash on its way down the gases of a chimney by a copper hoop or pair of hoops over its mouth. Mr. Goolden tells me that he has just applied this method to a new chimney at his works in the Harrow Road. For ordinary houses, a wire down each corner and along the gables is as much as can be expected. At many places even this will not be done; a couple of vertical wires from the highest chimney stacks on opposite sides must be held better than nothing, or than only one.

Earths will be made, but probably they will be simple ones, entailing no great expense. A deep damp hole for each conductor, with the wire led into it and twisted round an old harrow or a load of coke, may be held sufficient. And as to terminals: rudely sharpened projections, as numerous as is liked, may be arranged along ridges and chimney stacks; but I have at present no great faith in the effective discharging power of a few points, and should not be disposed to urge any considerable expense in erecting or maintaining them. Crowns of points on chimneys and steeples are certainly desirable, to ward off, as far as they can, the chance of a discharge, but a multitude of rude iron ones will be more effective than a few highly sharpened platinum cones. *I find that points do not discharge much till they begin to fizz and audibly spit; and when the tension is high enough for this, blunt and rough terminals are nearly as efficient as the finest needle points.* The latter, indeed, begin to act at comparatively low potentials, but the amount of electricity they can get rid of at such potentials is surprisingly trivial, and of no moment whatever when dealing with a thundercloud.

But the main change I look for in the direction of cheapness and greater universality of protection is in the size and material of the conducting rod itself. No longer will it be thought necessary to use a great thick conductor of inappreciable resistance; it will be perceived that very moderate thickness suffices to prevent fusion by simple current strength, and that excessive conducting power is useless.

In the days when the laws of common "divided circuits" were supposed to govern these matters, the lightning rod had to be of highly conducting copper, and of such dimensions that no other path to earth could hope to compete against it. But now it is known that low resistance is no particular advantage: it is not a question of resistance. The path of a flash is a question of impedance; and the impedance of a conductor to these sudden rushes depends very little on cross-section, and scarcely at all on material. A thin iron wire is nearly as good as a thick copper rod; and its extra resistance has actually an advantage in this respect, that it dissipates some of the energy, and tends to damp out the vibrations sooner. Owing to this cause a side flash from a thin iron wire is actually less likely to occur than from a stout copper rod.

The only limit is reached when the heat generated by the current fuses the wire, or runs the risk of fusing it. But in so far as oscillations are prevented, the mean square of current strength, on which its heating power depends, is diminished. Accordingly, a fairly thick iron wire runs no great risk of being melted. Its outer skin may, indeed, be considerably heated, for these sudden currents keep entirely to the outer skin, penetrating only a fraction of a millimetre into iron, and may make this skin intensely hot. But the central core keeps cool until conduction has time to act; and, consequently, unless the wire is so thin as to be bodily deflagrated by the discharge, its continuity is not likely to be interrupted. Thickness of wire is thus more needed in order to resist ordinary deterioration by chemical processes of the atmosphere than for any other reason.

But the liability to intense heating of the outer skin should not be forgotten, and care should be taken not to take the wire past readily inflammable substances for that reason. For instance, it would be madness to depend on Harris's notion that a lightning conductor through a barrel of gunpowder was perfectly safe, especially if said conductor were an iron wire or rod.

In the old days a lightning conductor of one or two hundred ohms resistance was considered dangerously obstructive, but the impedance really offered by the best conductor that ever was made to these sudden currents is much more like 1,000 ohms. A column of copper a foot thick may easily offer this obstruction, and the resistance of any reasonably good earth connection becomes negligible by comparison. A mere wire of copper or iron has an impedance not greatly more than a thick rod, and the difference between the impedance of copper and iron is not worth noticing.

But although, in respect of obstructing a flash, copper and iron and all other metals are on an approximate equality, it is far otherwise with their resistances, on which their powers of dissipating energy into heat depend. It is generally supposed that iron resists seven times more than copper of equal section, and so it does steady currents, but to these sudden flashes its resistance is often 100 times as great as copper, by reason of its magnetic properties. This statement is quite reconcilable with the previous statement, that in the matter of total obstruction there is very little to choose between them; the apparent paradox is explicable by the knowledge that rapidly varying currents are conveyed by the outer skin only of their conductor, and that the outer skin available in the case of magnetic metals is much thinner than in the case of non-magnetic.

Questions about shape of cross-section are rather barren. Thin tape is electrically better than round rod; but better than either is a bundle of detached and well-separated wires—for instance, a set of four, one down each cardinal point of a chimney; but it is easy to over-estimate the advantage of large surface as opposed to solid contents of a conductor. The problem is not a purely electrical one—it is rather mixed. The central portion or core of a solid rod is electrically neutral, but chemically and thermally and mechanically it may be very efficient. It confers permanence and strength; and the more electrically neutral it is, the less likely it is to be melted. Its skin may be gradually rusted and dissolved off, or it may be suddenly blistered off by a flash; but the tenacity of the cool and solid interior holds the thing together, and enables it to withstand many flashes more. Very thin ribbon or multiple wire, though electrically meritorious, is deficient in these commonplace advantages. Painting the surface has no electrical effect either way.

There were two functions attributed to high conducting power in the old days—first, the overpowering of all other paths to earth; second, the avoidance of destruction by heat. The first we have seen to be fallacious; on the second a few more explanations can be made. In so far as fusion by simple current strength is the thing dreaded, it must be noticed that a good conductor has no great advantage over a bad conductor. It is a thing known to junior classes that, when a given current has to be conveyed, less heat is developed in a good conductor; but that, when an electro-motive force is the given magnitude, less heat is developed in a bad conductor. The lightning problem is neither of these, but it has quite as much relationship to the second as to the first. There is a given store of energy to be got rid of, and accordingly the heat ultimately generated is a fixed quantity. But the rise of temperature caused by that heat will be less in proportion as the production of it is slow; and though by sudden discharge a quantity of the energy can be made to take the radiant form, and spread itself a great distance before final conversion into heat, instead of concentrating itself on the conductor, yet this cannot be thought an advantage. For, just as in the old days a lightning rod was expected to protect the neighbourhood at its own expense by conveying the whole of a given charge to earth, so now it must be expected to concentrate energy as far as possible on itself, and reduce it to a quiet thermal form at once; instead of, by defect of resistance and over-violent radiation, insisting on every other metallic mass in its neighbourhood taking part in the dissipation of energy.

The fact that an iron wire, such as No. 5 or even No. 8 B.W.G., is electrically sufficient for all ordinary flashes, and that resistance is not a thing to be objected to, renders a reasonable amount of protection for a dwelling-house much cheaper than it was when a half-inch copper rod or tape was thought necessary.

A recognition of all the dangers to which a struck neighbourhood is liable, doubtless prevents our feeling of confidence from being absolute in any simple system of dwelling-house protection; but at the same time an amount of protection superior to what has been in reality supplied in the past is attainable now at a far less outlay; while, for an expenditure comparable in amount to that at present bestowed, but quite otherwise distributed, a very adequate system of conductors can be erected.

Only one difficulty do I see. In coal-burning towns galvanized iron wire is, I fear, not very durable; and renewal expenditure is always unpleasant. It is quite possible that some alloy or coating able to avoid this objection will be forthcoming, now that inventors may know that the problem is a chemical one and that high conductivity is unnecessary.

PART II.

CHAPTER XXIX.

ON LIGHTNING GUARDS FOR TELEGRAPHIC PURPOSES, AND ON THE PROTECTION OF CABLES FROM LIGHTNING,

With Observations on the Effect of Conducting Enclosures.[1]

1.

In the paper which I read before the Institution last year, I spoke at length and showed some experiments on the subject of the protection of buildings from lightning; but although there were a few sections in that paper, as ultimately printed, dealing with lightning protectors for telegraphic instruments and cables, yet time prevented my calling attention to this portion of the subject at the meeting, or of showing any experiments in connection with it.

The present communication may be regarded as a development of this omitted but important branch of the subject.

I do not know whether I shall escape controversy this time, but I have no wish to provoke it; and I wish entirely to avoid decrying the merits of other specific protectors in order to emphasize the advantages of my own. All I have to say on this head will have a quite general application, and amounts to about this: that whereas all protectors proceed on the admitted fact that the greater portion of a sudden flash prefers

[1] Excerpt, "Journal of the Proceedings of the Institution of Electrical Engineers," Part 87, vol. xix.

to jump an air space, rather than traverse a moderate length of wire, the cause of this well-known fact was but imperfectly appreciated; could not, indeed, be perfectly appreciated without recognizing the rapid oscillatory character of sudden discharges, and the doctrine of impedance, or immense obstruction which good conducting wires offer to currents of this character: an impedance, it may be, of 100 or 1,000 ohms, whereas the resistance to steady currents may be but a ten-thousandth part of this amount.

In a word, the complete theory of lightning protectors has not till recently been known, and accordingly the many ingenious devices which are in use are naturally deficient in details which a completer recognition of theory would have suggested.

2. Long ago it was shown by the experiments of M. Guillemin and Professor Hughes that existing lightning protectors were not perfect safeguards, inasmuch as they could not protect a fine wire from deflagration by a Leyden jar discharge. At the same time it was admitted by these experimenters, and is, I understand, a matter of common experience, that almost any lightning protector is better than none at all, and that the best of those in use are by no means inefficient instruments. They do not afford perfect security, but the occasions when they partially fail are perhaps not very numerous. It is therefore the practice to establish some sort of proportion between the elaborateness of the protector and the value of the apparatus to be guarded. In telephone exchanges a simple double comb is used. In telegraph offices a pair of plates finely adjusted close together is the form employed. While at cable stations the most elaborate kind of protectors, and often a number of different ones in combination, are arranged, because of the enormous interests at stake.

3. In electric light installations, on the other hand, it is customary, I believe, at present to use no protector at all. But I cannot suppose that this will much longer be regarded as a reasonable procedure. When one considers the extensive

ramifications of conducting leads which are rapidly growing up, and the liability of the neighbourhood of house lightning conductors to some portion of such ramification, it is apparent, I think, that lightning arresters are just as important for electric light leads as for telegraph wires. However, there is no need for me to emphasize this assertion, for I expect that before long a flash from somebody's lightning conductor getting into some electric light leads and burrowing underground throughout a district will do an amount of damage sufficient to eloquently call attention to the danger. At Peterhouse, Cambridge, lightning got at the steam engine, thence to the dynamo, and thence into the leads; where, I believe, it did little more than fuse a large number of cut-outs. But it is not likely always to spare the lamps in so considerate a manner; while danger of an altogether more disastrous kind can by no means be considered absent.

4. Returning from this digression to the telegraphic protectors at present in use, my contention is that, in all except the very simplest and crudest form of protector, a perception of the true conditions may lead to more complete protection without necessarily much increase in cost. And, proceeding to cases where cost is altogether a secondary consideration, and supposing it possible that telegraph engineers are fairly satisfied with the quality of the protection at present supplied to submarine cables, I would ask them (not pretending to know anything at first hand about the matter) whether a number of obscure and unexplained faults do not develop themselves in the gutta-percha of cables; whether it is not possible that some of these are due to electric waves of high potential which have got into the core and punctured its dielectric at previously weak spots; whether, indeed, such faults are not found occasionally to develop in the wake of a storm.[1]

[1] Dr. Alexander Muirhead tells me that condensers have several times been returned to him perforated by lightning; and that he is of opinion that many cable faults arise from the same cause, for he

5. I would also point out that the fact that a lightning switch is found damaged, or its wire perhaps fused by a flash, is no proof that it has entirely protected the cable from hostile influence. It shows that it has done the best it can; but since that best falls short of perfection, the damage to the protector, proving that there has been occasion for its activity, proves likewise that the cable has been in a position of danger, has received some proportion of the damaging current, and in all probability has suffered some amount of deterioration. The voltage which plays about in the very feeblest form of statical discharge is so enormously greater than anything ever developed by ordinary voltaic batteries that one cannot regard with equanimity the entrance of any portion of such electromotive forces into the delicate vitals of a cable. Remember that a millimetre spark means 3,000 volts.

6. To illustrate by experiment the assertion I have virtually made—that no conceivable form of single air gap, whether it be between knobs or points or plates or wires, or between wire and tube, or coil and cylinder, or any other possible device, can possibly afford complete and adequate protection—I take the following simple instances:

finds that if the gutta-percha is punctured by a spark it does not show itself as a fault for three or four months afterwards.

Mr. Gott, chief electrician of the Commercial Cable Company, sent me last year an account of damage done to lightning guard and instruments at Canso, Nova Scotia, and I communicated it to "The Electrician" for July 12th, 1889.

The following is an extract from the Cable Company's superintendent at Canso:

"On the northern cable, the signalling coil (Thomson recorder) and lightning guard were fused, a one-microfarad section of condensers in cable-sending block destroyed (short-circuited), and a five-microfarad section artificial-line block damaged. Insulation reduced to 15,000 ohms. On the southern cable, the signalling coil and lightning guard were also fused, and two ten-microfarad sections of cable block destroyed. The fine platinum wires in lightning guards were entirely dissipated. Owing to violent kicks, cables had been earthed ten minutes before thunderstorm broke over the station."

Experiment No. 1.

Arrange a small air gap, as, for instance, between the knobs of a common "universal discharger;" connect the ends of a tangle or loop of thin covered wire, one on each side of the gap, and pass a Leyden jar discharge across it. The insulation of the thin wire is sparked through wherever two portions happen to come close enough together. [The wire actually employed had been the secondary of an old induction coil, and had been well soaked in paraffin.]

Experiment No. 2.

Diminish the air gap until the knobs actually touch. The sparking at the crossings of the tangle are rather less bright and numerous than before, but they still frequently occur.

Experiment No. 3.

Instead of an air gap, however short, use a foot or two of stout No. 0 copper wire or rod of highest conductivity, and shunt this with a thin wire tangle or coil. On passing a discharge along the rod, the crossings of the tangle sparkle as before, showing that the double insulation is still broken down.

7. These extremely simple observations establish the position I hold that no ordinary protector can switch the whole of a flash out of a coil; for a solid copper rod, no matter how thick, is unable to do it, although such a shunt as that would divert every appreciable vestige of signalling or other useful current. The fact is that to sudden discharges the impedance of a short copper rod may run as high as 100 ohms; and if the discharge current while it lasts is 100 amperes, which is a very moderate value, then we see at once that the E.M.F. needed to drive the current through the rod is 10,000 volts, and this is therefore for an instant the difference of potential between its ends. Such a difference of potential can leap three or four millimetres of air, and is amply sufficient to

burst through the insulation of silk-covered wire several times over.

8. The particular mode adopted in these and such-like experiments for sending a Leyden jar discharge through the rod or across the air gap is quite immaterial; any convenient plan serves, from simple hand discharging tongs upwards. But, once more, I may say that the handiest method of working is to use a couple of jars, to connect their internal coatings with the terminals of an electrical machine, the distance apart of whose terminals regulates the energy of the discharge employed; and to connect the outer coats of the jars to the two ends of the rod, or to the two rods of the universal discharger, or to whatever the flash is wanted to pass through.

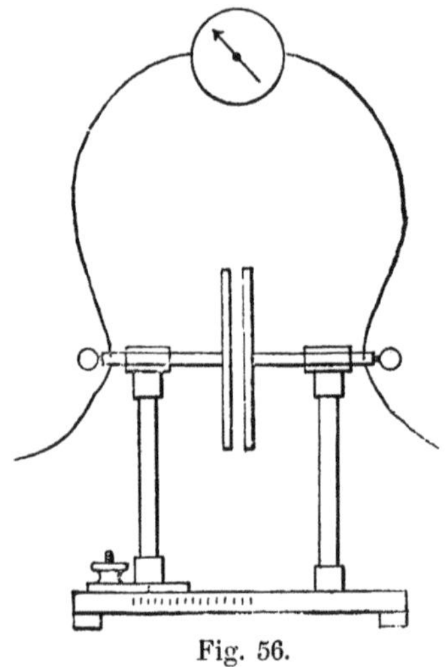

Fig. 56.

9. Instead of a wire tangle or loop, which only represents the coil of an instrument, and whose only advantage is that one has no compunction in spoiling it, I next proceed to employ an actual galvanometer, and take it as representing anything requiring protection, whether it be telegraph instrument or cable or electric light installation of any kind. A galvanometer serves pretty well, because it indicates whether any disturbing current passes round it or not. Notice, how ever, that it indicates properly only when the discharge really goes round the coil. If it jump across insulation, and, still more, if it jump from terminal to terminal, the needle indicates less or nothing. There are, in fact, two things to be aimed at and considered separately in the protection of instruments:

First: the current has to be prevented from passing round the coil, and thus disturbing the magnetism of the needle.

Second: it has to be prevented from jumping across from layer to layer, and so damaging the insulation.

Conduction protection, and insulation protection: both must be attended to. To carry out both these protections completely is not easy; but I apprehend that the protection of the insulation from permanent damage is usually more important than the elimination of temporary disturbing currents which interfere with signalling and make the needle kick while they last.

10. The galvanometer I use is a simple reflecting instrument, of a pattern devised by Professor Stuart at Cambridge some time ago. A copper iron junction momentarily touched with the finger demonstrates that it is in a sensitive condition, and could easily be made to furnish its effective constant were it worth while. Connecting its terminals to the outsides of the pair of Leyden jars whose knobs are attached to the machine, separating the machine terminals about the fiftieth of an inch, and turning slowly, we see an attempted steady deflection representing the charging current, interrupted by a series of reverse kicks which occur at every minute discharge; though the sparks of any discharge such as it is safe to send through the instrument are too faint to be heard.

Experiment No. 4.

Now insert between the wires leading to the galvanometer a sort of lightning guard: a pair of plates mounted on a sliding arrangement, so that their distance can be varied (Fig. 56). Directly the plates approach within sparking distance, the galvanometer is apparently protected, and its kicks cease, even though by further separating the machine terminals the energy of the discharges be increased.

Interpolated Experiment No. 5.

11. Although not immediately important, a little fact may here be noted. If the plates of the guard are pushed still nearer together, or lightly pinched together, so as to leave only a microscopic interval, and almost to obliterate the existence of a spark between them, the needle of the galvanometer again begins to kick at every discharge; but this time wildly and irregularly, and sometimes in the reverse direction. Occasionally these disturbances are very strong, and the spot of light disappears; only to be recovered by tapping the instrument. At the instant when these kicks occur, the plates are momentarily short-circuited, as may be proved by replacing the galvanometer by a Léclanché and electric bell. The bell is liable to ring at every discharge, and obviously for the same reason as the galvanometer kicks.

But whereas the bell only proves momentary conducting contact, the galvanometer proves this plus an electro-motive force of occasionally uncertain direction and always uncertain magnitude. This E.M.F. would seem possibly to have something to do with the infinitesimal spark which temporarily connects the plates, and suggests an E.M.F. like the E.M.F. in an arc.[1]

[1] When I first came across this effect some weeks ago it was with tinfoil inserted between a pair of sparking terminals, and after the kick the tinfoil was often found fused on to one or other terminal. I therefore put it down to a thermal junction between tin and copper, caused and excited by the heat of the spark; and since it was a toss-up which side of the tinfoil adhered (being the side which made a just imperfect enough contact), the fluctuations of direction were easily accounted for; and by special trial this explanation was found to hold good. But then in this case after one kick the galvanometer was quiescent at subsequent discharges, until the tinfoil was disturbed sufficiently to break the fused contact again.

The experiment described in the text differs, in that there are no two metals, the metal is not fusible, the infinitesimal spark occurs between a pair of similar brass plates, or knobs, and the short-circuiting is quite temporary. I surmised therefore, that it might

12. Returning now to Experiment 4, which illustrated the protection afforded to a galvanometer by a shunting air gap, it is natural to ask how it can be reconciled with our previous observation, that protection by a single air gap was impossible. But remember that complete protection involved two things, conduction protection and insulation protection; and the non-deflection of the needle merely shows that no important quantity of electricity now finds its way round the coil. But is the galvanometer therefore safe? By no means. Its insulation is in imminent danger, and if but moderately energetic flashes are employed we shall see sparks leaping across the coils or jumping from them to the metal work in a very pronounced manner.

Experiment No. 6.

As this is rather rough on the galvanometer, I prefer not to use strong flashes, but to show the existence of the tendency with smaller ones by arranging a safety-valve or supplementary minute air gap between the galvanometer terminals; said safety-valve being either a couple of pins brought close together or a chink cut across a narrow strip of tinfoil pasted on glass.

The high electro-motive forces which are endangering the insulation, and very likely already jumping in invisible places, can now demonstrate their existence by leaping this chink; and no matter how the lightning-guard plates are arranged, it is impossible to check the little sparks occurring at the

possibly be a phenomenon of greater interest than a mere thermo-electric one. But after the reading of the present paper, Professor Hughes informed me that he had come across the very same effect, and had satisfied himself that it was only a thermo-electric one. I now think it possible that he is correct, and that the junction is caused by a momentary heat-pimple after the fashion of a Trevelyan rocker or Gore's circular railway. And though this plausible explanation deprives the observation of any theoretical interest, I do not blot it out of the text, but leave it as a record in order to save the time of future experimenters who may easily come across the same thing.

safety-valve at every flash. Bringing the plates into absolute contact lessens the brightness of these sparks, but does not stop them; neither does replacing the plates by a solid bar of metal, as in Fig. 57.

13. But, directly the safety-valve is employed to filter off and render manifest the residual effects left by a lightning guard, the galvanometer needle begins to kick again whenever it acts; so that, singularly enough, whereas when no safety-valve was employed the galvanometer appeared protected by the lightning switch, inasmuch as its needle is stationary, directly the safety-valve is added and allowed to sparkle the needle kicks wildly with the very same flashes as before it ignored.

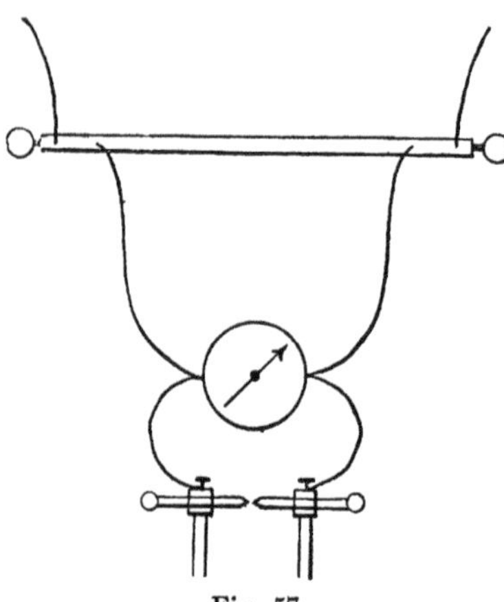

Fig. 57.

This behaviour appears mainly due to the same cause as produced the disturbance in the experiment interpolated above;[1] but it may be partly due to the weakening down of the flash by the safety-valve so much that a residue of it is able to make its way round the coil, whereas it had previously been too strong and preferred jumping across insulation in a manner ineffective for galvanometry. The possibility of the occasional truth of this latter explanation is by no means to be overlooked. It often happens that an unprotected galvanometer has its needle less strongly affected by sparks say an eighth of an inch long, than by sparks the fiftieth of an inch long; but the reason obviously is that the more

[1] I do not feel quite certain of the sufficiency of Prof, Hughes' thermo-electric explanation,

violent discharges are not really passing round the coil, but are taking all manner of short cuts.

14. Instead of employing a rigged-up model of a protector, for the purpose of calling attention to principles, an actual lightning guard may of course be used, and its behaviour studied in detail. I take, as an excellent instance, a Saunders protector, so much employed in connection with submarine cables. Its essential part is a fine wire, through which the useful currents have to pass, surrounded by an earth tube with points protruding towards the thin wire, which is stretched by a spring along its axis. The idea is that the tube will relieve the wire of dangerously high potential, while the wire itself will fuse if dangerously strong currents try to pass along it. A supplementary device is a short-circuiting contact, whereby a spring puts the cable to earth directly the wire is fused or in any other way broken. The diagram (Fig. 58) sufficiently represents the instrument. A Jamieson protector is another neat arrangement, involving a fine wire as well as an air gap; there is also some wire coiled on a metal cylinder, but the metal deprives it of all appreciable self-induction.

Experiment No. 7.

If I now send very small discharges down the line wire towards a galvanometer protected by this Saunders guard, they will, if small enough, escape it and pass through the galvanometer, either exciting its safety-valve or disturbing its needle, or both. But if stronger flashes are sent, the protector begins to act, sparks are seen between fine wire and surrounding tube, and the galvanometer disturbances diminish as already explained; but those at the safety-valve never wholly cease. One is not to suppose that the bigger the flash the less the effect; that would be a very desirable but somewhat impossible conjuncture. The smallest effect is got with the weakest flash which is just sufficient to spark to the protecting tube. Anything stronger or anything not greatly weaker than this gives larger effects.

15. The fine wire part of this guard is a good feature and one that may be advantageously introduced into any protector, on the principle of a safety-fuse or cut-out, to eliminate steady or slowly-varying currents of too great strength. But the short-circuiting of the protected terminal to earth by the terminal *D* as soon as the wire is destroyed, is perhaps not an unmixed good, for a subsequent earth-seeking disturbance

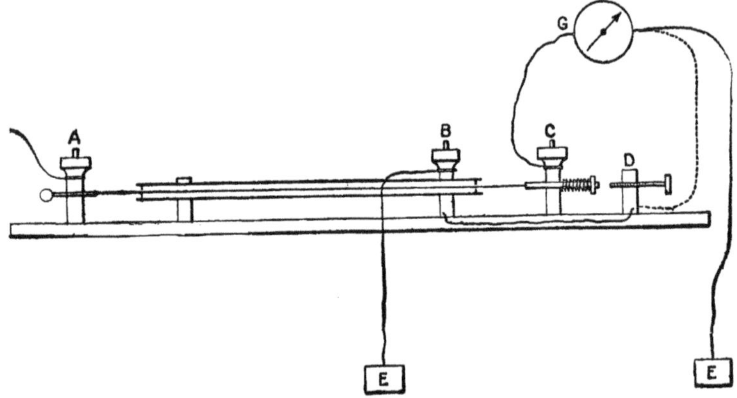

Fig. 58. MR. SAUNDERS' GUARD.

has then two paths between which it may divide—if the connections are made as in Fig. 58—one along the intended earth wire, and the other through the instrument intended to be protected.

True, this latter is a much longer route, but not so infinitely longer that it need convey none; especially if a few short cuts across insulation can be taken.

We see, at any rate, how important it is to make the earth lead as short and direct as possible, and how it is better to connect the thing to be protected to the two binding screws *C* and *D* of the instrument as indicated by the dotted line rather than to make an independent earth.[1] If, for instance, a cable be joined inside to C, outside to B or D, the spring short-circuiting is all good, the only objection being that however prompt the spring may be, there is ample time for damage before the contact can be made.

16. The statement concerning the importance of a short and direct earth also applies to the desirable mode of connecting up lightning guards in general. They should always be inserted direct into the line circuit between line and earth, never be simply led up to by side wires.

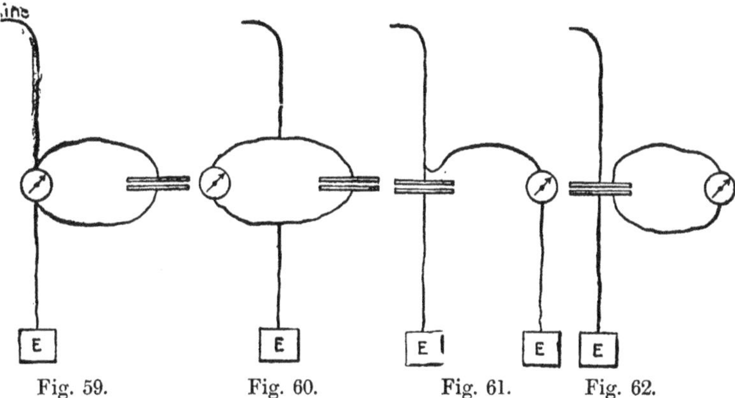

Fig. 59. Fig. 60. Fig. 61. Fig. 62.

Thus, of the various modes of connecting a plate or other lightning protector to a telegraph instrument, Fig. 59 is an altogether bad mode; Fig. 60 is but little better; Fig. 61 is the same thing, or even worse; while Fig. 62 is a good way. As I have said so often, even this is not perfection, but it is quite the best that can be done with a single air gap of whatever kind.

Throw the leads into the protected circuit; let nothing interfere with direct connection of lightning switch to line

[1] The dotted line connection to D in Fig. 58 is the most favourable possible, and it was the one used at the meeting. It is far better than an independent earth for the galvanometer terminal.

and earth respectively, and don't use an independent earth for your instrument's earth terminal. The time taken for a disturbance to travel even a foot of copper wire is by no means to be despised, notwithstanding that it travels with the speed of light; and the impedance of every inch tells.

17. I will now describe the principle of my own lightning guard, and will then connect it to the circuit in place of Saunders', and show that it affords practically complete protection for both small and big flashes.

The principle is one very easy to understand. It is merely to take the overflow from one protector and give it the chance of another, then to take the overflow from this and offer it

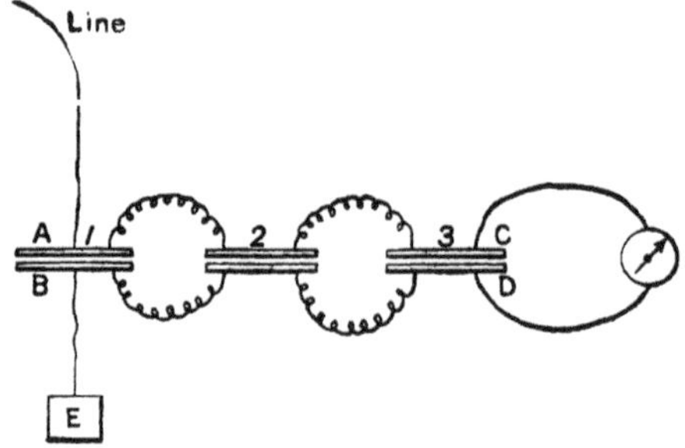

Fig. 63. PRINCIPLE OF LODGE'S GUARD.

another air gap, and so on till nothing is left; at the same time diminishing the overflow from each protector as much as possible by the use of highly insulated small self-induction coils, which impede the violently varying or alternating rushes by their electro-magnetic inertia; the best shape of these coils being the flat, or collar-box, shape employed by Hughes in his induction balance.

Thus, for instance, using a series of plate protectors, we may couple them up as shown in Fig. 63; the coils being only diagrammatically indicated.

Only a small fraction of a sudden shock will escape No. 1; say a thousandth part, since it is offered the inertia or impedance of the coils as the only alternative. A thousandth of this again may escape No. 2; and a thousandth of this, or a thousand-millionth of the whole, is all that is left for the galvanometer. By adding to the series, if it were necessary, it is manifest that a disturbance may be diluted down to any desired extent, with the rapidity of a geometrical progression.

18. I do not, indeed, purpose to use plates commonly; simply because they are more bulky than necessary, not so easy to adjust, and not so open to inspection as knobs or points. Moreover, the first or exposed pair of such a series is likely to be damaged by lightning; and when damaged, it may be permanently short-circuited or otherwise inconveniently altered in a troublesome and invisible manner. I prefer that the terminals of the first air gap shall be easy to examine, easy to remove, and cheap to replace.

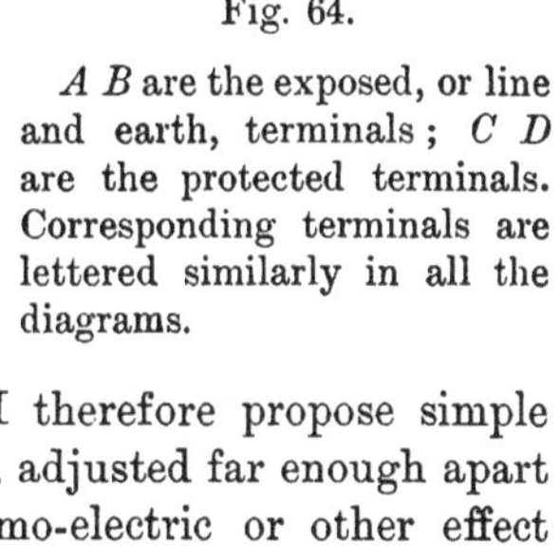

Fig. 64.

A B are the exposed, or line and earth, terminals; *C D* are the protected terminals. Corresponding terminals are lettered similarly in all the diagrams.

The last air gap of a series I prefer to be very finely adjustable indeed, with screw adjustment; and all of them should be open to inspection, so as to avoid accidental contact on the one hand, or undue air space on the other. I therefore propose simple brass rods for the exposed air gap, adjusted far enough apart to exclude that disturbing thermo-electric or other effect caused by a strong spark occurring between their surfaces. All rods must be short, so that heat-expansion may not short-circuit them. The theoretically best place to tap off the useful current is from near the tips in contact, so as to tap off a minimum of impedance with a maximum, as thus, Fig. 64:

19. Making a rough model of such an arrangement, with coils of a few yards of stout gutta-percha-covered wire, wound on cotton reels, I tested it by inserting a scrap of extremely

fine wire between *D* and *C*, by holding them with wet fingers, and so on; but was unable to fuse the finest wire, or to feel any disturbance, although great flashes were going to *A* and *B*, and the early air gaps were sparking properly. Large condensers, composed of great piles of window-glass—the same condensers as I had used for obtaining very slow oscillation, and the discharge of which had a powerful deflagrating effect, were used in this experiment, as well as more moderate

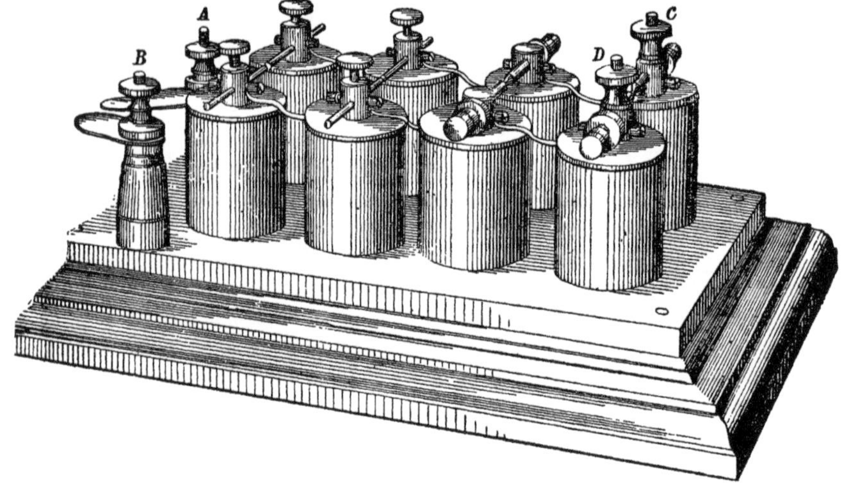

Fig. 65. DOUBLE FORM OF LODGE'S GUARD.

capacities, such as a Leyden jar battery and single jars; but still no effect at the protected terminals. This was the stage I had reached when I read my paper on Lightning Conductors to the Institution of Electrical Engineers, see p. 179 above.

20. I have now to report that Dr. Alex. Muirhead has kindly made me an actual and highly-finished instrument on this plan; designing it himself. This instrument I have here (Fig. 65); and I have also received from him this round and more compact and, I suppose, cheaper form of the same instrument (Fig. 66); with one-half the coils omitted, accord-

ing to a plan of mine which, though not in all respects so theoretically satisfactory, may answer well enough for many purposes—as I shall mention further on. The width of the air gaps I adjust to the thickness of millboard, cartridge-paper, note-paper, and tissue-paper respectively.

21. I now insert the first-mentioned protector in the path from Leyden jars to galvanometer, and I show that with all-sized flashes, from the smallest to the biggest here practicable, the galvanometer is protected; its needle does not move, neither does its safety-valve sparkle. Examining it still more strictly and rigorously, I find that the insulation protection is indeed quite perfect, but a faint trace of a wave does pass round the galvanometer wire and affects the needle slightly when in a sensitive condition. To see this clearly, one must allow the needle to recover from the disturbance due to the charging current before permitting the discharge. The galvanometer shows easily the charging current from a common frictional plate machine; in fact, the spot of light moves several inches with such a current. Connecting it up through the protector, with a small single-pair-of-plates inductive machine charging the jars and giving pretty long sparks, the usual occurrences observed are as follows: between each spark, while the jars are charging, the spot of light deflects considerably, gradually less as the jars get fuller, until they discharge; when, instead of a kick back, as one might expect, there is rather a leap forward, by reason of the suddenly restored strength of the now almost unopposed charging current.

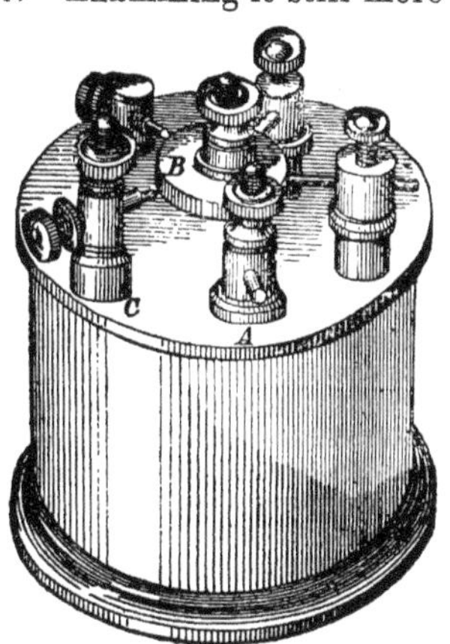

Fig. 66. SINGLE FORM OF LODGE'S GUARD.

22. Whenever we thus want to see the charging current, we

must of course make fair metallic contact with the outsides of the jars, so as to close the circuit through the galvanometer; but, in order to further test the instrument, I make a break at one or both of its terminals, and allow flashes to strike either A or B, or both. It is not really a more severe test than the other—not quite so severe, in fact; but it looks worse, perhaps, and at any rate it is the simplest plan of keeping the charging current out of the galvanometer, and so securing that the needle shall be ready to indicate the slightest effect which the unfiltered-off portion of the discharge is able to produce. But there is practically no effect to be seen.

23. You observe that my instrument is symmetrical, *i.e.*, that there are just the same coils in its earth-connected as in its insulated portion. It may easily be supposed that this doubling of the parts is unnecessary, or even deleterious. It may be plausibly argued that coils interposed in the earth connection are bad as well as useless, and that both the terminals on that side, B and D, should be agglomerated together: that, in fact, a preferable pattern would be that shown in Fig. 66, where there is only one earth terminal towards which all the others point. I think, indeed, that there are many cases where this will serve sufficiently well, but I do not regard it as so theoretically sound as the other, for this reason. It proceeds on the assumption that every disturbance arrives from above and is anxious to make its way to the earth. But we have no guarantee that such shall be always the case: disturbances are as likely to reach the apparatus from below, surging up from earth to line, and in that case the coils are wanted in the earthed half of the instrument. I return to this question later.

24. When I speak of a disturbance travelling from earth to sky, instead of from sky to earth, I do not mean that in one case the sky is negative and in the other positive. Questions about sign of charge have nothing to do with it. There is some amount of unnecessary haze abroad on this matter. Think of a sudden electrical disturbance imparted to a thin

isolated copper wire; it starts at some point and flashes along the wire with precisely the speed of light, and the electric wave or pulse reaches the different portions of the wire in successive epochs of time.

Instead of a single wire, think of what must always exist, viz., a closed circuit: two pulses of waves originating at some point of this circuit flash round it both ways, at a pace usually rather less but never more than the speed of light, and meet at the antipodes of the starting point. If the circuit is unclosed, each pulse will get reflected and return, surging to and fro perhaps several times, and in such cases any point of the wire is reached by pulses travelling first one way and then the other—a phenomenon very characteristic of disruptive disturbances; but the first pulse is likely to be the strongest. It must be clear, I think, that for such alternating currents, as well as for rushes of uncertain direction, a symmetrical protector is best.

25. To illustrate these things, make a few more experiments with Saunders' protector, which I choose as one of the best; any other will do.

Experiment No. 8.

Connect a Saunders protector to earth and to any line wire in the proper way, and attach a single wire to the "protected" terminal: like the wire *C G* in Fig. 58. Now send a discharge between line and earth either way, and the "protected" wire will be found ready to give off sparks at every flash. It will spark to anything: to the earth, to an insulated body, even to its own earth screw *B* or *D*. Connect to it a galvanometer, or better, a coil of wire one has no compunction in spoiling; then if the far end of the coil be attached to anything, either to the earth or to another line, or to an insulated body—even an insulated body of small size—sparks between the turns of wire will demonstrate the fact that the insulation is being broken down by the lateral waves rushing along the nominally "protected" wire, and being either reflected or absorbed according as it is connected to an insulated body or to the earth. The connection which permits least disturbance is to the screw B or D. But there must be no chance of the galvanometer being either purposely or accidentally connected to earth in some other way also, else even this partial protection has its virtue removed.

Experiment No. 9.

26. The "earth" in the previous experiment being the gas-pipes: instead of striking the instrument by a flash direct, let a flash be imparted to the gas-main at some other point—say in another room—to typify the possibility of a lightning flash striking the earth in the neighbourhood of a telegraph station. Immediately some of the charge splashes up from the earth, and the protected wire again emits a spark; or, if it be connected to anything by a scrap of fine wire, that wire may be deflagrated. Thus we see that earth connection is not so utterly safe as might be supposed: secondary surgings may rise up out of the ground and do damage to whatever is connected to it. I believe there are more instances of such occurrences than are usually recognized. But I prefer to leave the enumeration and discrimination of instances to persons of experience.

Experiment No. 10.

27. But now insert in the so-called protected wire an arrangement of air gaps and self-induction coils, after the fashion which constitutes my system of protection (Fig. 67). Then, inserting a fine wire, or a coil, or a galvanometer, or any other detector, in the interval, and connecting the far end of the wire to anything as before, it will be found that although every trace of signalling current is able to affect the galvanometer, no appreciable trace of a violent disturbance is felt there; it is now securely protected whichever way the disturbance comes.

28. The electro-magnetic inertia possessed by the coils

protects from sudden currents in the same manner as the inertia of a penny protects it from disturbance when it is balanced on a finger with a card under it, and the card smartly fillipped away. It is not, however, to be supposed that inertia alone, without successive air gaps, can exert this protective influence. The coil of a galvanometer has plenty of the required impedance, far more than the thick-wire coils on my instrument, but the only effect of that is to shunt violent disturbances *through the insulation*—by no means a satisfactory property. The *combination* of air gaps or escape valves along with obstructive inertia is essential to the device. Let me here interpolate the remark that the self-induction of my coils is quite small; a very small amount of wire thus distributed

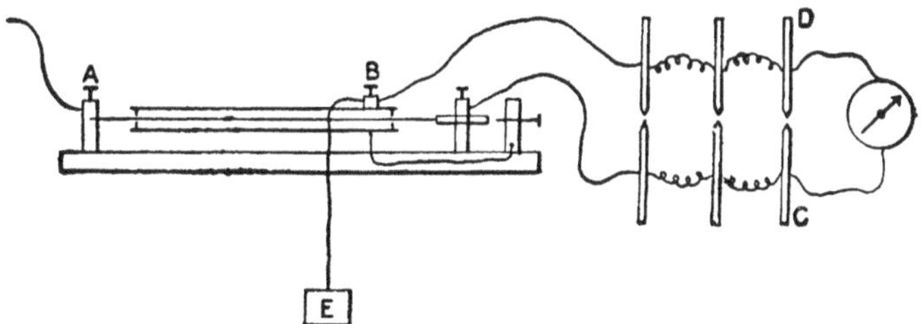

Fig. 67. SAUNDERS' AND LODGE'S COMBINED.

suffices. Two or three yards of No. 16 wire for each are all I need use. I am well aware of the objection to introducing great self-induction in circuits where rapidity of signalling is desired. Very thick gutta-percha insulation is used for the more exposed coils, to prevent any avoidance of impedance by jumping across from layer to layer.

29. With respect to the use of the one-sided pattern (Fig. 66), its effect will be represented in Fig. 64 or 65 if the right-hand set of coils are short-circuited out by a thick wire. In that case the galvanometer, though it is protected from disturbances arriving from the left, is exposed to those coming from the right. Moreover, it is possible for disturbances

arriving at *A* to jump into *C* across a couple of air gaps without going through the coils.

Experiment No. 11.

To illustrate this I take the round-pattern guard and connect it as indicated by Fig. 68 to a coil of wire and safety-valve, or to a scrap of fine wire, or anything convenient. On sending disturbances along the *A B* leads as usual, the safety-valve sparkles, even when the wires *C* and *G* are detached from its lower knob as shown in the figure. The rush into the central terminal of the guard is so strong as to cause a spitting off from every wire connected to it, even into such a little body as the insulated rod of the safety-valve. The sparks are not strong, but they cannot be prevented so long as this pattern is used with direct connection between one exposed (although earthed) and one intended-to-be-protected terminal. The sparks are able to fuse fine wire, and it is impossible to protect the finest wire from fusion with this pattern, and *B* merely earthed. Connecting up the *C* wire makes no difference, except that now its spark gap begins to sparkle too, whereas it might have been quiet: it manifestly receives a charge through the central terminal.

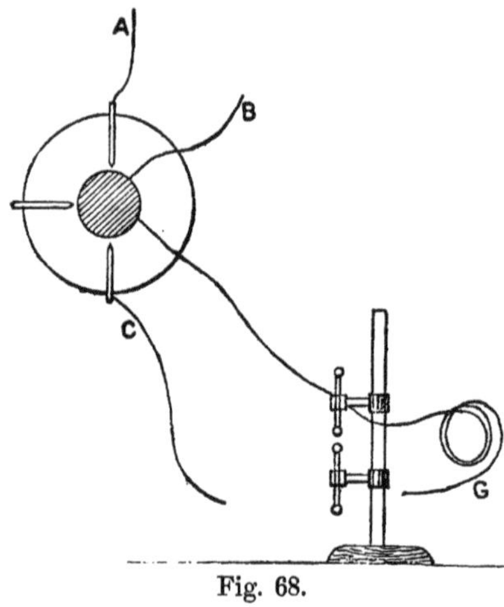

Fig. 68.

This dichotomized pattern is, therefore, only permissible when there is a fair presumption, or complete certainty, against disturbances arriving from the earth, and when there is a guarantee that nothing can splash across the earth terminal direct between *A* and *C*. Cases of this sort will be mentioned in connection with the protection of cables.

30. The patterns so far drawn correspond rather to the sort suitable for a terminal station; but for an intermediate station where an "up" and "down" line wire meet, but where there is no necessary "earth," except an earth adapted to filter off lightning disturbances from the wire instead of passing them from station to station, the following pattern is suitable (Fig. 69). It explains itself; it is virtually the round pattern doubled.

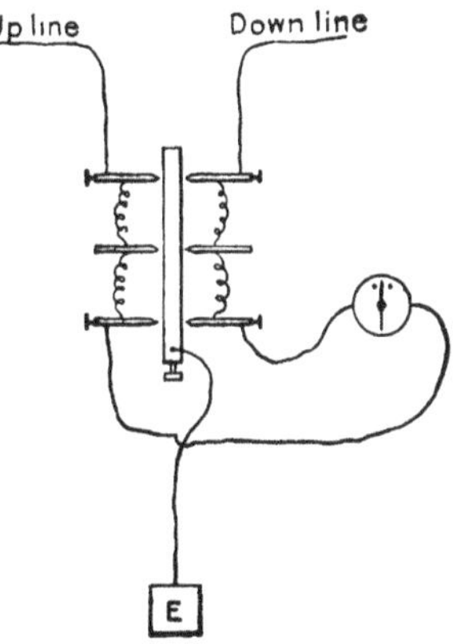

Fig. 69. PROTECTOR FOR INTERMEDIATE STATION.

31. There is one fact which, though fairly obvious, may be here explicitly mentioned, viz., that the air-gap method of protection only avails for very rapidly-varying currents. If the discharge from a large condenser be prolonged, and its oscillation made more leisurely by including a great self-induction coil in its discharge circuit, so that, for instance, the spark ceases to snap and approximates to a very short, shrill whistle, then my whole series of gaps fail to protect a fine wire from being deflagrated, or a galvanometer needle from being strongly disturbed. It is manifest that anything which could filter off currents varying with moderate rapidity would eliminate the very currents on which signalling depends; hence it is impossible to stop this kind of disturbing current unless it gets strong. So soon as it is much stronger than the telegraphic currents desired, it can be stopped by the fusion of a cut-out—a short piece of very fine wire interposed near the protected terminal. These leisurely currents do not endanger insulation, but they are more troublesome

to get rid of than the dangerous sudden ones. The fine-wire cut-out requires replacement, and I arrange for an easy supply of fresh ones by merely turning a button. But whereas the fine wire of ordinary protectors gets damaged by every kind of disturbance, sudden as well as slow, in mine the fuses only come into operation when absolutely necessary —*i.e.*, when no other means suffices. *The fine wire protects against amperes; the series of air gaps against volts.*

Submarine Cables.

32. All that has been said applies by implication to the protection of cables, as much as to any other sort of covered wire out of which it is desired to keep violent rushes of potential, but about cables there are a few special things to be said which we will proceed to say now. Not only is a cable a tremendously valuable piece of property in which a slight fault costs a large sum to repair, and hence the utmost precaution ought to be taken in their case (even the instruments employed in signalling being so expensive as to deserve a thorough protection if it can be given), but the fact that cables are always coated with a stout metallic sheathing is a peculiar circumstance not found in the lines of land telegraphs, whether overhead or underground, and it is a circumstance which, I wish to point out, renders their complete protection from lightning peculiarly definite and easy.

33. First, it is clear that risk is run wherever a cable is connected to a land line. I do not suppose this is ever done with the long ocean cables; but for short lengths, across gulfs, etc., I suppose transmission is usually immediate. Even with ocean cables I understand that the land line to the station is often led to the same switch-board as the cable instruments are connected to; and whenever there is any sort of proximity of this kind, a flash received by some distant part of the land line must be liable to spit across some of the terminals and flick off a bit of itself into the cable and its instruments.

Two lightning switches at least ought to be employed in every such station: one, a coarse one, at the place where the land line enters the building, to eliminate the grosser violence; and another, a fine one, at the mouth of the cable, to filter out the last traces of dangerous disturbance. An intermediate one between switch-board and instrument may occasionally be desirable.

34. A proper mode of connecting one of my protectors to a cable is shown in Fig. 70. Here the outside sheath of the cable is used as sole earth; as is, I believe, customary. Another proper mode is to have a subsidiary or local earth; connection being made as in Fig. 71.

I am not prepared to support one of these in strong pre-

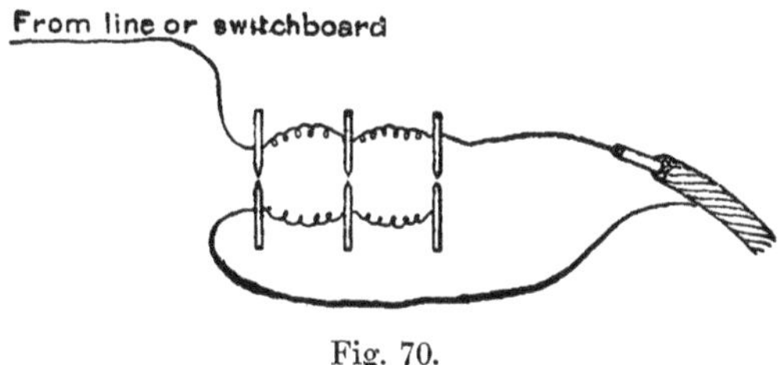

Fig. 70.

ference to the other. The second plan, however, is not to be adopted without the local earth. If there be no local earth, the first is the only proper plan, for it permits disturbances coming down the line to get to the outer sheath of the cable as directly as possible, the interior being protected by impedance; it also permits disturbances surging up the cable sheath, as they may when the shallow shore water is struck by a flash, to get to the line capacity directly, and not to easily enter the cable core.

35. But now suppose a case where no land line or connection is permitted to come within many yards of the cable station, though I suppose such a plan would be extremely inconvenient: let the cable be connected to nothing but its

own instruments--where is the need of a lightning switch then?

The only danger that can occur now is when these instruments are struck direct, either from the roof or walls, or from gas-pipes, or from the earth upon which they stand. A filter must therefore be arranged to protect the cable, even in this case. No fragment of cable core exposed outside its sheath can be considered safe. It would only be quite safe if it could be wholly put inside its metallic sheath and kept there.

But while so carefully contemplating the protection of the cable, why not protect its instruments as well?—for a siphon recorder and an artificial cable are no cheap toys. It can be

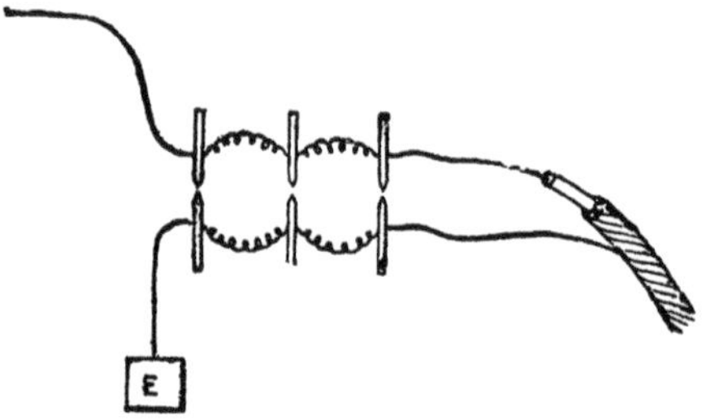

Fig. 71.

done perfectly well. *Put them all inside the sheath of the cable and they are safe.* Of course the sheath of the cable must be enlarged to receive them, but that is easy enough. Use a metal house, and at the point where the cable enters it attach thoroughly its sheathing wires to the house, making a good joint all round, and the thing is done. The cable station is now an expansion of the sheathing, and everything inside it is perfectly safe (Fig. 74).

36. It may be objected that a metal house in hot climates would be a nuisance. Very likely; there may be sufficient practical objections to the plan, but that is for others to judge. However, continuous sheet metal is unnecessary. Wire gauze, even with meshes so large as is used for poultry

yards, may serve sufficiently well, if a few stout wires be added to make effective contact with the cable sheathing.

Unless practical difficulties in so casing in a cable station are greater than any I foresee, I cannot help thinking that it would be worth while; because then, even during a thunderstorm, the operators might continue signalling in undisturbed security, instead of having to suspend operations, disconnect the instruments, and short-circuit the cable to its sheath. Slow trails of disturbing current might indeed render signalling difficult or even impossible, but there would be no *danger*.

37. Reliance is commonly placed on a short-circuiting of cable and sheath. If the short-circuit is very short indeed, the reliance is fairly justified; but if it be effected by a loop

Fig. 72. Fig. 73.

of any size (Fig. 72), then there is no absolute security; for a flash striking at *A* will bifurcate, and part of it rush into the cable. It may be said that that will not matter, because another part of it will be travelling down the outside, and that hence the G.P. between the two pulses (the internal and the external) will not be strained. But there is no guarantee that they will travel at the same rate. There is, indeed, every certainty that they will not. The outside pulse, moreover, will soon dissipate itself by leakage into sea water, leaving the internal one to work its way out through any weak place it can find. Several hundred volts is a very insignificant potential for such a pulse, but it is not customary to apply several hundred volts to a cable with equanimity.

But if instead of a mere loop we put a thimble or hollow

metal vessel over the end of the cable, connecting its sheathing to the walls, and connecting its core to the interior, then no sudden splashes can enter the cable at all; they will keep to the outside and do no harm (Fig. 73).

Such a magnified thimble is the proposed metallic or wire-netted cable station; and it is one that has the advantage of never needing to be removed; one which contains instruments and operators, and protects them all alike from dangerous disturbances (Fig. 74).

38. It may be asked whether such a house should depend

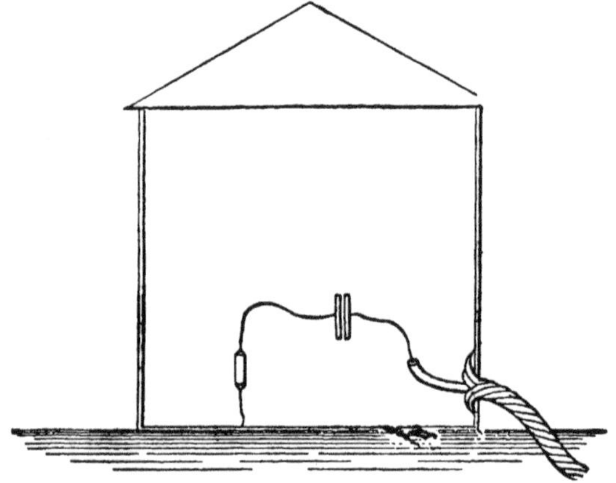

Fig. 74. DIAGRAM OF COMPLETELY PROTECTED STATION.

for its earth entirely on the cable sheathing, or whether a local earth should be provided.

It is not a vital point, but a local earth is to be recommended in order that even the outer sheath of the cable shall not have to carry flashes of exceptional violence, which might unduly heat it. Extra earthing the house can do no harm, and may thus occasionally do good. But do not think of sending out to any local earth a wire from the earth terminal of the recorder or any testing instrument inside. The inside of the metal house is their appropriate earth, and no other must be permitted. Again, gas-pipes, water-pipes and everything else may be permitted to enter the house, but only on the strict condition that they be effectively united to it at the point of entering—not at some other point. Connecting them or anything else to the house by a mere wire is not a bit of good. The cable, the gas-pipes, and every other conductor which penetrate the walls of the house must be *at that place* united to those walls. I have ample experimental as well as theoretical grounds for this assertion. See, for instance, the remarks below on Figs. 77 and 78. No insulated conductor must be allowed to protrude, unless it be enclosed in a metallic sheath connected to the walls, with the distant end of this sheath itself closed. Such an insulated conductor is, of course, the core of the cable itself; the completing closed sheathing being the distant cable station. If that be a mere wooden shanty, disturbances may there enter the cable, and not only do it damage but derange the signalling power of your instruments at this end also; all your fine protection being no good if not imitated at the far end of the cable.

39. Since no *insulated* conductor must on this plan be allowed to penetrate the walls, it is manifest that no land line can be permitted to enter a cable station. The land signalling station must be a separate chamber. It need not be a distinct building; it may be under the same roof, but it must be wire-fenced off from the cable compartment, and the messages must be put through non-electrically.

Whether such a separation is too inconvenient to be contemplated, I leave to those with more practical knowledge to say. I am only pointing out what is necessary for absolute protection. Of course a reasonable amount of protection can be obtained under less stringent conditions, and I am by no means laying down the law as to what ought to be done in practice; all I say is that unless all this is done the protection will not be absolute. Practical men are far the best judges of questions relating to temporizing and expediency. For instance, windows and doors are necessary in a cable station, and although they may be fenced over with wire netting, it is not to be supposed that in all conditions of practice they will never be left open, or that there will never be a break or a serious imperfection in the continuity of the metallic sheathing to the house; and in face of the liability to such accidents, it is not to be supposed that proper lightning guards can be dispensed with. They will, however, then become the last citadel of security, and not, as at present, the outer line of defence.

40. It is customary, I believe, to land the ends of cables in a cable hut on the beach, so as to be able to get testing instruments as close as possible to the real thing, undisturbed by insignificant shore leads. There is not the slightest difficulty

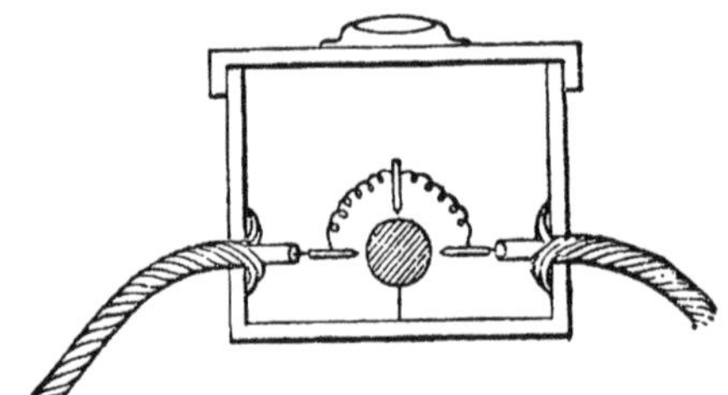

Fig. 75. SUGGESTED IRON BOX FOR CABLE HUT.

or danger introduced by this practice, provided the hut or an enclosure inside it is made of iron, and both cable and shore-lead are well connected to its walls by their outer sheathing as they enter it. Testing instruments and all inside the enclosure are then perfectly safe, except from disturbances conducted in through the central core of the lead; and this can only happen when there is some great defect at the cable station. It is this lead from hut to station which now represents the cable, and all the care described as proper to be taken of the cable must of course be taken of it.

The cable hut need not be itself of iron; if it contains an iron box, as shown in Fig. 75, all requirements will be satisfied, and such a plan may perhaps be convenient; the lid can be removed for occasional testing without appreciable risk. In case a trace of disturbance should by chance get into the lead, a lightning switch may, for a final appeal, be included in the box.

41. I said above that there were cases when the dichotomized pattern of Fig. 66 was sufficient. This is one of them. There is not the slightest objection to a single terminal here, connected with the inside of the box, because nothing violent can splash up into it.

The box may advantageously have some rude local earth of its own, merely in case its hut should happen to receive a direct flash of lightning.

Experiment No. 12.

42. To demonstrate the protecting action of cages I put a parrot cage over a thin-wire reflecting galvanometer standing on a copper plate through which a lead-sheathed cable passes, the lead being attached to the copper where it passes through the hole by a number of soldered wires. If this junction is defective the experiment is liable to fail; *i.e.*, the protection will not be thorough. The cage also should be well connected with the copper plate. The terminals of the galvanometer are connected, one to the inside of the cage at any point whatever, the other to the core of the cable. The distant end of the cable may be similarly treated, or for simplicity its core may be directly connected to its sheathing, and a tight-fitting metal thimble or hat put over it. Before doing this, however, I apply something representing a signalling E.M.F. to the distant end, to show the sensitiveness of the caged galvanometer: *e.g.*, touching the core and its sheath simultaneously with a pair of wet fingers immediately drives the spot of light off the scale.

All being properly arranged, flashes are sent from a jar to any point of the cable—to its far end, or to the cage of the galvanometer, etc.; some other point of the cable or cage being earthed. Not a wink does the needle show, at any rate to a cursory inspection; and if a safety-valve be supplied, no matter with how microscopic a spark gap, it cannot be made to overflow.

But, all the time these splashes are going on, an operator in the distant cage may be sending voltaic currents along the cable and signalling to the sensitive galvanometer inside the cage without any disturbance from the violent rushes outside.

43. It is not to be supposed that such cable houses afford any protection against earth currents, or any other steady or slowly varying effects. These will produce their full effect quite careless of the fact of metallic enclosure. Nothing but a *perfectly* conducting screen can entirely protect against them,

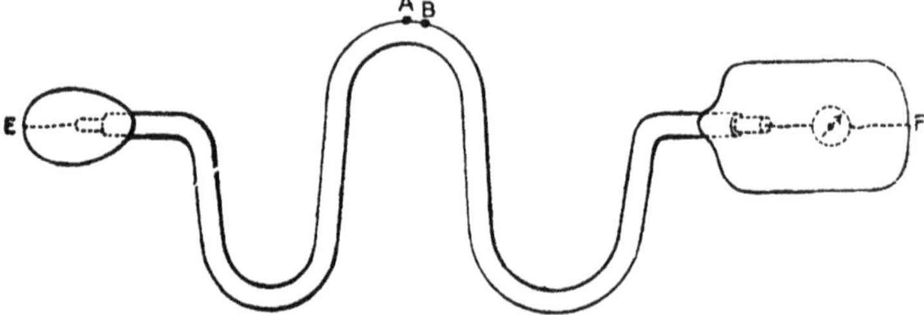

Fig. 76. DIAGRAM OF THOROUGHLY PROTECTED CABLE.

and that is not a likely invention. Atmospheric disturbances also, and even discharges *in so far as they are slow* or leave subsiding effects behind them, will be felt by the galvanometer; but then, although inconvenient to a reader of signals, slow disturbances are not dangerous. It is the violent and sudden rushes that are dangerous, and these are wholly excluded, occurring only in the outer metallic sheath. To exclude steady currents of too great strength a short length of very fine wire must be included in the circuit.

Experiment No. 13.

To emphasize the fact that a caged galvanometer is not in the smallest degree protected from steady currents, except in so far as the resistance of core and bobbin bears a large ratio to the resistance of sheath and cage, according to rudimentary laws of divided circuits, I take the terminals of a single storage cell and touch them to two points of the outer sheath of the cable an inch apart, like *A B* (Fig. 76). Instantly the spot of light is flung off the scale. Applying the terminals to points one-eighth inch apart only, the needle is still strongly disturbed; and indeed it is only by very careful feeling about and adjusting within the hundredth of an inch on the same sectional circumference of the cable-sheath that a neutral position for the second terminal can be found, and the galvanometer cease to tap off any fraction of the current.

Diagrammatically the arrangement is shewn in Fig. 76:

With battery terminals at *A B*, the galvanometer plainly receives a minute branch current, unless *A B* are coincident.

The simplest way to realize the matter is to recognize that when a current is passing in the sheath from *A* to *B* all the left-hand portion of the cable is at *A* potential, and all the right-hand portion is at *B* potential; hence, of course, there is a tendency to leak along any available path between the two portions; and the core is such a path.

But Leyden-jar flashes may be sent, not merely to points close together like *A* and *B*, but to points as far apart as you please, like *E* and *F*, and the galvanometer shall show nothing.

44. But now arises a very interesting question. The protection thus illustrated, is it theoretically perfect, or only perfect enough for all practical purposes? Does a metal cage protect a galvanometer absolutely from these sudden disturbances?

Put it in another way. Everybody knows Faraday's cage or metal-lined room, into which he went with electroscopes, birds, frogs, and other instruments, arranged that the outside should receive violent flashes, and detected nothing inside. Suppose he had added to his stock a galvanometer, connecting its terminals to two points of the walls, would he have got anything then? It naturally occurs to ask, Why did he not try it? What would he have expected the result to be? Did he perhaps not try it because he felt already sure of the result? If so, the feeling of certainty was premature. Whether his guess might happen to turn out right or no, he certainly had not all the facts before him on which to base an opinion.

45. Let us go into the question. First of all, Mr. Chattock and I obtained the following result a year or two ago, viz.: that between two wires wholly inside a wire-gauze house not a trace of spark can be obtained, whatever flashes pass along the outside. The only way we got sparks inside a metal

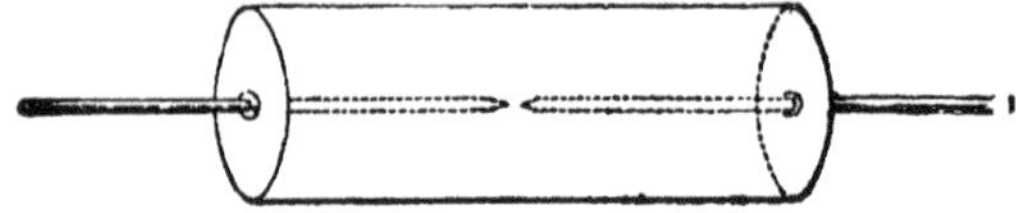

Fig. 77.

enclosure was by allowing one of the wires to protrude a little through a hole in the walls without touching the sides of the hole.

Thus, for instance, the arrangement of Fig. 77, with the wires soldered to the metal case where they enter it, gives no sparks inside; but modifying it as shown in Fig. 78, where

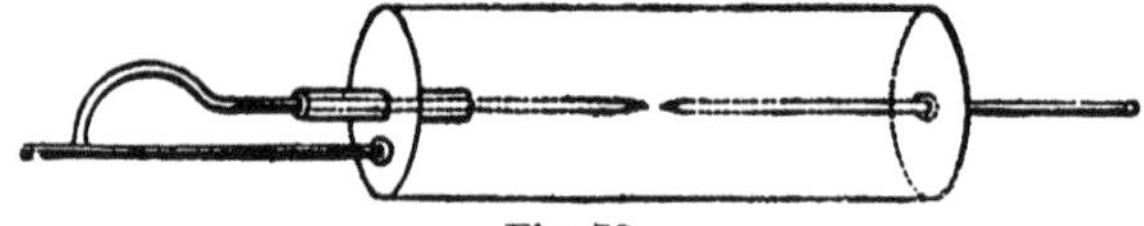

Fig. 78.

one of the entering wires is soldered to the cage elsewhere than where it enters it, may give some very small ones.

But this last is hardly a fair device: it is equivalent to turning a portion of the enclosure inside out; the protruding wire can be regarded as part of the interior surface. Inside a thorough and fair metal enclosure sparks cannot be got between its walls.

46. Recently Hertz tried what is practically the same experiment, his object being to demonstrate that rapidly varying currents traverse only the outside of a conductor ("Phil.

Mag.," August, 1889). He took a sort of mouse mill of wires (Fig. 79), with two rods entering it and joined to it—one of them through a silvered glass tube—and, inserting it in a place where sparks would naturally occur between the axial wires, he found it not possible to obtain any until many ribs of the mouse mill had been removed, so as to leave a great gap for the penetration of electro-magnetic waves, or unless he coated the glass tube portion of his enclosure with only a film of silver so thin as to become incipiently transparent to such waves—*i.e.*, transparent to light. All this shows clearly why a cage protects a cable: no sparks are possible inside, and therefore its insulation cannot get damaged.

47. But what about a galvanometer or telegraph instrument?—*i.e.*, what about the conduction test? Insulation is

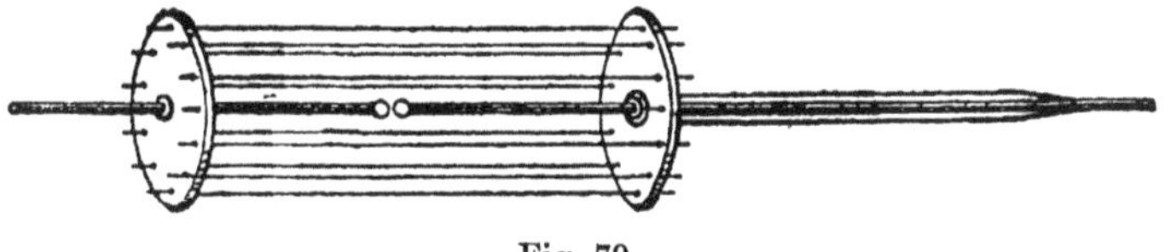

Fig. 79.

safe enough, but will the signalling be interfered with during the continuance of a storm?

Well, if we try the experiment by putting a galvanometer inside a good conducting vessel and attaching its terminals to the walls—a small patch of gauze being, of course, provided for the ray of light to get in and out—nothing will be seen when ordinary flashes are given to the cage; and that must certainly have been the result of the experiment if it had been tried in Faraday's time with the galvanometers of that day.

48. But a galvanometer is in this arrangement so severely handicapped by short-circuiting that it is perhaps unfair. It may be better to use an elongated form of enclosure and attach a low-resistance galvanometer to points of it a good distance apart. The metallic sheathing of a cable is such an elongated enclosure; hence modify the simple cage experiment thus:

Experiment No. 14.

Instead of the thin-wire galvanometer, which we find easy to protect, encage a thick-wire galvanometer and join it up to the lead-covered cable as already described, with good connection between cage and sheath. Sheathe also the distant end, and connect up as in Fig. 76. Now send flashes to *E*, or to *F*, or to both; or spark to *E*, and let *F* be earthed, or *vice versâ*. The galvanometer, if sensitive, gives a slight kick at every flash. Its safety-valve shows nothing, but a feeble pulse does pass round its wire.

49. When I first got this effect the cage employed was a common parrot cage; but, since its wire junctions could not be depended on, I had one made of fine copper gauze, soldered to a disc of copper at top and to a flat ring at bottom, suitable for clamping down to a copper plate. The same disturbance was still found when it was used; so a solid sheet-copper water-tight hat was made and soldered down to the plate, with the galvanometer inside; the gauze cage being placed over all as a supplemental and outer covering. Still the residual effect remained, apparently unaltered in strength.

50. The protection of a galvanometer from these sudden discharges is therefore not theoretically complete. No amount of covering-in can absolutely eliminate all disturbance. No matter how sudden a pulse may be, so long as there is an integral passage of electricity in one direction along a conductor, the central portion of that conductor will convey some trace of it. The integral passage of electricity along the axis is, in fact, that determined simply by its relative conductivity as compared with its surrounding sheath, irrespective of more complicated considerations. But whereas in the outermost layer of the sheath there are violent surgings, the current-strength being very great and alternating, nothing of the sort occurs in the central portions; the violence is all damped out, and there remains nothing there but a quiet and sluggish flow. The rush in the outer layer of a conductor induces

currents in the layer next below, these again in the layer below that, and so on; a sort of diffusion of currents occurring towards the axis, like the diffusion of heat in a body to whose surface heat, or an alternation of heat and cold, has been suddenly applied. Such diffusion is accompanied by a flattening out of the waves, a decay of all their suddenness, so that the axial disturbance is a mere peaceful flow, analogous to that given by a voltaic cell, not having a trace of jump left in it. The very smallest breach of continuity stops it altogether: it is impossible to get a spark in the axial wire; but if this wire be a completed conductor it takes its share of integral current along with the rest. Not simultaneously,

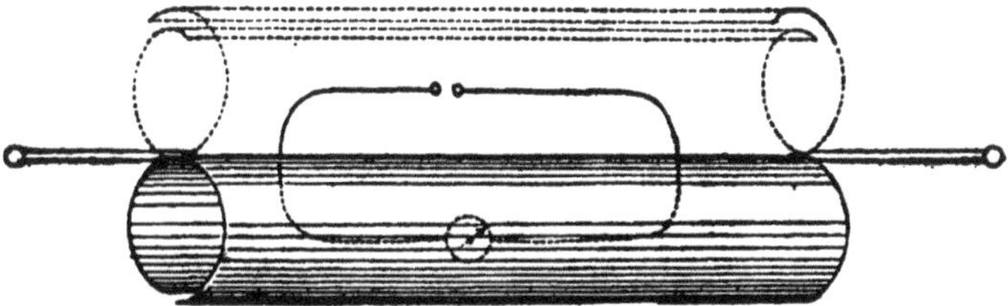

Fig. 80. EFFECT OF ENCLOSURE IN PREVENTING SPARKS.

however, the diffusion inwards occupies time, but the delay has no particular effect to call for notice here.

51. Another way of putting the matter is to use the mechanical analogue of slipping wheel-work as representative of a conductor; then, if the outer layer of wheels be spun to and fro violently and left with a certain residual spin, to represent the impact of electro-magnetic waves from the dielectric on the metal, that residual spin will penetrate to all parts of the geared mechanism, except where there is absolutely perfect slip—*i.e.*, except into space bounded by a perfect conductor.

52. To suppose that the axial part of a conductor (whether solid all through or hollow with a wire along its centre makes no difference in principle) takes no part in conveying a momentary transfer of electricity, is to make the same error as is published so uniquely and interestingly by Sir William

Thomson in the March, 1890, "Phil. Mag.," wherein he corrects his original idea that the ordinary resistance of a ballistic galvanometer wire would not be the right resistance to introduce into its formula, because the outer layer of the wire conducts most of the current; and shows that, examining the matter still more completely, every part of the wire is ultimately effective, and equally effective as regards integral flow. All the rush does go by the outside at first, and all the *violence* is there expended anyhow, but every part of the section does its full share of conduction ultimately.

It is worth entering thus fully into the matter, because these are things about which it is easy to get bothered if one does not happen to get hold of them right way up. And this matter of protection by cages is one on which there has long been some uncertainty or hesitation.

Now that I see clearly how they act, their behaviour seems natural enough, and what one might have expected—what I think Maxwell (doubtless others also) would have expected; certainly it all comes out clearly enough on his principles. But a little time ago the matter was by no means so clear in my mind, and I rather gather that several others felt a similar sort of temporary uncertainty.

53. One more way of putting the result may be permitted. It is not given as an experiment, because I have not tried it. Take a sheet of metal or gauze, tap it along a certain line with a galvanometer on one side and with a spark gap on the other, arrange to send flashes along the same line, and then fold the sheet about this line into a cylinder, either upward or downward, so as to enclose at pleasure the galvanometer or the spark gap, but not both (see Fig. 80).

The indications of the galvanometer will be wholly unaffected by the way the sheet is folded, or whether it be left flat; provided always the insulation of its wire is and remains perfect. But to the air gap the folding of the sheet makes all the difference. When enclosed it will be quiescent, when exposed it may sparkle.

CHAPTER XXX.

REPLY TO CRITICISMS.

IN the discussion Sir WILLIAM THOMSON made the following remark: "Dr. Lodge's new principle of inductive quasi-inertia interposed in the line of conduction between a series of points separated from an earthed conductor supplies what is, I believe, a practically convenient and a thoroughly effective protection against the electrification to too high a potential of even the shore end of the insulated wire."[1]

Mr. C. E. SPAGNOLETTI gave the following results of experience: "Bearing on some remarks that have been made in this discussion, I should like to bring before the Institution one or two cases of the effects of lightning on telegraph wires, which will show that its vagaries depend entirely on the potential of the flash. In one case, between Worcester and Wolverhampton, a severe storm visited that neighbourhood, and ten poles were split down, the cross arms broken, insulators smashed and scattered about, and wires fused. This must have been a flash of extremely high potential, as it would have to jump an air space of 8 or 10 inches.

"In another case, which happened on the Shrewsbury and Hereford Railway, a lineman was up amongst the wires at Shrewsbury. It was a calm, fine summer's eve there, but a severe storm was raging at Hereford, 50 miles away, and the wires were struck and the man at Shrewsbury rendered insensible; he was burnt under the arm and on the leg by the entry and exit of the current through his body: he never recovered his vigour, and died three or four months after-

[1] Remarks of other speakers are reported in the Journal of the Institution of Electrical Engineers for 1890, pp. 382, *et seq.*

wards. Now this flash, although strong enough to have this serious effect on the body, did not damage a single pole along the route of the 50 miles it travelled.

"Another case happened on the Metropolitan Railway. There was a continuous tunnel from Edgware Road to King's Cross, and five intermediate stations were in this length; the wires were insulated, and attached to gas and water pipes for earth, so that there were not any exposed wires. In a violent storm passing over London the instrument coils at some of these intermediate (and, one would think protected) stations were fused. The only way the lightning could have got to them must have been through the gas or water pipes, or there must have been an upward flash from the earth.

"On another occasion, in North Wales, a Bell instrument was struck, and the wire on the coils was cut into nearly equal lengths and scattered with such force that the wooden case was lined like a wire brush with bits of wire sticking in by one of the ends; the core was stripped of wire, and it seemed as if the wire on the coil had been cut like a cross, taking the lines from top to bottom and from right to left, or *vice versâ.*

"On another occasion, the men reported that while working in the tunnel on the Clifton line, near Bristol, running wires in casing, the lightning during a violent storm ran into the tunnel along the rails, flashing and frisking along them and darting up to the wires they were then putting up against the wall, lighting the tunnel in brilliant flashes.

"Lightning protectors and conductors are unfortunate things; they get little or no credit for what good they do, as we know not the danger they avert."

Dr. Lodge: Several persons of great practical experience seem to think that the present instruments do their work well, and that no new one is necessary. Other authorities, I know, are not of the same opinion, and some of the reasoning which has led to this idea is rather fallacious. Thus one of the arguments is based upon the fact that existing protectors are sometimes found damaged, the idea being that therefore they have acted and saved whatever is attached to them. Mr. Saunders says that very often his fine wire is fused. Mr. Preece says that the plates of the Siemens or other plate protector were often found burnt or damaged in some way, showing that sparks had occurred. Now I say that experiments made in the laboratory demonstrate that whenever a protector has been damaged hitherto something else has been liable to be damaged also—not necessarily very much, not necessarily visibly, and possibly not at all, but it has run the risk of being damaged. I say that from the point of view of complete protection there is no cable and there is no instrument completely protected from all possible danger from lightning, and the damage of the instruments and the fusion of the fine wires show that a great amount of disturbance has entered the cable and may have done damage.[1]

With regard to the use of fine wire I entirely agree; fine wire is a most useful addition to any spark-gap lightning protector. There are two things to be guarded against—high potential and strong current. High potential is kept out by a judicious arrangement of spark gaps and self-induction; strong currents by a fine-wire fuse. High potential is by far the most dangerous; but strong currents are not to be desired. Strong steady currents cannot be kept out by spark gaps. Steady or slowly varying currents can only be kept out by fine wire. Fine wire has long been used in protectors. I have all along contemplated using it as an additional safeguard wherever convenient, but the essence of my instrument is its arrangement of coils and spark gaps. In some places

[1] That land signalling instruments do get damaged notwithstanding their lightning guards, at any rate in France, is proved by the official returns, a copy of which for the year 1883 has been kindly given me by Professor Hughes. It shows that 90 sounders, 24 Morse instruments, and 477 lightning guards were damaged in that year. And Dr. Muirhead tells me of cases where the condensers of cables have been burst by atmospheric electricity with a loud report.

fine wire is objectionable, since it entails replacement when damaged. As Sir William Thomson says, the fine wire ought to be at the protected terminals, not at the exposed terminals, or else it may be deflagrated unnecessarily by discharges which the spark gaps alone are well able to tackle. That is one defect in Mr. Jamieson's protector: the flash coming in from the line deflagrates its wire with extreme ease; it is not, however, the only defect. I may say here, in connection also with Professor Hughes's remarks, that no thin silk-covered wire coiled on a metal bobbin can be satisfactory; for, besides the inductive neutralization of self-induction, it acts very much as if the wire were wholly uncovered; the coil is practically shunted out by the metal. Thus in Jamieson's instrument the cable terminal receives discharges from the earthed bobbin by its sparking through to the wire upon it at the nearest point, or often at several points simultaneously. And no discharge thinks of going round the coil: it jumps from it at beginning and to it at end, travelling by the reel.

Mr. Preece and Sir Henry Mance contend that the self-induction in telegraph instruments is sufficient to cause sparks to jump the air gaps of their protectors. But lightning ignores altogether the self-induction of a silk-covered coil; it jumps right across it from layer to layer, and may easily fuse it. Moreover, a self-induction coil alone, without an overflow to get rid of the current which has got into it, is incomplete.

Corkscrew spirals of wire used for connection, although they have some self-induction, are by no means of the right shape if that were their object. The shape of coil which gives maximum self-induction with a given length of wire is a flat shape, very like that employed by Professor Hughes in his induction balance.

I am surprised that it is not thought desirable to protect telephones, considering their frequent proximity to the human ear; but I suppose the British Isles are exceptionally free from violent storms.

Sir William Thomson instructively said that a fine wire in

conjunction with a condenser must be of service in protecting a cable even from rapidly-varying currents; for in order to charge the condenser to high potential a certain Quantity is necessary, and a fine wire might exclude that. This may be the reason why Saunders's guard has acted as well as it has. The fine wire is probably the most important part of the protection in that instrument.

Major Cardew spoke about my experiments last time, and said I had to be careful to strike the lightning protector at the right end, and that if I had struck it with flashes at the wrong end it would have told a different tale; but what I did really was not to represent the lightning as striking the protector direct—of course it may strike the protector direct, and it may strike the end of the cable direct; you cannot prevent that, except by a metal house—but what I did was to imitate a flash or disturbance coming along a line wire to the instrument; what I struck was a line wire terminal, a line wire which might have been struck a mile away for that matter, and of course the disturbance reaches the instrument at the terminal to which the line wire is attached, and not the other way.

With regard to what General Webber said about leads—danger from overhead leads as compared with underground leads, and so on—I did not say anything about the danger from overhead wires, because it is obvious; but I wanted to point out that even underground leads were not safe. If they were entirely underground they might be—except from the neighbourhood of gas and water pipes, as Mr. Spagnoletti instructively reminds us—but that is absurd: if they are completely underground you cannot get the lines into your house. Danger arises from the above-ground or inside-house portion. Buried leads covered with insulation are plainly more susceptible to permanent damage than bare copper strips enclosed in an air-trough, because the latter would be none the worse if a spark or two did jump from them to earth.

Mr. Mordey has mentioned that sparks are liable to strike

arcs. I have had arcs struck like that in the course of these lightning experiments: some flashes put into the storage battery wires struck an arc across them, making a great flare. That is a thing very much to be taken into account in protecting electric light leads, but that does not mean that my lightning protector won't do for houses and electric lighting; it only means that a special pattern must be devised in order to stop the arc the instant it is formed. (See p. 423.)

Mr. Crompton wished to know how I thought that house leads should be protected. My suggestion is that a number of houses wired together should be disconnected as regards lightning flashes on the principle of fire-proof doors. An arrester should be placed wherever overhead leads enter the ground; it should also be placed between different houses, so as to isolate a struck house and not let damage spread through a district.

CHAPTER XXXI.

CONSTRUCTION AND USE OF INSTRUMENTS.

The following are the Directions for Lodge's Lightning Guards, as constructed and issued by Muirhead and Co., Cowley Street, Westminster, S.W.

THESE are of two main patterns, the double pattern and the single pattern.

The double pattern has four terminals, symmetrically situated. One pair of these may be called the exposed ter-

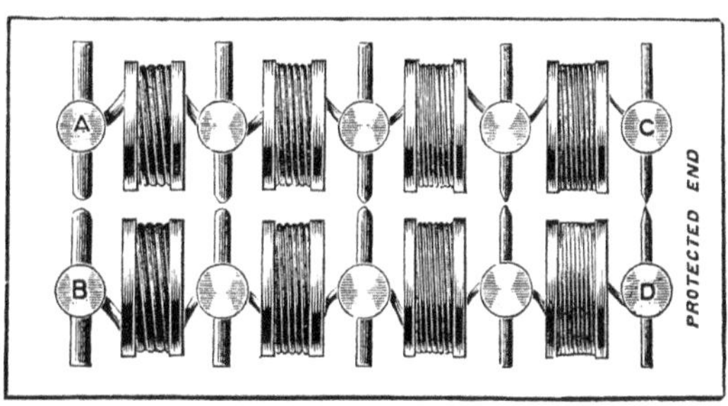

Fig. 81. DIAGRAM OF DOUBLE FORM OF GUARD FOR INSTRUMENTS.

(For actual instrument, see p. 390.)

minals, and are labelled *A* and *B*. The other pair are the protected terminals, and are labelled *C* and *D*.

The ordinary use of this pattern is to protect a telegraphic instrument, say a siphon recorder.

To do this, the recorder terminals are connected to *C* and *D*, while the switch board terminals, "line" and "earth," are connected to A and B.

The instrument acts as follows:—By its construction A and C are always electrically connected for signalling currents; and so are B and D. But for lightning and sudden discharges, both these pairs are practically disconnected, while A and B are connected instead.

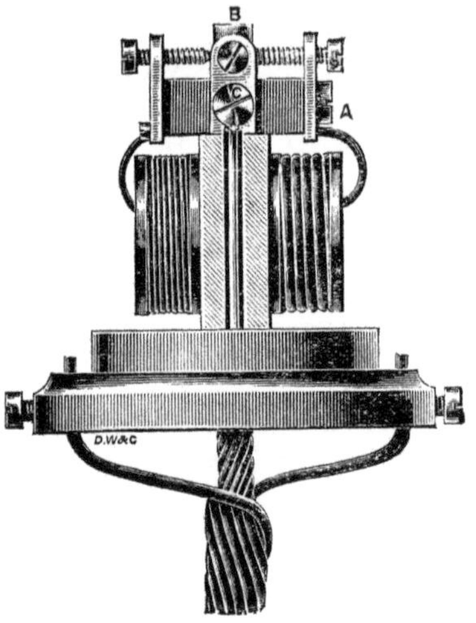

Fig. 82. DRAWING OF SINGLE PATTERN FORM OF GUARD FOR CABLES: SHOWING CONNEXION OF CABLE-SHEATH TO BASE. METAL COVER TO BE ADDED.

Thus lightning is shunted out, while the full signalling current proceeds on its way to the recorder.

Another use of the instrument is to switch out lightning flashes from a lead, and take them to earth. To this end the lead is cut and connected to A and C. B is connected to earth by the shortest available route, and D is not connected to anything.

But the single pattern is frequently used for this purpose, and *except when disturbances arrive from the earth instead of from above*, it is equally effective.

The single pattern has three terminals, *A*, *B* and *C*. It is simply the double pattern with *B* and *D* short-circuited together.

A and *C* are put in the lead. *B* is connected to earth direct.

The single pattern, when enclosed in a metallic case, is perfect for the protection of *all things enclosed or virtually enclosed in the same sheath.*

Fig. 83. VIEW OF SINGLE PATTERN FORM FOR CABLES WITH METAL COVER.

Thus, for instance, it is the pattern used to protect submarine cables.

The core of the cable is taken inside the case and attached to *C*, its metallic sheathing being at the same time connected closely with the metallic case, in such a way as to leave no part of the core exposed (fig. 82).

The sheath of the cable is thus virtually a prolongation or

enlargement of the metal case, and its core is entirely protected.

The wire bringing the signals (from the outlet board) is taken inside the case and attached to *A*.

The *B* terminal is permanently connected with the case and is thus already earthed by the cable sheathing.

Fig. 84. ANOTHER FORM OF DOUBLE-PATTERN LIGHTNING GUARD.

It may with some slight possible advantage be separately earthed as well, but only if the flashes feared are likely to be strong enough to damage the cable-sheathing when going to earth by its means alone.

Anything that can without inconvenience be put inside a metal house or cage, may be completely protected by one of these encased single pattern or "hat" protectors, provided the connecting leads be taken through a metal tube connecting the two cases.

Whenever a delicate instrument cannot conveniently be thus caged, a double pattern instrument must be used, as above described.

Lightning Guards for Electric Light Installations.

The same principle as has been applied in all other forms of the Lodge Lightning Guard is applicable also to Electric

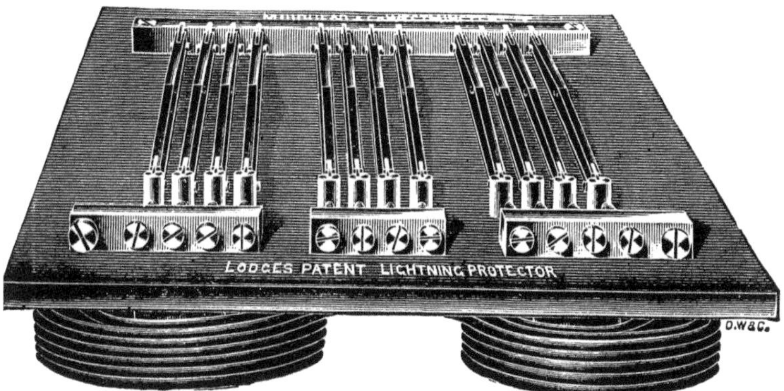

Fig. 85. SPECIAL PATTERN FOR HIGH VOLTAGE ALTERNATING CURRENTS.

Light Installations, viz. : a succession of air-gap paths to earth, connected up by coils of well insulated wire, across the turns of which the lightning, weakened as it is by the first air-gap to earth, is not able to break. The only extra difficulty which occurs in protecting electric light leads instead of telegraph lines and cables is that the lightning spark is able to start an arc across the air-gaps to earth, and thereby to divert the main current out of its proper channel.

To check this diversion without a moment's delay the air-gaps are led up to through a fine wire or tinfoil fuse ; this is

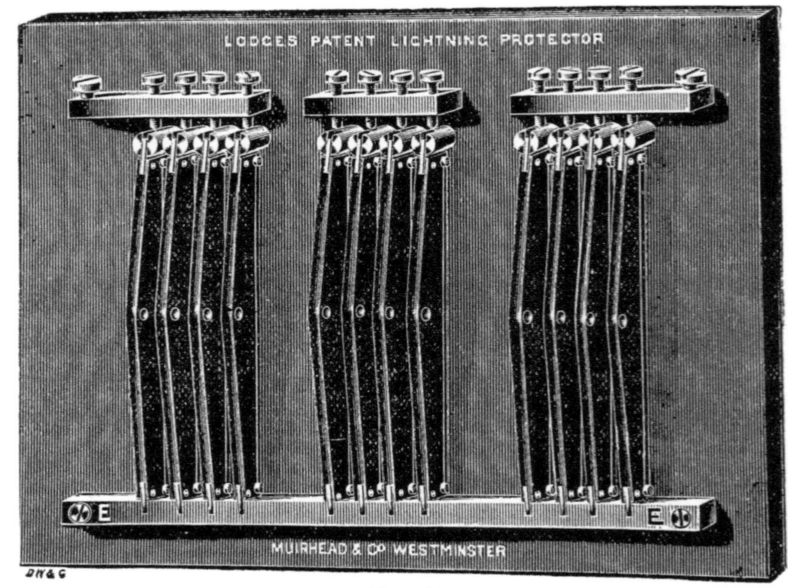
LODGES PATENT LIGHTNING PROTECTOR
E
E
MUIRHEAD & Cº WESTMINSTER
DW&G

Fig. 86.

able to guide the flash, but is destroyed either by it or by the main current, whose path to earth is thereby instantly stopped.

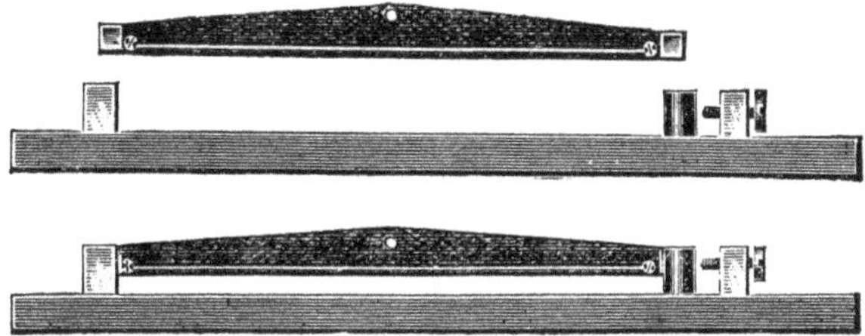

Fig. 87.

For high voltage installations the length of these fusesor cut-outs can be adapted to the demands of the circuit. They are now made to interrupt any arc up to one foot long (fig. 86).

Fig. 88. ANOTHER FORM OF LIGHTNING GUARD FOR LOW TENSION ELECTRIC LIGHT LEADS, WITH TINFOIL ARC-STOPPERS INSIDE CYLINDRICAL METAL CASE.

In order to enable the same instrument to take a succession of flashes without attention, a dozen fuses are provided in parallel; one of which is liable to go at every stroke. Fresh ones are supplied, and can be inserted with the greatest ease, as shown in the two figures (fig. 87).

Each instrument has three obvious terminals. One terminal *A* for the line (especially if it be an overhead line) at the point where it enters a building, another terminal *B* for the main inside leading wire; while the third, *E*, is to be connected as thoroughly and directly with earth as possible.

For low tension installations the same form of instrument is made, but the lengths of fuse are much less (fig. 88).

APPENDICES.

LIGHTNING CONDUCTORS AND PRACTICAL EXPERIENCE.

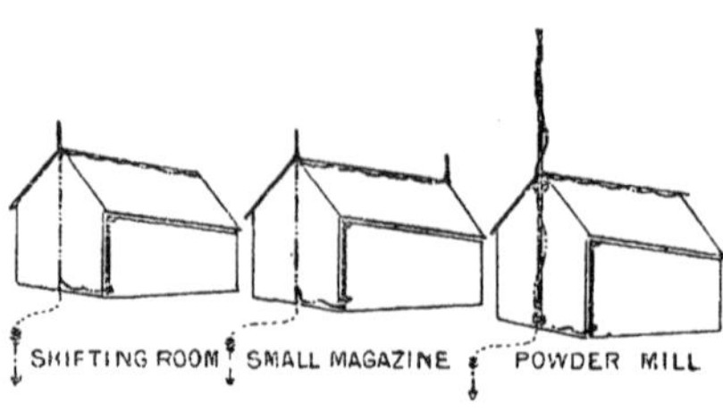

Fig. 1. Fig. 2. Fig. 3.

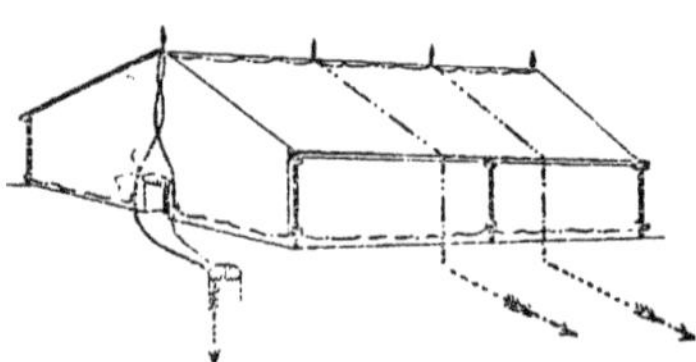

Fig. 4.

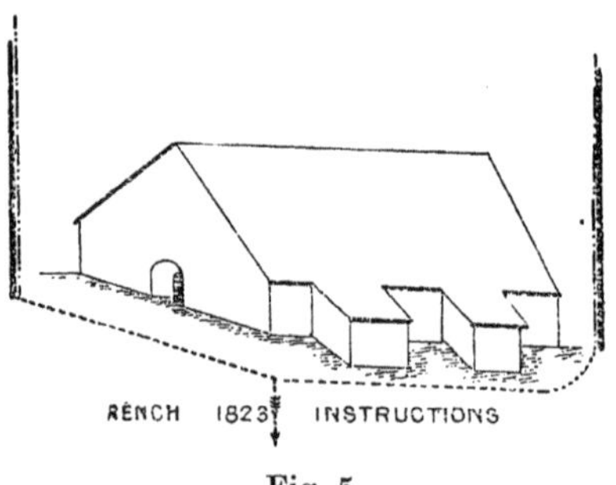

Fig. 5.

APPENDIX I.

Concerning Army and Navy Regulations for the Protection of Powder Magazines.

COLONEL BUCKNILL has been good enough to send me various documents relating to the protection of powder magazines and other important buildings, with which at one time he was closely connected. He also wrote me one or two letters which he asks me to reproduce in full. I accordingly do so, in amplification of his remarks at the official discussion at the Institute of Electrical Engineers (see above, p. 225). I also reproduce, at his request, a leading article which he wrote to the journal "Engineering," March 10th, 1882, on the occasion of the appearance of the Report of the Lightning Rod Conference.

Thornfield, Bitterne, Hants, 5th May, 1889.

PROFESSOR OLIVER LODGE.

"Dear Sir,—Your paper on Lightning and on Conductors, etc., for our protection from its ravages, clears away in a most satisfactory and convincing manner many of the difficulties which have for so long enshrouded the subject. I fear that I shall not be able to take part in the discussion upon it, but your remark numbered 46—first two lines—has given me the greatest satisfaction, as for so many years I have borne the odium of, I think, some very undeserved sarcasms in an abstract of my paper read at the Royal United Service Institution on 6th May, 1881—such abstract being published in the report of the Lightning Rod Conference as an appendix.

"My views were accepted at the time by the W. O. authorities, and I was appointed to re-write the Instructions regulating the application and inspection of lightning conductors, such instructions being issued as part of the W. O. Circulars of 1st Sept., 1881. The L. R. C., for some reason best known to the members, ignored those instructions, and quoted in full the W. O. Instructions of 1875! The Admiralty Instructions of 1880-1 were also ignored.

"When the L. R. C. Report came out, I wrote a leading article on it, which was published in "Engineering," 10th March, 1882 (reprinted p. 432).

"At the time I was writing the W. O. Instructions of 1881 I wished to insert a paragraph, that probably a piece of No. 8 iron wire would carry off any stroke of lightning, and gave some reasons in support of this view, but I was advised not to publish the same. I have the MS. now before me.

"I hope and trust that your powerful paper and convincing arguments will sweep away much of the quackery connected with the subject; a quackery that is only equalled (not surpassed) by the vendors of many patent medicines.

"I am, dear Sir, yours faithfully,

"J. T. Bucknill.

"P.S.—I cannot concur in some of your practical applications at the end of your paper, and they do not seem to me to be logical sequences from your theories, which are apparently incontestable.—J. T. B."

Thornfield, near Southampton,
Wednesday Evening, 8th May, 1889.

"MY DEAR SIR,—Your letter just to hand. I have sent a few friendly remarks on your paper, which, in my humble opinion, will be the cause of an entirely new departure in L. R. practice.

"You will have a quite sufficiently uphill battle to fight to establish the truth of your *principles;* the correct practical applications will follow as a matter of course: but each case will, I fancy, require much sagacity, and, as you truly imply, the problem becomes *more difficult* by your important discoveries.

"I have always maintained that it is *quite impossible* to make yourself perfectly secure from the effects of lightning, and your views emphasize this opinion.

"With regard to your sections 51 to end:

"I quarrel with sections numbered 59, 63, and 65—see my notes to the Institution of Electrical Engineers on p. 226.

"I also add now to you that I think the words—*but each should have a direct route to earth*—should be added to *section* 66.

"*As regards* 69, I cannot believe in "complete security."

"*As regards* 70, I don't like it; better connect the stoves, etc., if only by a small wire.

"*As regards* 71, I should much prefer to connect the conductor by branches through brickwork to roof inside in several places.

"72 *is first-rate.*

"73—I should prefer to have two or more conductors connected to a flat roof at each end.

"74—*Quite so.*

"75—I prefer the orthodox rule in *majority* of cases.

"76—I don't believe it. I believe that the earth charge is generally collected there, and *will* flash to conductor unless you connect them to it.

"78—I always recommend a double conductor for a tall chimney : one on each side, joined by a cross conductor at top to attract the attention of the stroke, and prevent it flashing down the INSIDE of chimney.

"79—I think there are exceptions ; notably on ships of war.

"80—French Instruction, 1823.

"81—Clerk-Maxwell nearly; but very difficult to carry out.

"82—A big IF.

"83—Difficult—difficult.

"84—Not understood.

"85, 86, 87—VERY difficult to carry out, except as scientific experiments.

"88—What is an 'EFFICIENT' Lightning Protector?

"89—I have no experience in such circumstances.

"Yours very faithfully,

"J. T. BUCKNILL."

ARTICLE FROM "ENGINEERING," 1882, BY COL. BUCKNILL.

The Lightning Rod Conference.

"The devastation produced by lightning is fortunately not so considerable in England as in many other countries, and this may account for the fact that no rules have ever been drawn up to guide the general public in the erection of lightning conductors,—except by private individuals. In France, the Academy were requested by Government on several occasions to draw up such rules, and the practice in France is consequently very much more uniform than in this country. The want being felt, the Meteorological Society issued invitations in May, 1878, to the Royal Institute of British Architects, to the Physical Society, and to the Society of Telegraph Engineers, and delegates were chosen by each society to co-operate in drawing up a report on the subject. Had the matter been taken up by Government it is probable that the Royal Society would have been asked to report; but inasmuch as there were no less than seven Fellows of this society on the Conference, it may be conceded that the report almost carries upon it the stamp of our great scientific society. For this reason we must confess to a feeling of disappointment in perusing the report, which occupies only nineteen pages of large letter-press, which contains several suggestions of a doubtful character, without giving any reasons for them, and which omits many matters of scientific interest, which were certainly to be expected from so distinguished a Conference. The principal labours and the greatest success lie in the appendices, which occupy 260 pages of closely-printed matter, and which give in a condensed form a vast amount of historical information on the subject. The members of the Conference divided this labour, and a most valuable book of reference is the result. The small advance in the knowledge of the lightning discharge is due to the impossibility of carrying out experiments except with electricity of a much lower potential. Authentic history is therefore of the greatest importance, and all records written by scientific and careful observers become of the utmost value. The Appendices of the Conference enable the student to easily get at any desired information, and any further investigation is assisted by the excellent catalogue of works upon lightning conductors in Appendix G.

"The report recommends that each terminal rod should be provided with a multiple point, the central point being a few inches higher than the others, and thereby the opinion of Mr. Preece, Appendix B, that "each conductor should end in one fine platinum point," and that "he sees no advantage whatever in multiplying these points," is not endorsed. The report next recommends copper as the material for a conductor, although giving a comparison with iron which seems to be in favour of the latter metal. As regards the size of a conductor, a sectional area of 0.11 of a square inch for copper and of 0.64 of a square inch for iron are recommended. Concerning the sectional shape of a conductor, and the desirability of its being free from any joints and the encasing of the foot of the conductor to protect it from the thief, and the painting, &c., there is nothing new. In reference to the attachment of a conductor to a building, it is stated that it shall not compress the rod but that it shall hold it firmly, and yet shall allow play for its expansion and contraction. The form of attachment which possesses these opposite virtues is not explained.

" As regards the earth connection the usual precautions are recommended, but we note that a connection to a gas main is not only permitted when other things fail, but is advocated as an efficient arrangement at all times, although an accident in Halifax, N.S., is on record in which the Provincial buildings narrowly escaped destruction by fire occasioned by a lightning rod being struck which had been taken to earth in this way. We also note that in a rocky and dry situation it is recommended to bury 3 cwt. or 4 cwt. of iron at the foot of the conductor in addition to the ordinary earth plate. What action the 3 cwt. or 4 cwt. of iron will have upon the electric discharge is not explained, although it would be very interesting to learn.

"The space protected by a lightning rod receives considerable attention, this being a hobby of one of the members of the Conference. His views, however, are not fully endorsed, and the more guarded opinions quoted from the War Office instructions of 1875 (which are printed in one of the Appendices) are practically adopted. And here we would observe that the more recent instructions issued by the War Office last year are not quoted or mentioned, nor are the last instructions issued by the Admiralty in 1880. The recommendations of the Conference on the height of the upper terminal are illogical, as the matter is first stated to be one which may be left entirely to the option of architects and engineers, whereas in the following sentence the lofty rods often used in France, and the low rods usual in England, are regarded as 'opposite errors.'

"Concerning the testing of conductors, an annual examination both visual and electrical is recommended, and various apparatus are noted. 'The simplest and best is effective as regards testing the efficiency of the conductor, but not that of the earth connection;' but inasmuch as the earth connection is exactly that part of a conductor which most frequently requires repair, and is also the part which being under ground can only be examined electrically, it would not have been anticipated that a testing apparatus which could be used only for the conductor and not for the earth connection, would have been recommended as the 'best,' although it might easily be the 'simplest.' This apparatus is made after designs which were furnished by Mr. Preece, who saw a similar arrangement in France.

"The Conference recommends that all masses of metal in a building (church bells in a well-protected spire excepted, reason not stated), whether internal or external, should be connected to each other and to a conductor, or to the earth. Clerk-Maxwell, who followed up theoretically by such beautiful mathematical processes so many of the discoveries due to the experimental researches of Faraday, considered that masses of metal should be treated differently according to their position; the internal masses being kept separate from the system of conductors, but the external masses being connected therewith.

"The report concludes with the code of rules, which seem to err on the side of brevity; indeed some details of importance are not mentioned at all. For instance, in the case of large buildings with flat roofs, the spacing of the points and of the conductors, as well as the number and the best positions of earth connections, are not mentioned, nor whether surface as well as deep earth connections are to be provided as recommended in the instructions drawn up by the French Academy.

"Some of the rules appear to be drawn up without much care; for instance, where it is stated that 'the lower end of a conductor should be buried in permamently damp soil,' and it adds, 'hence proximity to drains is desirable.' We were always under the impression that drains produce dryness in soils. Again, in the same rule we find, 'a strip of copper tape may be led from the bottom of the rod to the nearest gas or water main, not merely to a lead pipe.' The use of the word *merely* is probably not intended, as any connection to a soft metal pipe is deprecated previously in the report.

"The line of reasoning which appears in some of the rules does not carry conviction. Thus, in the rule about '*collieries*,' it is stated that 'undoubted evidence exists of the explosion of firedamp in collieries through sparks from atmospheric electricity being led into the mine by the wire ropes of the shafts and the iron rails of the galleries. Hence, the head gear of all shafts should be protected by proper lightning conductors.' As the damage, however, is not occasioned above ground but in the mine, it would, perhaps, be preferable to 'earth' the head gear by a wire rope down the shaft, the iron rails of the galleries being connected *en route*, and earth plates provided both near to bank and at the various seams, and at the bottom of the pit. At all events, some reason should be given for the belief that protecting the head gear with a proper lightning conductor, and 'earthing' this conductor as laid down in the rule for 'earth connection' in the previous paragraph, would rectify the evil. The head gear, the winding engines, and the boilers and pipes, &c., already afford a metallic system to 'earth' which would appear to require no special lightning conductor.

"It is most unfortunate that the rules and the report have not been drawn up in a fuller and more convincing manner, as something of the kind of unquestioned authority was and, we are compelled to think, still is much required.

On the Protection of Buildings from Lightning.

By Captain J. T. Bucknill, R.E.[1]

"In whatever manner the electricity is produced, the thunder clouds act as collectors; and more than this,

[1] Extracts from a paper read at the United Service Institution, Friday, May 6th, 1881. Reprinted by permission of the Institution.

when the surface of the earth beneath them is not far distant, and is composed of fairly good conducting media, the earth, the clouds, and the intervening air form huge condensers—the electrified clouds acting by induction upon the earth, and the latter reacting upon the cloud.

"Now the amount of electricity of a given potential which a cloud is capable of receiving, depends firstly upon its size, the amount varying directly as the linear dimensions of the cloud; and, secondly, upon the intensity of inductive action of the earth's surface, the cloud's power of receiving electricity being greatly increased thereby.

"For example, a cloud of given dimensions at an altitude of 300 feet, could be charged by 80 times the electricity that would charge it were its altitude increased to four sea miles.

"For a similar reason a cloud over a conducting area could be charged much more highly than the same cloud at the same height over a non-conducting area.[1]

"Now it generally happens that the thunder clouds in a storm are sufficiently numerous to cover both favourable and unfavourable areas of the earth's surface, and, as little or no inductive action occurs over the latter, but very considerable action over the former, the electrostatic capacities of the clouds become greatly altered, and lightning plays from cloud to cloud, until those which are situated over the earth's conducting surfaces become so highly charged that the electricities are able to overcome the resistance of the intervening air and to unite across it by what is termed the disruptive discharge. This is lightning.

"I have been thus particular in describing the actions produced by the earth's surface upon thunder clouds,

[1] It is questionable whether any actual area is sufficiently non-conducting to make this true.—O. J. L.

because the somewhat important conclusion must be arrived at, that lightning is most to be feared by those who live on well conducting areas, even of low elevation; and that lightning is least to be feared by those who live on non-conducting areas. This is shown on plate, Fig. 9, where the distribution of the electrical charge is shaded in. The cloud over the Portsdown Hill, although nearer to the ground, is much less highly charged than the cloud over Portsmouth and Spithead, because the former

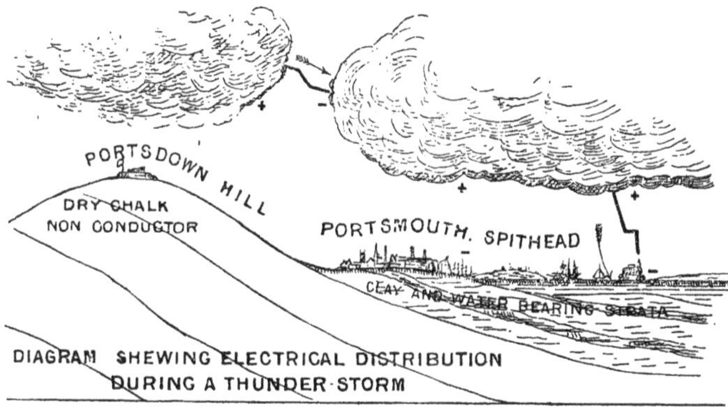

Fig. 6.

presents a non-conducting area and the latter a conducting area. This electrical distribution is of considerable importance, and it shows that it is much more necessary to provide lightning conductors for buildings situated upon a damp clay or boggy bottom than for those on a chalk down.[1] This is very convenient, for it is almost impossible to make an efficient earth connection in the latter situation.

[1] "At Portsmouth it has been noticed that although severe thunderstorms often occur in the vicinity, the clouds move round and seem to avoid the Portsdown Hills, which are of chalk and possess few trees."

"As before stated, disruptive discharge constitutes a lightning flash. Immediately before the stroke the particles of air are subjected to a high strain by static induction, producing a polar tension which is proportional to the square of the potential. Faraday's experiments proved this, as well as the fact that the stroke tends to traverse the air in the direction of such polarity. The tendency of lightning is therefore to strike in a direction normal to the earth's surface.

"But there is another mode by which thunder-clouds are discharged, viz., by the brush discharge.

"Electricity of high potential leaks, as it were, from conductors which are provided with projections in the nature of points, where the distribution of electrical density is greatest, a stream of electrified air being thrown from each point, and the charged conductor robbed by continuous streams of its electricity in this manner.

"Although the brush discharge is frequently so intense as to be luminous to a height of 6 or 8 inches, it is not attended with any appreciable heat. Its action should therefore be fostered, as it often wards off a dangerous stroke of lightning by neutralizing the opposing electricities in a harmless manner.

"It has been observed so long ago as 1758 by a Mr. Wilcke, that a thunder-cloud in sweeping at low elevation over a forest, not unfrequently appears to lose charge without the occurrence of lightning. The under surfaces of such clouds at first present a serrated or tooth-like appearance, which gradually disappears, the teeth retreating into the cloud, and finally the cloud itself rising away from the forest.

"In such cases the numerous points on the branches of the trees present facilities for the brush discharge on an extended scale.

"To illustrate this action, an experiment was made by Franklin, as follows:—A very fine lock of cotton was suspended from the conductor of an electric machine by a thread, and other locks were hung below it; on turning the machine the locks of cotton spread forth their fine filaments like the lower surface of the before-mentioned thunder-cloud; on presenting a point which was connected to earth below them, they shrank back upon each other, and finally upon the conductor.

"But to return to the lightning. Just as a certain amount of water falling through a difference of level produces a definite amount of energy, so a certain amount of electricity falling through a difference of electrical potential produces a definite amount of energy. It is known that if p be the potential and q the quantity of electricity in a flash, the work done during the stroke is $\frac{1}{2}qp$. Now the duration of the illumination of a stroke is rather less than the 10,000th part of a second,[1] and although q is small (Faraday said not more than would decompose a single drop of water), p is so enormous that the flash is often capable of decomposing a million drops of water in series. The potential can be calculated approximately, because it is known that 10,000 volts will spark across a little more than half an inch at ordinary atmospheric pressure; and, as the sparking distance varies as the square of the potential, a flash of lightning 1,000 feet long must be impelled by an elec-

[1] "This fact has been distinctly proved by experiments with revolving chequered discs. Wheatstone's classical experiment proved that the duration of the luminosity of the spark from an electrical machine is about the 24,000th of a second. The longer duration of luminosity in the case of lightning is probably due to the higher temperature to which the particles of the dielectric are raised by the stroke, and their consequent more tardy return to a non-luminous condition." (See also Rood, p. 85.)

trical potential of $1\frac{1}{2}$ millions of volts or thereabouts. This is only approximately accurate, because the mean atmopheric pressure would be less than that at the earth's surface, and therefore a correction should be made, as the pressure of the atmosphere decreases very rapidly with altitude, and the sparking distance increases very rapidly with decrease of atmospheric pressure. The work $\frac{1}{2}qp$ done by a flash of lightning is used up in the disruption of the air, in the destruction of non-conducting solids that obstruct its path, in heat, light, and in chemical decomposition. Ozone is always produced during thunderstorms. All that can be done to protect buildings from its destructive action is (first) to attract the lightning to another spot if possible, and (second) to arrange that even if the building be struck, the work shall be given out at other portions of the path of the stroke. To do this it is necessary to provide a sufficient conducting channel or channels to convey the electricity past the buildings from the air to the ground.

"Firstly, let us examine the methods which have been pursued for attracting lightning away from the building which it may be desired to protect. The French Académie des Sciences has issued information concerning lightning conductors on different occasions, the several instructions having been the results of the labours of various commissions of celebrated physicists.

"In the first instruction, 1823, with Gay Lussac as reporter, the rule is laid down that *a conductor will effectually protect a circular space whose radius is twice the height of the rod,* and it is stated to be in accordance with calculations made by M. Charles.

"Accordingly we afterwards find in the same instructions that magazines should be protected in the manner shown on Fig. 5, the wording being: 'The conductors should not be placed on the magazines, but on poles at from 6 to 8 feet distance. The terminal rods should be about 7 feet long, and the poles be of such a height that the rod may project from 15 to 20 feet above the top of the building. It is also advisable to have several conductors round each magazine.'

"In 1854, however, the next commission, with M. Pouillet as reporter, no longer supported this rule. The report says:

"'At the end of the last century it was a generally accepted opinion that the circle protected by a conductor possessed a radius equal to twice the height of the point. The Instruction of 1829 (Gay Lussac, *rapporteur*) having found that practice established, adopted it with certain reservations. . . . These rules . . . rest on much that is arbitrary, . . . and they cannot be laid down with any pretence to accuracy, since the extent of the area of protection in each case is dependent on a multitude of circumstances.'

"It is the more necessary to make this quotation, because an attempt has recently been made by Mr. Preece to revive the theory in a modified form. In a paper which he read before the British Association last year he attempted to prove that—

"'*A lightning rod protects a conic space whose height is the length of the rod, whose base is a circle having its radius equal to the height of the rod, and whose side is the quadrant of a circle whose radius is equal to the height of the rod.*'

"His argument was similar to, but not of such general application as, that used by M. Lacoine in a somewhat remarkable paper read 20th June, 1879, before the French Société de Physique, from which the following is extracted:

"'Experience shows that a thunderbolt has a tendency to fall on the metallic portions of a building. If then, by the assistance of a lightning conductor we are enabled to protect a certain metallic surface, much more therefore will the same conductor protect the same surface if non-metallic.

"'Let *N*, Fig. 7, represent a thunder-cloud situated over the surface *AC* to be protected. Assume that the cloud is at such a distance from the point *P* of the lightning conductor *PO*, that the circle described from *N* as centre with *NP* as radius will be tangential to the surface *AC*. Then the cloud will be equally attracted by the points *P* and *E*,[1] because these points are at the same potential, this rule having always been admitted in all the instructions of the Académie

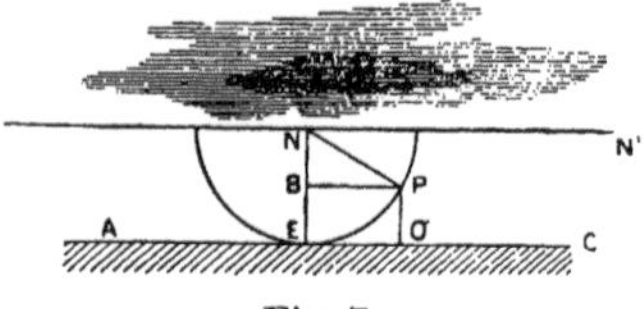

Fig. 7.

[1] "This is open to doubt; the electrical charge on the cloud is attracted by the induction of an opposing *surface*, the total attraction being proportional to the sum of the tubes of force existing between the two opposing surfaces, charged by inductive action. To assume that the charge on a thunder-cloud is concentrated at a single point is not in accordance with the circumstances of the case in nature.

"Faraday's experiments have conclusively proved that static induction polarizes the particles or molecules of the interposing dielectric, and that dynamic currents tend to traverse the same by disruptive discharge in the direction of the said polarity.

"Assuming therefore that a lightning flash from the charged surface *NN* occur at *N*, it will have a tendency to follow the direction *NE* rather than the alternative route *NP*, because polarity exists between *NE* to a greater extent than between *NP*.

"This consideration will cause the theoretical circle of protection advocated by M. Lacoine to be considerably diminished when the charged cloud lies low, but when the cloud is at a considerable altitude *NP* becomes more nearly normal to the surface *AC*, and more nearly parallel to the direction of polarity of the atmospheric particles."

Française. Consequently every point on the surface AC within the circle with radius OE will be protected, but every point outside E towards A would be unprotected.

"'Hence the radius of protection $r = \sqrt{NE^2 - NB^2}$, NE being the height of cloud above the ground, NB being the height of cloud above the conductor.

"'It is enough then, to know the height of the thunder-cloud, to know the radius of action of a certain conductor.

"'By several years' observation, and by direct measurement, the average height of thunder-clouds could be obtained, and the mean value of r for any given conductor deduced therefrom.'[1]

"Mr. Preece does not work out any such formula, but bases his rule on an assumption that a thunder-cloud would never be nearer to the earth than the height of the lightning rod. This is open to question, as very low-lying thunder-clouds may be driven by the wind into the neighbourhood of lofty conductors that command the clouds, and this is corroborated by a case recorded in Mr. Anderson's excellent book on lightning conductors, page 67, where the belfry of an edifice, 115 feet high, 'remained standing out clear above the electric cloud' whence issued lightning that killed two priests near the altar of the church. As a single application Mr Preece's rule comes at once from M. Lacoine's formula.

"It is perhaps important to bear in mind these theories concerning the area of protection given by conductors,

[1] "As the height of thunder-clouds varies enormously, the values for r would range between proportionately wide limits, and the *mean* value of r obtained by M. Lacoine would seem to possess no definite or practical utility. If, however, the observations were directed to observing the minimum altitudes of thunder-clouds in each locality (the altitudes will be found to vary with the locality), the smallest areas of protection given to conductors there situated could be approximately established."

when it is necessary to fix a few conductors on buildings of considerable extent, such as barracks, hospitals, etc., but sufficient reliance cannot be placed upon the rule to enable us to consider the protection to magazines, as shown on Fig. 1, p. 428, and already alluded to, as efficient.

Fig. 8.

"The area of protection afforded by a conductor depends much more upon the efficiency of the earth connections than upon the height of the terminal point, and in proof thereof many instances might be cited. For example, in the case of Shelton Church, in the Potteries, which was struck on the 10th June, 1880, the tower, about 16 feet square, is surrounded by four pinnacles 16 feet above the roof, which is nearly flat and covered with slates, with lead guttering and ridges. From the centre of the roof springs a large flagstaff, about 40 feet high, see Fig. 8, secured to the tower in the upper chamber 20 feet below the roof by large cross beams unconnected, except by stonework, with the clock-works, bells, and gas-pipes, in the chambers of the tower. A copper wire rope $\frac{5}{8}$ inch diameter is fitted to one pinnacle and taken direct to earth. Although the flagstaff projects some 20 feet above the conductor, and is distant only 10 feet, a very heavy stroke of lightning, which caused much alarm, and which was seen to fall upon the tower, struck the conductor, knocked the point slightly out of the perpendicular, and passed off by it innocuously. In this case a good conductor, well connected to earth, protected something higher than itself, but not well connected to earth.

"Again, Sir William Snow Harris mentions a chimney at Devonport which, although provided with a conductor, was struck on the other side, and shattered down to the level of a metal roof below. Here the conductor must[1] have been badly connected to earth, and was useless.

"Moreover the safe area rule may be upset in practice by all sorts of accidental circumstances. Thus, a house within the theoretical circle of protection given by a church spire close at hand might be struck if the line of least resistance from cloud to earth were afforded by a column of rising smoke from the kitchen fire, and the shorter of the two chimneys in Fig. 8 would most assuredly be struck, for a similar reason, although it is within the theoretical cone of safety of the taller chimney as fixed by Mr. Preece.

"In short, if thorough protection be desired for any

[1] Query.—O. J. L.

building, it is necessary to put a conductor or conductors upon it.[1]

"Let us now examine the manner in which conductors should be applied.

"Churches and dwelling houses of ordinary dimensions, factory chimneys, monumental columns, etc., need but one conductor led from the most lofty point to the ground, to which a thoroughly efficient earth connection (to be described presently) must be given. As a rule it is the best plan to fix the conductor externally, in which case it should be connected to all *external* metal surfaces, but *not* to any masses of metal wholly within the building. It should be fixed to the exterior by strong cramps of iron or other metal, and provision should be made for its expansion and contraction due to differences in tem-

[1] "A lamentable result of the practice of placing lightning conductors distant from a building occurred at Compton Lodge, in Jamaica, the residence of J. Senior, Esq. A lightning rod, of small dimensions, of iron, had been set up within 10 feet of the south-east angle of the building, as used to be the practice with gunpowder magazines, on the assumption that the rod would attract the lightning and secure the building. So far from this, the building itself was struck in a heavy thunderstorm, 28th July, 1857. The south-east angle was shattered in pieces; the escape of the family appears to have been miraculous; whilst the lightning rod, 10 feet distant, remained untouched. If this building had been a deposit of gunpowder, it would certainly have blown up.

"Sir Wm. Snow Harris said:—'To detach or insulate the conductors is to run away from our principle, which is, that the conductor is the channel of communication with the ground, in which the electrical discharge will move in *preference to any other course.* To detach or insulate the conductor is to provide for a contingency at once subversive of our principle. Is it possible to conceive that an agency which can rend large rocks and trees, break down perhaps a mile of dense air, and lay the mast of a ship weighing 18 tons in ruins, is to be arrested in its course by a ring of glass or pitch, an inch thick or less, supposing its course were from any cause determined in that direction?'"

perature. It should be continuous from top to toe. It should possess a proper amount of conducting power per unit of length.

As regards the last mentioned and most important matter of conductivity, the last French instructions, dated 14th February, 1867, state that there is no case on record where lightning has fused a square bar of iron having a side of 0·6 inch, or a section of 0·36 □″—and square iron conductors 0·8-inch side are recommended, which gives a section of 0·64 □″. Also Sir William Thomson considers that a round iron bar 1″ diameter would form a very safe protection for magazines; this would be about 0·77 □″ sectional area. It would appear that continuous iron conductors weighing 6 lbs. per yard would be quite safe, as shown in the following table:

TABLE A.

	Iron Conductors.		
	Side.	□″	lb. per yard.
Limits of safety—French instructions	□ 0·6″	0·36	3·6
Conductors recommended by ditto—from	□ 0·75″	0·56	5·6
to	□ 0·8″	0·64	6·4
Sir William Thomson recommended	○ 1·0″	0·77	7·7
New W. O. Instructions	...	0·8	8·0
Now proposed for general purposes.	...	0·6	6·0

Now iron has about one-seventh, and good commercial copper about four-fifths of the conductivity of *pure* copper. Hence iron has about one-sixth conductivity of good commercial copper. A safe conductor in good copper must therefore weigh 1 lb. per yard.[1]

[1] This portion, relating to conductivity, is now admittedly erroneous.—O. J. L.

It is, however, inconvenient to specify for a conductor either by sectional area or by weight per yard, because different samples of metal, and especially of copper, vary considerably in their conducting power. See Table.

Table of conducting power of different descriptions of copper:

TABLE B.

Pure Copper	100
Lake Superior	98·8
„ Commercial	92·6
Burra Burra	88·7
Best selected	81·3
Bright wire	72·2
Tough	71·0
Demidoff	59·3
Rio Tinto	14·2

Temp. about 15° C. or 60° F.

Imagine a conductor made of Rio Tinto copper (!) No doubt many exist.

A limit of electrical resistance per unit of length should therefore figure in any contract for a lightning conductor, and for the conductors already recommended this limit would be 0·3 ohm per 1,000 yards, or 0·03 ohm per 100 yards, at 60° Fahrenheit or 15° C.

This would be obtained from iron wire rigging ropes weighing 6 lb. per yard, or from copper (equal to 80 per cent. pure in conductivity) ropes weighing 1 lb. per yard.

When two "earths" are used, and the conductor is carried up one side and along the ridge and down the other side of the building to be protected, it is evident that the conductor may be reduced in power by one-half, but no further reduction can be made when a still greater number of "earths" are used, because the lightning may strike the system of conductors at any point. A 3-lb.

iron (or a half-pound copper) rope is therefore the smallest that should ever be used in any situation.

There is much difference of opinion as to whether iron or copper is the better material for lightning conductors.

The French use iron almost exclusively, and Sir W. Thomson prefers it to copper.

For the same money the same conductivity can be purchased in either metal (iron being one-sixth of the price and one-sixth of the conductivity of copper), and iron has the following advantages:

(*a.*) The mass of an iron conductor being greater than that of a copper conductor of equal conductivity, it is heated less by a given current of electricity.

(*b.*) The fusing point of iron (2,786° F.) is much higher than that of copper (1,994° F.).

(*c.*) Iron is more constant in its conductory power than copper of different samples.

(*d.*) A conductor made of iron is not so liable to be stolen as copper, and being so much the stronger is therefore less liable to be broken, accidentally or otherwise.

(*e.*) A copper conductor if connected to a cast-iron water supply pipe (to form an "earth") produces galvanic action, to the damage of the pipe.

On the other hand, a copper conductor lasts longer in smoky towns or near the sea shore, where the air rusts iron quickly, and being of much *smaller* size it does not interfere so much with architectural effects. But Sir W. Thomson has suggested that iron conductors should be treated boldly by architects, and brought into prominence purposely and artistically, and the late Professor Clerk-Maxwell recommended that in the case of new buildings the conductors should be built into the walls.

They would then not only be hidden but protected from the weather, from the British workman carrying out repairs, and from the thief.

As regards the liability of iron to rust, galvanizing is in most situations a sufficient protection, and in smoky towns an iron conductor should be painted periodically.

On the whole, therefore, the advantages of iron outweigh those of copper so considerably, that the employment of copper in lightning conductors should be the exception instead of the rule.

Those who make, supply, and apply lightning conductors in this country, nevertheless, invariably recommend copper; and it is quite difficult to convince them to the contrary.

Another point I notice is that large conductors are always recommended for lofty buildings, and smaller conductors for smaller buildings, and the same for masts of ships. This is unscientific and wrong. The stroke of lightning falling on a short conductor is no less powerful than the stroke that falls on a lofty conductor; indeed the chances are in favour of the shortest conductors receiving the heaviest strokes, if they are struck at all. On costly and important buildings, the proper course to pursue is to increase the number of conductors, and of the earth connections, the limit of electrical resistance between any possible striking point and earth being kept below what is fixed upon as the point of safety, viz., 0·3 ohm per 1,000 yards.

We will now examine the question as to the best *form* of conductor. Mr. Preece has investigated this subject, and by permission of Dr. Warren de la Rue carried out in that gentleman's splendid laboratory, a series of experiments on the best sectional form for lightning conductors. The results were communicated to the British

Association at Swansea last year. He found that ribbons, rods, and tubes, of the same weight per foot, were equally efficient.

The application of rods and tubes necessitate frequent joints, generally made by means of screw collars. I have found by electrical tests that these joints after long exposure to weather offer very high resistances; especially so in copper conductors. For instance, at Tipner magazine

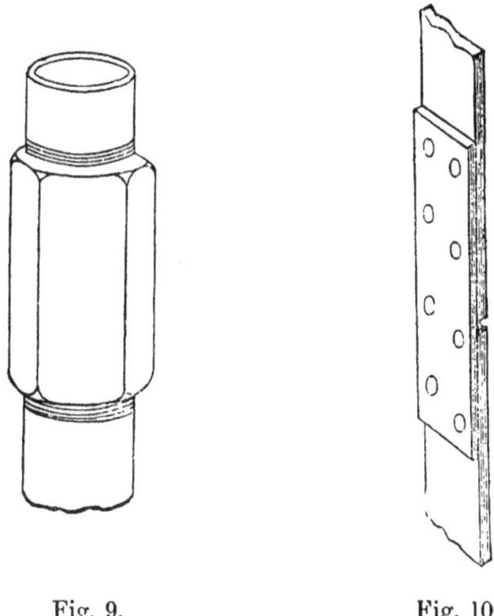

Fig. 9. Fig. 10.

a screwed joint in a large tubular copper conductor tested 10,000 ohms, and a riveted joint in a ribbon conductor on a battery in the Isle of Wight 700 ohms. These joints could not be moved by hand, and were apparently quite tight.

Ribbons of *copper* are now made in long continuous pieces (as much as 70 or 80 feet in one length), and can be applied to irregular architectural outlines, but the joints, although less frequent than with rods and tubes, are open to the same objections. The copper ribbon, however, possesses one decided advantage, viz., that by the introduction of suitable bends, the expansion and contraction from heat and cold can be allowed for. Iron conductors, when in the form of tubes, rods, or ribbons, are difficult to apply, and must possess a number of joints. Moreover, in long conductors, compensators to allow for expansion and contraction by heat and cold have to be introduced. In order therefore to obtain with iron the necessary continuity and pliability, it is best to resort to the wire rope, which form is already very generally employed for copper conductors. Pliability can be obtained in several ways—

1. By using small wires.
2. By making the rope *flat*.
3. By using a hemp core with the round rope.

It is not advisable to make the iron wire ropes with very small wires, because oxidation destroys such a rope rapidly if through carelessness the conductor be left unpainted. A fair amount of pliability can be obtained with a round iron rope 6 lb. per yard if the wires are about No. 11 B.W. gauge, and arranged in six strands of seven wires each round a hemp core, thus producing a rope about $3\frac{3}{4}$ inches in circumference.

But there are few situations in which two ropes of half the size could not be more readily applied; and I think the double rope, if taken up on one side of a tower and down on the other, in one continuous length, has many advantages.

Where a single conductor is desired, the best for general purposes is probably a flat iron wire rope about $2\frac{1}{4}'' \times \frac{1}{2}''$ (11 lb. per fathom), or $2\frac{1}{2}'' \times \frac{1}{2}''$ (13 lb. per fathom). The round ropes cost from 21*s*. to 24*s*. a cwt., or about 2*s*. 6*d*. per fathom for a 12 lb. rope; and the flat ropes 33 per cent. more, or add one-third.

The next question that presents itself is concerning the terminal point, and a good deal of nonsense has been written about it. Points made of silver or of copper, points covered with platinum or with gold, points of so many millimetres in height and diameter, and possessing certain exact forms, have been proposed, and rejected or adopted as the case may be.

The height of the points above the surrounding roof or tower to be protected has also been much debated with very little profit, for to this day many of the rods erected on the continent are made much longer than is necessary.

It is a good plan to carry conductors on lofty rods high above powder mills, flour mills, and petroleum oil wells; but these are exceptional cases, the air close to the buildings being frequently charged so as to be dangerously explosive.

The English practice of using a short rod in most situations is a reasonable plan, the rod being placed on the highest part of the building. The rod should be made of the same metal as the conductor, and the connection formed with bolts and afterwards run in with molten zinc or solder. The weight of the rod per foot should be the same as the conductor. The top of each rod should be provided with several points, (*a*) because the gathering power is increased thereby, and the chance of lightning striking other things in the immediate vicinity of the conductor is proportionately diminished; (*b*) because the top of the rod is less likely to be fused when struck, the stroke being divided between the various points; and finally (*c*) because the brush discharge is facilitated.

Another plan is to carry the wire rope up the side of

the rod, which in this case might have one point, the wires being opened out to form a brush-like arrangement just under the point. The wire rope and the rod should be bound together with wire and connected with molten zinc.

We must now pass to the foot of the conductor, and here we enter upon the most difficult part of our subject. The earth connections of a lightning conductor constitute the most important portion of the whole arrangement. If the electrical resistance of the earth connections be high, a conductor, perfect in all other respects, may fail, some alternative and perhaps dangerous route being taken by the lightning discharge. It is difficult to fix the limit of maximum resistance of the earth connections.

The Académie des Sciences recommends an iron earth plate, consisting of four arms on a central bar, or five arms in all, each 2 feet long and of square section 0·8 inch side, thus presenting a combined surface of 2·6 square feet, to be immersed in water in a well that never dries.

Again, Mr. Anderson, in his book before referred to, says that: "When a conductor is taken deep enough into the ground to reach permanent moisture, the single rope touching it will be quite sufficient. But when the permanency of the moisture is doubtful, it will certainly be advisable to spread out the rope like the fibres in the root of a tree." Here a few square inches touching permanent moisture is considered sufficient.

Again, Professor Melsens used three earths for the Hôtel de Ville at Brussels—one the gas-main, another the water-main, and the third a cast-iron pipe, nearly 2 feet diameter, sunk in a well and giving 100 square feet of surface to the water, which was rendered alkaline with lime to prevent oxidation. The total surface of these three earth connections amounts to more than $2\frac{1}{2}$ millions of square feet!

As opinions differ so greatly concerning the surface required for the earth connections, it will be necessary, before laying down any rule, to give some of the reasons upon which it is based.

The electrical resistance offered by a cylinder of spring water 1 yard long is as great as the resistance offered by a cylinder of copper of equal diameter, but seven times longer than the distance of the moon.

Now the practice in the War Department has always been to give joints in conductors a surface of about six times the sectional area of the conductor. This is a very good rule, and is borne out by the French practice, where even with soldered joints, 6 square inches of surface is laid down as necessary at each joint in an iron conductor. An obvious corollary to this rule is that when a conductor is made of two metals (end to end) the joint must have a surface equal to six times the efficient section of that conductor of the two joined which possesses the lowest conductivity. The efficient section of the better conductor ought not in any way to govern the amount of surface of the joint. Thus copper to iron requires a joint of 6 square inches, the same as would be required by iron to iron. In short, the joints should be made of such a size as to prevent the conductors of lower conductivity being damaged by the lightning.

A copper to copper joint only requires 1 square inch of surface, but it is generally convenient to give more.

Now the earth connection is really a joint, a very difficult joint to make well, and one that should follow the rules of other joints, *unless we can show good reason to the contrary.*

It is found that increasing the size of an earth plate

does not proportionately decrease the electrical resistance. A limit of size is soon arrived at, beyond which it is useless to go. "In the sea this limit is quickly reached."—(Culley.)

Culley states that if a plate containing 1 square foot of surface gives a resistance of 174 ohms, a plate of 4 square feet will give 140 ohms, and so on, a reduction of only 20 per cent. in resistance being obtained by quadrupling the earth plate surface.

The explanation that suggests itself as probable is that the electric current is distributed through the humid ground by an ever-increasing sectional area (often by an hemispherical surface), thus arriving at the efficient section for a water conductor of 2 millions of square feet (see Table C), at the small distance of 200 yards, or thereabouts,[1] from the earth plate; and this is borne out by the fact, noted by Culley, that the resistance depends to a certain extent upon the depth at which the plate is buried. Thus, a deep plate would disperse its charge in all directions by an ever-increasing spherical surface up to the limit of a sphere whose radius is equal to the depth of the plate underground, and afterwards by a segment of an ever-increasing sphere, which segment would always in this case be larger than, but would gradually approximate, the hemisphere. These sections are roughly shown on Fig. 11:

Culley states that the resistance alters with the depth at which the earth plate is buried, as follows:

4 inches . . .	100 ohms.
10 ,, . . .	90 ,,
40 ,, . . .	80 ,,
80 ,, . . .	77 ,,

[1] In an arid plain with a dry subsoil, the surface of which was wet by rain only to a depth of 1 inch, the efficient section of a water conductor would not be reached at a less distance than fifty miles,

It would appear, therefore, that little is to be gained by increasing the surface of junction between the earth plate and the earth (1) beyond the amount required to insure that the resistance to earth at foot of conductor is less than the resistance to earth through possible alternative routes in the vicinity of the conductor, and (2) beyond the amount required to prevent damage to the conductor by the flash of lightning when it leaves for earth. It is evidently impracticable to give a surface of some millions of square feet to the earth connections, and if it were practicable, the foregoing considerations prove, I think, that it is not necessary to do so.

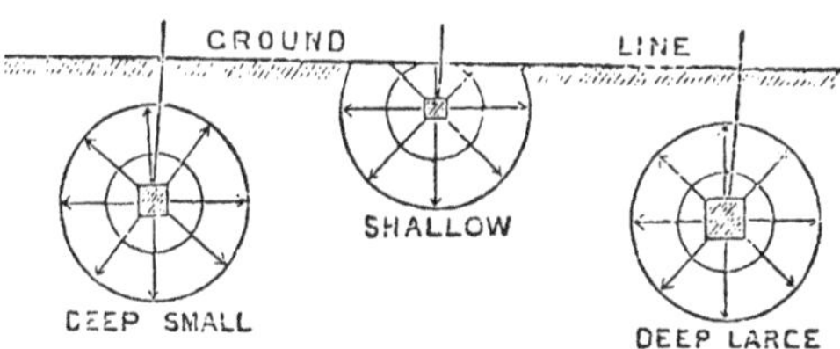

Fig. 11.

The difference in the conductivity of iron and water is so enormous that an intermediary appears to be very desirable. Carbon is eminently suited to act in this manner, especially if used in the cheap form of coke or ashes. The minimum effective section for coke is about 4 square feet, the iron which is surrounded by coke should, therefore, have a surface of 24 square feet. Moreover, inasmuch as the contact between an iron plate, of whatever form, and coke loosely surrounding it must frequently be discontinuous, and as the conductivity of coke in a mass composed of loose particles must be very much lower than that of a solid piece, the above surface should in practice be a minimum.

The total surface may, however, be divided if a number of earths be used.

The outer surface which should be given to the coke, must depend very much upon the nature of the ground; when the conductor is led into soil which cannot be regarded as permanently damp, the surface of the carbon "earths" must be increased.

As the surface of the earth connection should vary directly as the resistance per unit of area, an intermediary of coke becomes unnecessary where a conductor is led into salt water; but the conductor should still present a total surface to earth of from 20 to 30 square feet, the amount being divided between the "earths" if several conductors be connected.[1]

Professor Pouillet's committee, which reported upon the application of conductors to the Louvre in 1854-55 (the said report being adopted by the Académie des Sciences), recommended that when permanent water is not found near the surface, two descriptions of "earth" are necessary; first, the deep earth connections to permanent water, and secondly, the shallow earth connection to the surface water. This for the following reasons: After a long drought, the "terminating plane of action" (to use Sir William Snow Harris's term) is situated on the upper surface of the deep water-bearing strata, the induced charge being consequently collected there. After a heavy rain, however, which thoroughly impregnates the upper strata with water, the "terminating plane of action" is raised to the surface of the ground, and the induced charge is accordingly collected there. It is evident, therefore, that a perfect arrangement should in many situations provide both for *surface earths* and for

[1] A 3″ (circumference) wire rope offers about 1 □′ surface per 4′ run.

deep earths. In some situations, however, such as the top of a chalk hill, deep earths would be of little value; whereas in other situations surface earths would be inefficient—in a well-paved town, for instance, where the surface water is at once carried off by gutters and drains.

A deep earth connection can be effected in the manner shown in Fig. 12, the well being carried down 10 feet below water level in the driest seasons. The diameter of the well may be fixed at 3 feet. It should be rendered alkaline with lime, so as to protect the iron from rust.

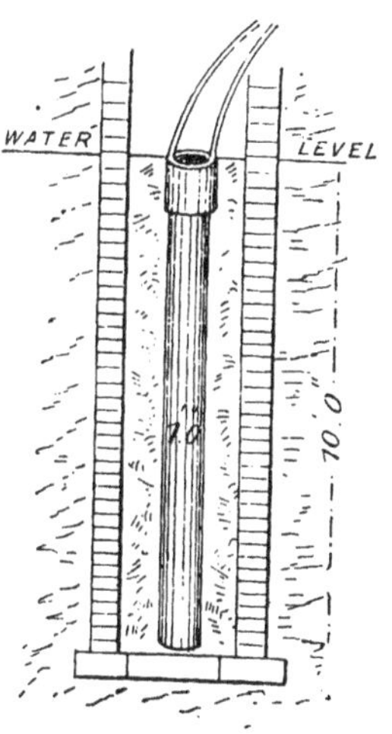

Fig. 12.

The bottom 10 feet should have no mortar or cement in the walls, and should be filled in with blocks of coke. The iron conductors should terminate in cast-iron pipes, offering together 24 square feet of outside surface. The pipe should be galvanized to preserve it from oxidation. The dimensions of the pipe may be, length 10 feet, diameter 1 foot. The pipe may rest on the bottom of the well, in a vertical position. The best way to connect the pipe with the conductor is to have a flange at the top (all ordinary gas or water pipes have such flanges), and to rivet a small cylinder to the inside of the pipe at the upper end, thus forming a ring or annulus, into which the end of the conductor can be introduced, and the space filled in with molten zinc, the surfaces of the conductor and of the pipe having first been cleaned and painted with hydrochloric acid.

In situations where iron water supply pipes are at hand, they can be employed in place of the deep earth connections already described, but great care must be devoted to the connections. The conductor must be laid along the iron pipe for a distance of 4 feet (if an iron wire rope it should be unlaid for this distance), it must then be bound to the pipe with wire, and a metallic connection formed by means of lead, zinc, or solder. The connection should then be tarred and covered with tarred tape to prevent galvanic action.

Surface "earths" should consist of a trench filled with coke and ashes, and carried away from the walls. Clay and other soils which keep the rain-water near to the surface, require shallow trenches about 1 foot deep; whereas gravel, sand, or shingle, through which the water penetrates easily, require deeper trenches, say 2 feet deep.

In each case, however, the top surface should be kept on the ground level.

The end of the metal conductor should be carried along the bottom and through the whole length of each trench. This length may in ordinary soils be fixed at 25 feet, and in very porous soils at 50 feet.

The water-pipes from the roof of the magazine or building may with advantage be caused to deliver into gutters which lead to the surface "earth" trenches.

The shallow trenches, 1 foot deep, recommended for stiff soils, may conveniently be split into a V shape on plan (the conductor being split also), so that the total side surface may be equal to that given by the same length of deeper trench used with porous soils.

Important buildings and magazines provided with several conductors, may have a few deep "earths," and several shallow "earths," an "earth" of one or the other description being provided at the foot of each vertical conductor, and in order to connect the whole it is advisable to employ a horizontal conductor near the foot of the wall, but above ground in order that it may be open to inspection, such conductor being carefully connected to all the vertical conductors, and to all the metal water-pipes. By this means not only is the cage principle advocated by the late Professor Clerk-Maxwell and other physicists embodied, but the earth connections are connected in an efficient and reliable manner. (Figs. on p. 428.)

Sir W. Thomson considers that conductors on magazines should be spaced at intervals of about 50 feet, by which plan no portion of the building would be more than 25 feet from a conductor. This rule has been adopted by the War Department for all large magazines, and a conductor of power equal to an iron rod weighing 8 lb. per yard has been adopted for single conductors, and of half that weight for all others. A wire rope of 4 lb. per yard, applied *as shown on diagram*, Fig. 4, page 428, is now considered the best arrangement.

It will be seen that wherever the lightning falls a conductivity equal to, or more than, that of a single large conductor will carry the stroke off to earth.

Small magazines can be protected by one rope led to a deep "earth" at one end and to a shallow "earth" at the other, as shown on diagram, page 428.

Powder mills must be provided with lofty conductors, to guard as much as possible against powder dust in the air being ignited by the stroke.

As regards the inspection of lightning conductors, opinions vary greatly, and it was mainly in order to obtain a report on this matter that I was ordered last summer to inspect a number of conductors on magazines in the Portsmouth district. I will read a few extracts from my report. (See Appendix A.)

Before concluding this paper, I may observe that the principal object has been to prove the following points:

1. That iron is the best metal to use in conductors.
2. That wire ropes are more easily applied than rods, ribbons, tubes, etc.
3. That conductors should be continuous, and that all unavoidable joints should be soldered.
4. That conductors should be specified in terms of electrical units.
5. That lofty conductors require no additional conductivity per unit of length.
6. That high lightning rods are only required in exceptional situations.
7. That several points are preferable to a single point.
8. That greater surface than is usual with present practice should be given to earth connections.
9. That both deep and shallow earths are required.
10. That periodical inspection is most important.
11. That the history of conductors and of former tests should be carefully recorded.
12. That electrical tests may then be of value."

Appendix A.

"I have to report that, in accordance with instructions, I have made nearly 500 tests, and have inspected the whole of the lightning conductors on fortifications in the Portsmouth and Gosport Divisions of the southern district, and have come to the deliberate conclusion, after a careful study of the subject, that *with the lightning conductors erected as they are at present by W.D.*, electrical testing is of small value.

The fact that the conductors on one building test

lower than the conductors on another building, certainly points to the inference that the earth connections in the former case are of superior efficiency; but it does not prove it. Moreover, although the tests are sometimes of value to the inspector *when he knows the details of the earth connections from office records,*[1] the tests taken by themselves are frequently positively misleading, so far as the earth connections are concerned. As regards the conductors themselves, above ground, high resistance tests do not prove inefficiency when the W.O. rule that the surface of the joint shall be at least six times the sectional area of the conductor is strictly adhered to; and in this view I am borne out by Sir William Thomson's opinion, which now lies before me, viz., "that although it would be desirable that the joints should be soldered and run in with lead, so as to make sure of absolute contact, at the same time it is to be remarked that the great resistance at imperfect joints is not detrimental to the lightning conductor, because, when a discharge takes place, the imperfect joint is bridged across, and the resistance, which is very great when tested by a feeble current, becomes practically annulled in the electric arc during discharge."

Dr. De la Rue also writes to me and says:—"The resistance of many megohms would offer an insignificant obstacle to a lightning discharge, on account of the extremely high potential of a thunder-cloud. Consequently, a conductor would be quite efficient, although offering a megohm resistance."

The opinion that lightning conductors with large surface joints are efficient, although offering high resistance at the joints, is also substantiated by the well-known action of plate paratonnères, as applied on the flanks of

[1] Which are but seldom obtainable.—J. T. B., 1891.

electric telegraph stations, to protect the instruments therein from the effects of strokes of lightning upon any portion of the line. These paratonnères consist of plates, in most patterns smaller than the flat joints of lightning conductors, and paraffined paper is interposed between the plates the more thoroughly to insulate the lower plate from "line." A number of these paratonnères are in store at Woolwich, and they each test from 3 to 40 megohms of resistance; yet in practice a flash of lightning is always found to pass across them to good "earth," in preference to the alternative path offered through the telegraph instrument, usually of less than 2,000 ohms. It is therefore quite erroneous to suppose that lightning always passes to earth by those paths, which, *to the ordinary voltaic current,* test lowest. It, however, does pass to earth by those paths which, *to a current of its own potential,* would test lowest. . . . .

With regard to the conductors now existing on our magazines and fortifications, and which have been erected for the most part on sound principles, and which have never yet failed, it would appear that the periodical inspection should be performed by a thoroughly competent inspector who has studied the subject. He should be provided with drawings and record plans, and every information that can be afforded of each and every conductor in the district to be inspected. The information concerning the earth connections should be most minute and exact. He should also be provided with a light equipment for making such electrical tests as he may find necessary. If this were done, my recent experience would point to the conclusion that the electrical tests would form the least important portions of his periodical reports. . . . .

As far as my own experience has gone, it would seem

that our conductors are, with few exceptions, as efficient now as when they were first put up ; but the earth connections of most of the conductors are, and always were, considerably below the standard. . . . . .

Although the lightning conductors at present on our magazines and forts are no doubt, so far as the conductors themselves are concerned, efficient, their efficiency could nevertheless be guaranteed with greater certainty if more modern practice were followed. . . . .

The adoption of modern practice would at once make electrical testing of considerable value, because with *unbroken continuity* and the *best earth connection,* all conductors would test at a very low figure, unless out of order. An economy would also be effected on all new works, because the metal pipes and rods with costly sliding joints, to allow for expansion and contraction, would no longer be required.

As regards the testing of conductors: a few tests were taken with the three-coil galvanometer, but with no satisfactory results, as the instrument is not sufficiently accurate when used as a measurer of electrical resistance. An attempt was then made to test by means of the "earth" cells produced by the earth of the lightning conductor, which was always either of copper or iron, and a test earth of iron or copper. This gave promise at first of becoming a good test, the astatic galvanometer being employed, but the method was soon discarded from want of accuracy. It is, however, useful for the tester sometimes to discover the metal of the earth connection of a conductor, and the above method can then be resorted to. . . . .

A quarter of a mile of the light insulated wire for Engineer mountain equipment (60 lbs. per mile) was cut up into three pieces, each 110 yards long and 4 ohms resistance, and two pieces each 55 yards long and 2 ohms resistance. This wire was found to answer well, and being so light, could be carried over a man's shoulder without any difficulty for considerable distances. . . . .

Two small plates (one copper and one iron) were used, their dimensions being 7 inches wide and 8½ inches long; they were of oval shape, and made of quite thin metal. A lip was formed at the top, and a hole punched in the plate 2 inches below it; a 2-foot piece of Navy demolition cable was then brought through the lip, passed through the hole, the wires cleared of insulation for 1½ inches, and the ends spread out like a fan and soldered to the plate. The lip at the top was then firmly hammered over the covered wire until it held the wire tightly. The other end of the piece of core was then stripped and the wires sweated together ready for insertion into a brass connector when required.

A number of resistance tests having been taken with the P.O. pattern resistance coils, an astatic, and service six-celled portable test battery, it was found that the tests usually ranged below 200 ohms; and I designed an instrument to test these resistances with approximate accuracy up to 200 ohms, and to measure roughly up to 2,000 ohms, the bottom plug being placed in the "× *TEN*" hole when measuring the higher resistances. The whole arrangement weighs less than 6 lbs. when the battery is charged; its dimensions, moreover, are only 9″ × 5¼″ × 6″ over all, and the method of using it can be taught to any intelligent man in a few minutes. The instrument shown on Fig. 13 is the latest and improved pattern, and has a range up to 1,110 ohms, when testing direct by steps of 1 ohm; and to 11,100 ohms by steps of 10 ohms, when using the multiplying hole marked × *TEN*. In testing a conductor's "earth" the wire to the conductor would be taken to terminal *L'*, one pole of the battery and the wire to the test earth plate to terminal *BL*, and the other pole of the battery to terminal *B'*; the plugs on the upper row of brasses would then be moved about until no deflection is produced upon the galvanoscope on the battery key being pressed down, the bottom plug being placed in the "*EQUAL*" hole. If, however, the resistance to be found is more than 1,110 (shown by above trial) the bottom plug is moved to the "× *TEN*" hole, and a balance obtained and recorded.

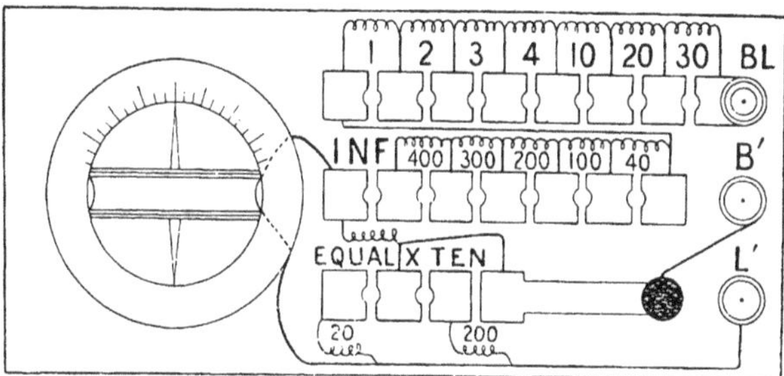

Fig. 13.

A special clamp was found to be useful in connecting the test wire to the conductors, a small clean spot being produced by a file for the end of the screw to seat upon. When the leads had to be connected for long stretches the naval pattern brass connectors were used."

APPENDIX B.

Extracts from a Memorandum by Colonel H. Schaw, R.E., 1879, on Lightning Conductors.

The testing of the electrical resistance of a system of lightning conductors will generally present great difficulties, because the ordinary means of allowing for expansion and contraction by slotted joints destroys the metallic continuity of the conductors, and introduces a variable resistance of oxides and foreign substances between the slipping surfaces.

This resistance will generally be very much in excess of that of the whole length of the conductors; it is, however, of little or no consequence when opposed to electromotive force of such high tension as a lightning discharge, which will easily pass the obstruction as exemplified in the form of lightning protector used by Messrs. Siemens for electric telegraph stations, which is formed by two brass plates with roughened surfaces placed face to face, but prevented from coming into contact by a thin strip of mica.

If the line wire is struck by lightning, the discharge takes place to earth through the protector, the two plates becoming oppositely charged by induction, and a spark passing between them. . . .

The ordinary currents have not a sufficient tension to pass the air space in the lightning protector, but go to earth through the more circuitous route of the instrument.

The test by simple inspection would seem to be the best for the conductors above ground. A resistance test could only be applied with advantage where there were no slip joints, and where the conductors were difficult of access.

As regards the earth connection, simple inspection may frequently be the easiest and most satisfactory test also. It is known by experience that 10 superficial feet of metallic conductor in contact with *wet earth or water* is sufficient to carry off safely any discharge of lightning. If then we can by inspection ascertain that in *dry summer weather* we have such a connection we may be satisfied. Should it be difficult to inspect, then the electrical test should be used, and I should prefer the Wheatstone balance test. . . .

It might happen that the connection between the conductor and the plate, or tube, or mass of metal forming the earth, was imperfect, owing to oxidation. In such a case the resistance would appear considerable, yet in reality the connections might be practically good as regards lightning, as a spark would pass from the conductor to the plate, etc., and from its large surface in contact with water it would escape freely and harmlessly. . . .

Hence I consider that in all possible cases inspection is the best test, but that electricity carefully used may assist the inspection in cases where the earth connection is difficult to get at.

It is most necessary that tests or inspections of earth connections should be made at the driest time of the year. In wet weather they must always be unreliable.

In rocky or very dry sites good earth connections are most difficult of attainment. . . .

I do not think that tests made by weak currents are of any very great value in deciding on the resistance of earth connections intended to carry off a great charge of electricity at one instant of time, as in the case of a lightning discharge.

24th January, 1879. H. SCHAW, Colonel, R.E.

P.S. — Were all systems of lightning conductors arranged so that expansion and contraction might be allowed for by S bands of flat iron instead of by slip joints, and all other joints welded or soldered, electrical resistance tests could be applied without difficulty, and I consider this would be very desirable.—H. S.

Extract from Discussion.

Major HAMILTON TOVEY, R.E.: I have had to superintend the arrangements for the protection of extensive buildings at Waltham Abbey, in connection with the powder works, and there naturally we had to be very careful. During the time I have been there, there have been four distinct cases in which buildings have been struck, and I had the opportunity of seeing them immediately afterwards. This circumstance, taken in connection with the situation in which Waltham Abbey lies, is a striking illustration of the truth of what Captain Bucknill says as to lightning being particularly liable to strike on damp soils, for, of course, four distinct cases within a short period and within a limited area is very far above the average. The first case was that of an entirely new building—a range of new mills for incorporating powder. The centre building was about 50 feet high, and on each side of it extended about 60 or 70 feet of lower buildings. There were lightning points over the centre and also at the extreme end of each wing. During a thunderstorm, the point at the end of one of the wings was struck, but no damage was done, excepting that the stroke seemed to have a sort of shaking effect, loosening part of the iron roof trusses with which the conductor was connected, and shaking all the mortar out of some of the joints of the ironwork and brickwork. The stroke passed away without doing any other harm to the building. That is rather a striking illustration of how the lower part of a building can be struck when the high part is not, because the tower, which is considerably higher than the point struck, escaped.

The second case was rather a striking one. A small low wooden building, fitted with a copper conductor leading into water, was struck, although it was within 220 feet of a very high chimney—150 feet high—and was also surrounded by trees, a most unlikely place to be struck. The building was situated alongside a stream of considerable size, and the conductor led directly into the water.[1] That, possibly, might have led to its being struck. There again the passage of the lightning had a shaking effect. The building was a low wooden one, and there was an arrangement by which a large copper basin full of water was balanced over the mill, so that in case of an explosion it would be upset over the powder, and this was shaken down. At the same time the lightning conductor was shaken away from the woodwork in places.

In another case the lightning struck a bell wire which was carried along upon several posts, and was used for ringing a bell at the works. The lightning seemed to have struck a tree to which the wire was fastened, and then ran along the wire and passed down the posts. It was curious to see the way in which the electricity passed from one copper nail to another on the posts. After passing down the copper as far as it extended, it seemed to have jumped from one nail to the other, tearing the intervening wood out as it went.

[1] The conductor was of copper band, $1\frac{1}{2}$ in. by $\frac{1}{8}$ in. in section; it was immersed for 3 feet of its length in water, the last 2 feet of the band being split up, forked apart, and covered with stones to keep it in place.

The last case was not in the powder factory, but about one and a half miles off, in a private house, close to some works that we were executing—a row of villas. The chimney of one was struck. There was no lightning conductor, and the lightning passed down the chimney, probably attracted by the warm air from the fires, and then went from room to room down three floors, shaking the iron grates out of place, and in one case throwing it right out into the room, but fortunately no one was injured. That case showed how very difficult it is to know when you are near a flash of lightning, exactly how near you are, because a number of workmen were about the buildings, and although they must have been at least 100 yards away, they were terrified at the flash, and were all ready to swear that it struck the place they were in.

After this paper, Colonel Bucknill was asked to draw up a report on the subject for the War Office. From this draft, which contains many of the suggestions adopted in the last War Office Rules, 1887, and is too long to quote in full, I extract the following important appendices, viz., statements by Sir William Thomson, and the account of an observation made by Mr. Brough :

Appendix C.

Sir W. Thomson, F.R.S., etc., in answer to a letter from Colonel Stotherd, R.E., on the subject of lightning conductors, made the following remarks, 11. 2. 74 :

"I have always maintained that iron is better than copper for three reasons:

"First—for the same value of metal, roughly speaking, as much conducting power can be obtained in iron as copper.

"Second—with the same conducting power, the mass of iron is greater than the mass of copper, and the specific heats being nearly the same, the elevation of temperature produced by the same electric energy dissipated is less in iron than in copper.

"Third—the melting temperature of iron is much higher than that of copper, and, therefore, even were the elevation of temperature the same, the copper conductor would melt before the iron.

"On the other hand, some, even taking the advantages into account, prefer copper to iron because of its less liability to rust. You on the other hand point out that the copper is more liable to suffer from the thief. . . . I think it is quite clear that for all powder magazines, powerful iron conductors are preferable to copper. . . . .

I believe that a solid iron rod of one inch diameter will always be sufficient. I believe that no conductor of the same conducting power as this, whether of iron or copper, has ever been melted by a lightning discharge. . . . . The merit of a tubular conductor in comparison with a solid one of the same mass must not be rashly decided. So far as mere conducting power is concerned, one is as good as the other, but there is a *quasi inertia* due to mutual electro-magnetic induction between parallel conductors, in virtue of which the solid iron rod will be somewhat less effective in permitting a very sudden discharge through it. . . . .

I should think that a solid iron rod of an inch diameter every fifteen yards would be very safe . . . . and a solid pointed iron rod with the point gilt will certainly give in any case, for the same expense, much better protection, than a solid pointed copper rod.

I would certainly have all the joints soldered, as much resistance is added by any unsoldered joint that can be made, and there is danger of fire at any unsoldered joint. The expansion and contraction might be allowed for by suitable bends introduced at intervals in any horizontal lines of conductor. . . . .

Periodic inspection of every joint, and of the earth connections, ought to be ordered and regularly carried out. . . . . The ease with which copper wire rope can be placed on buildings with complicated architectural forms recommends it strongly to architects. For example, it was chosen on this account for our own University Buildings. . . . .

For protecting powder mills the external pointed conductors ought to be at a sufficient height above the buildings, or to be so placed that a lightning discharge to the point may have no chance of igniting dust of powder in the air. A fork or brush of three or four points at the top of a lightning rod is probably in general preferable to a single point; but of what practical value this preference may be I cannot tell for certain, although I think it may be considerable.

Believe me, etc.,

WILLIAM THOMSON.

Sir William Thomson was asked unofficially, 30th September, 1880, the following questions:

(1.) Are joints in lightning conductors objectionable when the said joints offer high resistances to the passage of voltaic currents?

N.B. All W.O. conductors have the joints (if any) outside the buildings, so that the danger of fire at the joints is reduced to a minimum, and the rule of the service is to make the surfaces of junction equal to six times the sectional area of the conductor, which itself is always equal in conductivity to, at least, a rod of copper half an inch in diameter; say ·00012 ohm per yard run.

(2.) Assuming, with Sir Wm. Snow Harris, that such a conductor is sufficient to carry off the largest stroke of lightning, is it necessary, when several vertical conductors are united at the top of the building or magazine by horizontal conductors, to make them all of such large dimensions? In principal magazines containing, say 700 tons of gunpowder, the present practice might seem desirable, but in positions of secondary importance and in the case of buildings (always assuming the earth connection or connections to be good), it would appear that a number of small conductors somewhat exceeding *in joint conductivity* the aforesaid limit of ·00012 ohm per yard of height, or $\frac{1}{4}$-ohm per knot, ought to be efficient. This idea is taken from the manner in which the Hotel de Ville, at Brussels, has been protected by a large number of small conducting wires joined at the top and joined again at the bottom, great care being taken to provide a good "earth" and good joints at the top and near the ground.

(3.) How much surface (in square feet) do you think is necessary for the earth connection?

(*a.*) In dry soil, as in the case of a fort on the top of a chalk hill, the only water supply being rain in cemented tanks.

(*b.*) In damp soil, such as water-bearing strata a few feet down.

(*c.*) In salt water, for sea forts.

(4.) Concerning the advisability of connecting all masses of metal with the system of conductors, for instance, the iron doors or copper-covered shutters of the windows of magazines. In such positions the metal-lined cases full of gunpowder may be close inside, and to me it would appear advisable to keep the conductors away from such openings and not to connect them, as is now the custom."

The following was received in reply:

The University, Glasgow,
October 19, 1880.

Dear Sir,

The following is from Sir Wm. Thomson in answer to your letter to him, of date September 30th:

"I think it would be desirable that the joints should be soldered and run in with lead so as to make sure of absolute metallic contact. At the same time it is to be remarked that the great resistance at imperfect joints is not detrimental to the lightning conductor, because when a discharge takes place, the imperfect joint is bridged across and the resistance, which is very great when tested by a feeble current, becomes practically annulled in the electric arc during the discharge.

"I quite assent to your answer to question No. 2.

"The third question with reference to earth connection is much the most difficult. For case (*a*) I think it would be scarcely possible to obtain a thoroughly safe earth connection. As to cases (*b*) and (*c*) I am not able to give any definite information, although, no doubt, some rules for them have been derived from practical experience.

"Perfect security would be obtained by having sheet-metal over the floor and walls and roof of the whole building. Thus a building of galvanized iron with sheet-iron floor would be perfectly safe for a powder magazine without any earth connection whatever. Even windows of some considerable size, and wooden doors would not,

I believe, impair its safety. Indeed, it seems to me that in all cases gunpowder magazines should be made of (? galvanized) sheet-iron. The gunpowder itself might be stored on stone slabs laid upon the sheet-iron floor or on stone shelves fixed to the iron walls. The metal floor and walls and roof should all be in thorough metallic connection. I should be glad to hear what you think of this from the practical point of view, that is to say whether sheet-iron buildings could or could not be used as gunpowder magazines.

"All the conductors should be as thoroughly connected as possible by solder and lead connections.

WILLIAM THOMSON.

Per A. GRAY.

APPENDIX D.

"*On a Case of Lightning*," *from the Proceedings of the Asiatic Society of Bengal for February*, 1877, *by R. S. Brough.*

The south-west monsoon of 1871, in the neighbourhood of Calcutta, may be considered to have been characterized no less by its copious and protracted rainfall than by the violence and frequency of its thunderstorms. During the progress of one of these storms in the early part of the monsoon, one of the trees standing near the gate of the compound building, then occupied by the Sadr Diwáni Adalat, and now used as the European Military Hospital, in Lower Circular Road, was struck by lightning. The branches of this tree overhung the wires of the telegraph line, from which they were only about a foot distant. The discharge passed from the tree to the wires (of which there are four), broke fourteen double-cup porcelain insulators, and passed to earth through the iron standards on which the wires are supported.

The one ends of all the four wires were connected to earth through instruments in the Calcutta Telegraph Office, at a distance of about $5\frac{1}{2}$ miles from the locality of the accident. The other ends were connected, as follows, to earth through instruments: the first at the telegraph workshops, a distance of less than a quarter of a mile; the second at the Lieutenant-Governor's residence, less than half a mile; the third at Atchipur, less than 14 miles; and the fourth at Diamond Harbour, less than 25 miles. At the moment of discharge nothing extraordinary was noticed at any of these offices.

It is often far too generally stated in text-books that lightning invariably follows the best conductor to earth. This statement is misleading at the best; and is absolutely untrue if the word "conductor" be employed in the sense to which it is usually restricted in electrical science. In this instance, for example, we find that the lightning broke 14 insulators, each having probably electrical resistance of several thousand megohms, in preference to traversing a resistance of not more than 500 ohms to earth through the receiving instrument in the telegraph workshops. The writers appear to overlook the fact (experimentally illustrated long ago by Faraday) that there is exerted a mechanical stress proportional to the square of the potential, tending to produce disruptive discharge, as well as an electromotive force proportional to the simple potential, tending to produce a conductive discharge. Thus the discharge may occur either along a path of minimum mechanical resistance, or along a path of minimum electrical resistance. Which form of discharge will occur in any particular instance depends of course on the special circumstances of the case; but generally speaking, as the potential increases the tendency naturally is for the disruptive to predominate over the conductive. In the case of lightning, the potential is so great that for any form of "lightning protector" to be efficient the conductive facilities offered must be correspondingly great, that is, the protector must offer no sensible resistance to earth, otherwise a disruptive discharge may take place from the protector itself, which under these circumstances becomes merely a source of danger. This tendency to disruptive discharge is taken advantage of to protect telegraph instruments from lightning.

Another assertion of the text-books is that the metallic rods now employed as lightning protectors on buildings do not "attract" lightning. This statement is literally true, according to the meaning of the word "attract," but it is untrue in fact. For such a rod lightning protector determines a line of maximum induction, and a discharge is more likely to occur at the place than if the protector were not there. Professor Clerk-Maxwell does not appear to hold this opinion, but it seems to me unquestionable that if a charged thunder-cloud is carried over a building furnished with a lofty metallic rod, discharge is more likely to occur than if the rod were away. Professor Clerk-Maxwell observed in his paper recently read before the British Association at Glasgow, that such lightning protectors are designed rather to relieve the charged cloud than to protect the threatened building. In fact lightning rods are legitimately employed for this very purpose in the vineyards, where the object in view is to relieve charged clouds and prevent disruptive discharges and the consequent showers of hail.

[The calculations then entered into in the paper prove

the electromotive force of the discharge under examination to have been at least 216,810 volts.]

Assuming the sparking distance to increase as the square of the potential, it can be calculated from the experimental results obtained by Messrs. Warren de la Rue and Muller (Proc. Roy. Soc., January, 1876), viz., that 1000-rod chloride of silver cells give a spark 0·009166 inch, that a difference of potential of 216,810 volts would produce a spark in air between two electrodes at a distance of about 36 feet apart. This is of course a relatively very short distance, but it must be remembered that we have only taken into consideration that portion of the energy of discharge which was employed in breaking the 14 insulators, and have neglected all that was spent in heat, light, etc.

Appendix E.

During the discussion on Mr. W. H. Preece's paper on Lightning and Lightning Conductors, read before the Society of Telegraphic Engineers, 11. 12. 72, Sir William Thomson observed:

There is no reason to suppose that clouds are essential to electrical discharge in the atmosphere. On the contrary, instances are recorded, both in ancient and modern times, of lightning flashes occurring in a perfectly clear sky. Clear air is generally of as much importance as cloud, possibly in general of greater importance than cloud, in the theory of atmospheric electricity, and clouds must not be spoken of and studied to the exclusion of the rest of the atmosphere. The electric potential of any point in the air can be measured relatively to the earth without the slightest ambiguity. One of the methods of doing this is by means of a water-dropping apparatus. This consists either of an insulated vessel from which a stream of water flows in the form of a fine jet, and breaks into drops at a definite point in the air, or of an insulated receiver in which drops from an insulated jet are collected. . . . .

I have learnt from this and other unmistakable experiments that the lower stratum of air is, in fine weather, in general negatively electrified.

The potential of the air out of doors in fine weather is always, but with very rare exceptions, found to be positive. In showery or wet weather it is sometimes positive and sometimes negative, occasionally altering with extreme rapidity. As the lower portions of the atmosphere are, in fine weather, negatively electrified; if a large quantity of air spread out in a horizontal sheet—say about a quarter of a mile thick, and extending over an area of several square miles—were to be raised by a current, so as to form a vertical column, this would produce exactly the effect that a negatively electrified cloud or other body would do if placed over the earth at that point. . . . .

Mr. Latimer Clark, during the same discussion, said that he witnessed in 1869, in the Persian Gulf, a most interesting electrical storm, of which he made memoranda at the time, and he would, with permission, read extracts from them :

The thermometer had fallen nearly thirty degrees, and a tempest of thunder and lightning burst over the two vessels on a scale of great grandeur and beauty, which, as the vessel's masts and riggings were all of iron, could be enjoyed without apprehension; the flashes averaged thirty or forty per minute, and the roll of thunder was incessant. Many of the flashes appeared to drop into the ocean perpendicularly as a single stream of fire, which enlarged at the point where it struck the water. From their distance and apparent height, many of these flashes were estimated to have fallen from a height of 1,000 feet. They were followed by rapid interchanges of electricity among the clouds above, as if the disturbed equilibrium were re-adjusting itself. . . . .

It was noticed that the thunder caused by those flashes of lightning which struck the vessels did not follow the flash instantaneously, but after a very perceptible interval of time, showing that from some cause the lightning travelled the last three or four hundred feet in silence. Another circumstance of a technical character, still more unexpected, was that the electrical instruments connected with the cable were not in any way affected during the storm, although they were of the most sensitive construction, and were arranged in a manner well suited to show any effects if they had existed. The vessel and rigging were of iron, and the cable was coiled in iron tanks riveted to the sides of the vessel, yet even when the discharges were sufficient to burn pieces of canvas on the rigging none of the electricity appeared inclined to enter the cable, but the whole escaped silently to the sea, without causing even a quiver of the galvanic needle; thus recalling to recollection Faraday's celebrated observation, that the whole quantity of electricity in a flash of lightning is not greater than that caused by the decomposition of a single drop of water.[1]

[1] Grove, in his "Corollation of Physical Forces," however, truly remarks on the above observation that the potential of a flash of lightning is sufficient to decompose a million drops of water in series.

APPENDIX II.

RULES OF WAR OFFICE.

The first War Office circular on the "Protection of Powder Magazines and other Buildings," seems to have been drawn up, in 1875, on the lines of Snow Harris's papers, and insists strongly on the importance of conductivity. It is an admirable summary of practice based on the drain-pipe view of the function of lightning conductors.

The next edition of the circular, issued in 1881, was largely modified on the lines of Col. Bucknill's paper quoted above.

The last circular, that of 1887, introduces some fresh modifications, and here and there seems to return to something more like the rules of 1875.

Since this circular may be taken as embodying the best existing practice, and as it was not referred to by the Lightning Rod Conference, it may be convenient and permissible to quote it in full. (I have obtained permission from the War Office and from H.M. Stationery Office.)

I would not, however, be understood as endorsing the whole of its statements by any means.

Regulations for the Royal Engineer Department.

Instructions as to the application of Lightning Conductors for the Protection of Powder Magazines and other Buildings.

General Principles.

(*a.*) A thunder-cloud is a mass of vapour charged with electricity at extremely high pressure or potential. The origin of this charge has been variously ascribed to evaporation, to friction of air currents, as well as to all the changes in the physical condition of the earth's surface which are incessantly occurring.

(*b.*) A thunder-cloud acts by induction on the land or water beneath, or on clouds near it, and draws a charge of electricity of opposite kind to the surface. This induced charge re-acts upon the cloud in a similar manner, thereby forming a huge electrical condenser.

(*c.*) When the difference of electrical pressure between the oppositely electrified cloud and earth, or cloud and cloud, is sufficiently strong to break across the air space which separates them, an electric discharge of a disruptive nature, with consequent disengagement of heat, takes place.

(*d.*) Clouds are imperfect conductors, and therefore do not part with all their charge at once. Hence a single discharge does not necessarily deprive a cloud of the whole of its charge; there may be several successive discharges.

(*e.*) The surface of the earth is formed of fairly good conducting media, but there are some portions, such as sandy deserts and chalk downs, which, after long droughts, form non-conducting areas. Lightning is least to be feared in such situations, the induced charge not being so easily drawn to the surface.

(*f.*) The lightning discharge between cloud and earth follows the line of least resistance, or, in other words, selects the easiest path. Objects which project above the general level are, therefore, *cœteris paribus*, most frequently struck. Dry air is practically a non-conductor of electricity, but moist or hot air possesses a certain conducting capacity. Hence rain or hail, and columns of rising smoke or steam, sometimes determine the direction of the discharge, which does not therefore always strike the highest points. Metals, which are the best conductors of all, generally determine the path of the electrical discharge.

(*g.*) The lightning discharge does not always follow a single track; it frequently divides into several lines. When alternative routes are offered for its passage, the electric discharge will divide itself among them in direct proportion to their several conducting capacities, *i.e.*, inversely as their respective resistances. In its passage to earth, however, the discharge will not leave a line of good conductors for an inferior one, with which it is unconnected, except when the latter offers a much more direct path to earth, in which case a portion of the discharge may leave it. Again, when a conductor is bent abruptly through a considerable angle, the discharge may seek a shorter path to earth by bridging the air space connecting the nearest portions of the conductor, and a portion of the discharge is then very liable to be diverted to any alternative route which may present itself. This action is due to the attractive effect of induction.

(*h.*) Atmospheric electricity is only destructive when it is overcoming high resistances. If the conductivity of its path to earth be sufficient, the discharge passes off harmlessly.

When the electric discharge bridges a gap or sharp bend in a conductor, or jumps from one conductor to another, a considerable mechanical effect is produced. The conductor may be broken, bent, or melted, both at the point where the discharge leaves and at that on which it jumps, and the effect is greater in proportion to the distance across which the discharge jumps. Hence the importance of insuring perfect metallic continuity in the joints of all lightning conductors, of leading them to earth by the most direct route, of avoiding sharp bends, and of connecting all masses of metal in the line of probable discharge with the lightning conductor, so as to avoid the danger of lateral discharges across the air, or it may be through the building.

(*i.*) A lightning rod is a pointed conductor, in intimate connection with the earth, fixed on the salient feature of a building with the object of protecting it from the destructive action of lightning. It fulfils two functions :

1st. A lightning rod tends to *prevent* a disruptive discharge occurring by silently [1] neutralizing the conditions which determine the formation of an induced charge in its neighbourhood.

2nd. It *protects* the building to which it is attached by offering a path of high conductivity by which the discharge may be carried off harmlessly to earth.

(*j.*) The *preventive* action of a lightning rod depends on the power of its pointed end. A thunder-cloud in the vicinity of a building draws a charge of electricity to the

[1] The word silent is the conventional term used to describe a continuous brush discharge which generally is not audible. The brush discharge when very rapid is, however, accompanied by considerable noise, and is frequently visible at night.

lightning rod, and this charge will escape from the point as fast as it can be induced, provided there be a sufficient number of sharp points all well connected with the earth. This power of points to dissipate a charge is due to the self-repulsive action of electricity of the same kind, and to the law of distribution of electricity on a surface that the density is greatest at points or on portions of the surface of greatest curvature. It is essential, therefore, to foster this gradual or brush discharge by providing a sufficient number of sharp lightning rods in intimate connection with the earth by means of continuous metallic conductors, so as to collect or tap the induced charge, and to oppose the least possible resistance to its escape from the points.

Further, the flow of electricity from the points being directed towards the charged cloud, some of the inducing charge may thereby become neutralized. Hence not only does a lightning rod tend to prevent the accumulation of electricity on the surface of the earth within its sphere of action, but it also tends to restore the clouds to their natural state, both of which concur in preventing lightning discharges.

(*k.*) Should, however, this brush discharge of electricity from the points of the lightning rods be insufficient to prevent the accumulation of a charge, and a lightning discharge take place, it would pass to the points, because the density and consequent attraction are greater there than anywhere in the neighbourhood, and also because the flow of electricity from the points reduces the mechanical resistance of the intervening air. The discharge in this case would, in all probability, be greatly modified by the previous escape of electricity from the points, and being conveyed to earth by a continuous conductor of ample capacity, would leave no trace of its passage.

This may be termed the *protective* function of a lightning rod.

(*l.*) A lightning rod in imperfect connection with the earth, due either to insufficient surface of conductor buried in the ground, or to defective joints, although it may save a building from actual damage by determining the path of the electric discharge to earth, is generally regarded as a source of danger, inasmuch as the sudden and disruptive discharge would be liable to fuse or scatter some portion of the conductor in its passage, and so leave the building unprotected from further strokes. Again, a lightning rod in such a condition, by allowing the comparatively slow accumulation of an induced charge, tends to attract or invite an electric discharge, while the resistance offered at the defective portions of the conductor may cause a portion of it to seek another path disruptively through the building. A faulty lightning conductor may thus prove worse than useless.

(*m.*) The lightning rod terminal should be designed so as to combine, as far as possible, both its preventive and protective action (*vide* paragraph *i*). The requirements are somewhat antagonistic, because the sharper the point the more rapid is the brush discharge, but at the same time the more liable is it to be fused should a heavy disruptive discharge fall upon it. Attempts have been made to obtain a resisting point by the use of platinum tips, and silver or other alloys, but they enormously increase the cost, and have not proved reliable. The system which has been adopted is to separate the double function of a lightning rod by prolonging the upper terminal, and bevelling it off to a blunt right-angled cone of the effective section of metal capable of safely carrying off any disruptive discharge, and with a view to facilitate as much as possible the brush discharge, to add three or four very sharp-tapered points projecting upwards from a ring fixed about a foot below the top of the lightning rod.

(*n.*) As regards the earth connections, it is most important that the electrical resistance which they offer shall be very far less than that offered by any alternative route in the line of probable discharge, such as the rain-water or gas pipes outside a building. The earth connections should be the best which the nature of the soil will admit of, and all available means which will assist in tapping a large extent of moist earth in the immediate vicinity of the building should be utilized.

(*o.*) Except where the permanent water level is very near the surface, both deep and shallow earth connections are required, because after a long period of dry weather the induced charge may be collected on a damp substratum, whilst after rain it may be collected on the surface. The deep earths should be carried down to water-bearing strata or to permanently moist soil, and the shallow earths should be arranged so as to offer a considerable surface of connection with the soil around the building.

(*p.*) Earth connections should be buried throughout in small coke, which is a fairly good conductor of electricity, readily absorbing and retaining moisture. The contact surfaces between the metal conductor and the soil in which it is laid are thereby much increased, and the tapping of any induced charge or the transmission of any discharge to earth facilitated. A layer of coke also tends to preserve copper from corrosion.

(*q.*) The metals employed in lightning rod construction are iron, plain or galvanized, and copper. Roughly speaking, they cost the same for equal conductivity, but conductors made of iron are stronger, less easily fused, and less liable to be stolen. Copper conductors, on the other hand, have the advantage of being far more durable, and, being smaller and lighter, they interfere less with architectural features, and are much cheaper to erect. Copper tape of high conductivity, which is now manufactured in long lengths, thereby obviating the necessity of numerous joints, has been adopted for all conductors on War Department buildings.

(*r.*) The size of conductor required for lightning rods is based on recorded instances of metal bars and rods which have been fused. The Lightning Rod Conference of 1881 recommended the following as the minimum sizes of conductors to be employed, viz :

Material.	Section.	Area sq. in.	Weight per ft.
Copper tape . .	$\frac{3}{4}'' \times \frac{1}{8}''$	0·09	6 oz.
„ rope . .	$\frac{1}{2}''$ diameter.	0·10	7 „
„ rod . .	$\frac{3}{8}''$ „	0·11	7 „
Iron rod	$\frac{9}{10}''$ „	0·64	35 „

The recorded instances of lightning rods which have been fused or damaged can invariably be traced either to faulty earth connections, defective joints, want of continuity or adequate sectional area in the conductor, or to the propinquity of neighbouring masses of metal unconnected with the conductor which have invited a portion of the discharge to leave the main conductor by providing an alternative route. There is no authentic record of a properly constructed lightning rod having been injured or having failed to do its work, and there is every reason to infer that a smaller size of conductor would suffice to carry off any electrical discharge, provided perfect continuity and efficient earth connections could be permanently maintained. The size of conductor required for lightning conductors is, however, practically ruled more by considerations of strength and surface for making good connections than by that of electrical resistance. The smallest size of copper tape to be used for the main conductors on War Department buildings is $1'' \times \frac{1}{8}''$.

(*s*.) It may be accepted that a lightning rod will protect a space included in a cone having the point for its apex, and a base whose radius equals the height from the ground. Buildings protected on this principle would require very lofty lightning rods. It is considered that a number of smaller rods well connected together by conductors, carried along the salient features of a building, provide a more reliable protection than an equal amount of metal in higher rods spaced at greater intervals, and the former is the system which has been adopted for the protection of all War Department buildings.

Rules.

I. A complete system of lightning conductors should be provided for all overground magazines, and for all buildings in which the manufacture or manipulation of explosives is carried on.

II. Important underground magazines, although less exposed to lightning than overground buildings, should nevertheless be provided with conductors, because magazines are now so frequently filled with gunpowder in metal cases that a line of smaller electrical resistance than through the surrounding earth might be offered to the lightning through the body of the magazine.

III. Expense small-arm ammunition magazines need not be fitted with lightning conductors, except in cases where they occupy very exposed sites, or have much metal connected with them. Steps should be taken to remove forthwith defective lightning conductors on these magazines, or to make them efficient should it be decided to retain them.

IV. The salient features of barrack buildings should be provided with lightning rods when experience has shown that the locality is attractive to lightning, more especially when any considerable mass of metal enters into their construction.

V. It is advisable to fix a lightning rod on any flag-staff that may be near a magazine, and also on all high chimney shafts.

VI. All lightning conductors on War Department buildings should be brought as far as possible into conformity with these rules, but existing arrangements may generally stand, provided care be taken to improve defective earth connections and to make good all joints.

VII. Whenever lightning conductors are to be erected or reconstructed on any War Department magazine or building, the Commanding Royal Engineer of the District should invariably forward a special report, accompanied by descriptive drawings, to the Inspector-General of Fortifications, in order that the details and general arrangements may be approved before any steps are taken to carry out the work. In cases where a doubt may exist as to the necessity of erecting lightning rods (*vide* paragraphs III. and IV.) a report should be made specifying any peculiarities of the site; its height as compared with the neighbouring ground; liability of locality to thunderstorms; nature of soil and substrata, and depth of permanent water level; and full particulars of building, and of all masses of metal entering into its construction or placed near it.

VIII. The angles and prominent features of a building being the most liable to be struck, lightning rods should be fixed on gable ends, chimneys, turrets, etc., and they should be connected together by continuous conductors along the ridges.

IX. Lightning rods should be about 4 feet high, and spaced at intervals not exceeding 50 feet, so that no point on the building is more than 25 feet horizontally distant from a lightning rod.

X. The material to be exclusively employed for the construction of new lightning conductors is copper tape. It is manufactured in lengths of 300 feet and upwards. A conductivity of at least 95 per cent. of that of pure copper should be specified, and the tape should be soft and flexible, so as to admit of its following closely the outlines of a building. Copper tape $1'' \times \frac{1}{8}''$, weighing about $\frac{1}{2}$ lb. per foot, is the most suitable size for all ordinary conductors, and $1\frac{1}{2}'' \times \frac{1}{8}''$ may be used for very high chimney shafts.

XI. In situations where copper tape is liable to be stolen, it may be let into the walls of the building and cemented over, or otherwise concealed where it is accessible. The practice of protecting the lower portion of copper conductors on buildings by enclosing them in iron pipes at the base of a building is questionable.

XII. In order to guard against those accidental defects and disarrangements to which conductors are liable, buildings provided with lightning conductors should have, as a rule, at least two earth connections, the conductors leading to them being connected at the base of the building, either above the ground line, or by a conductor underground forming a "surface" earth (see paragraph XV. Each lightning rod should be connected direct to earth by the shortest path outside the building, and, where practicable, it is well to carry the conductor down that face of the building which is most exposed to prevailing wet. In the case of gabled buildings, the conductors should be taken down the barge courses in preference to the gable, so as to protect the angle of the building, and at the same time secure the advantage of the additional moisture in the ground near the rain-water down-pipes, and facilitate their connection thereto.

XIII. When the level of water or permanently wet soil lies within a few feet of the surface, the conductors should terminate in earths offering each about 18 square feet of external surface. These may consist of copper plates, about $3' \times 3' \times \frac{1}{16}''$, riveted to the ends of the conductors, and buried in water or wet soil from 15 to 25 feet from the building. A better plan, however, is to coil the end of the conductor spirally on a wooden frame, the external diameter of the coil being 4 feet, with 6-inch intervals between the turns. About 33 feet of tape are required for this earth connection; it obviates the necessity of any underground joint, taps a larger surface of earth, and, the tape being twice the thickness of the plate, it is more durable, besides being a cheaper arrangement.

XIV. When the permanent water level is deep, it may be necessary to sink special wells for the earth plates, which should be of the same size as those specified in the previous paragraph. The wells should be carried down several feet below the water level in the driest seasons, and the lower portion of the well should be built dry.

If coils of copper tape, however, be employed for earths, special wells are not necessary, because, there being no joints underground, the same necessity for periodical examination no longer exists. In this case the earth coils may be buried at the bottom of a pit, sunk below the level of permanently wet soil.

Where the depth is considerable, two or more conductors may be connected to the same earth plate, and in the case of a coil of tape the inner, as well as the outer end, may be brought up to the surface of the ground so as to form two earth connections. In both cases the size of earths should be made proportionately larger.

XV. In addition to these deep earths it is necessary to provide surface "earths" laid in trenches, from 1 foot deep in clay soils to 2 feet deep in sand or shingle, through which the rain percolates more freely.

These surface earths may consist of that portion of the conductor which leads from the base of the building to the well or deep earth, or they may be arranged as separate conductors led in trenches away from the building. In the latter case, the deep and surface earths should be connected together by a conductor carried round the base of the building.

The length of each surface earth trench may be from 25 feet in ordinary soil to 50 feet in dry soil, and the width at bottom should be about 9 inches. A few inches of powdered coke should be spread both above and below the conductor, and the trench filled in with light soil. The rain-water down-pipes from the roofs may with advantage be led into these trenches.

XVI. In the case of forts and magazines near the sea, good earths can be obtained by laying a length of tape so that at least 5 square feet of it shall always be under water; or a coil of tape may be buried in permanently wet sand.

When the distance to the sea is considerable, these earths should be supplemented by surface ones round the building, so as efficiently to tap any induced charge in its vicinity.

XVII. Iron water mains form good earth connections.

Soft metal pipes and gas mains should not be used, but when they run close to the conductors from a building, they should be connected to the lightning conductor system. There are many recorded instances of both water and gas pipes which have been damaged by lightning springing on to them from neighbouring conductors, which would have been obviated had they been connected thereto. (*Vide* General Principle *h.*)

XVIII. In extremely dry or rocky situations it is often impossible to obtain good earth connections except at a great distance. In such cases the best plan to adopt is to bury several hundredweight of old iron at the foot of the earth coil or plate in a mass of coke, leading the rain-water pipes so as to discharge into it.

XIX. Coke, suitable for improving the earth connections of lightning rods (*vide* General Principle *p*), is procurable as a waste product of gas-works. Clean smiths' ashes may also be used. A layer of about 3 inches should be spread both below and above the conductors in the trenches, and also round the earth plates or coils.

XX. The earth connections of flagstaffs near magazines should be led in a direction away from the building. Should, however, the horizontal distance between the flagstaff and the nearest lightning rod on the magazine be within 50 feet, or should any portion of the building lie within the cone protected by the flagstaff rod (General Principle *s*), then the magazine and flagstaff earths may be connected, or have an "earth" common to both. This rule is also applicable to shafts of powder mills, etc.

XXI. The rain-water pipes and gutters should never be utilized as a portion of the system of lightning conductors, to which, however, they should be connected.

All external masses of metal, such as copper sheeting

on magazine doors and ventilators, etc., should also be connected to the nearest conductors by lengths of copper tape.

In the construction of magazines and other buildings in which the manipulation of explosives is carried on, the employment of external masses of metal should be avoided as far as possible.

XXII. Lines of rail near buildings protected by lightning rods should be connected to earth direct on both sides of the building, and when the line is carried inside the building it should be connected also to the system of lightning conductors. Iron railings round magazines should be connected direct to earth at intervals of about 50 feet.

XXIII. All large and long masses of metal, such as beams, girders, pipes, hot-water systems, and large ventilators fixed in the interior of buildings, should be electrically connected with the earth as well as with the conductor; but the soft metal gas-pipes should never be used as conductors; and the lightning conductors should be kept as far as possible from them, and also from all internal gas-pipes.

XXIV. Lightning conductors should not be insulated from the buildings to which they are attached. The copper tape should be laid on the ridges and walls, and secured by suitable fastenings screwed or nailed to the building. The holdfasts should be of gun-metal fixed by nails of hard copper, and they should allow free expansion or contraction, at the same time preventing all the weight falling on any one bearing.

XXV. Powder mills, etc., with zinc or galvanized roofs, should be protected by copper conductors laid over them, but protected from actual contact by strips of wood, paint, or tarred felt. The zinc roof itself should be treated like any other external mass of metal, and should be thoroughly well connected to the conductors in several places. The system sometimes adopted of having sheet zinc lightning rods, and copper tapes from the eaves of the roof to earth, is very unreliable, because the lightning rods are of ineffective section, the zinc sheets are insulated from one another by a layer of oxide, and the joint between the zinc and copper is liable to failure.

XXVI. Chimney shafts should be protected by a ring of copper tape $1\frac{1}{2}'' \times \frac{3}{16}''$, placed round the outside of the top of the cap and a few inches below it, having stout copper points, projecting 1 foot above the top of the shaft, at intervals of 3 or 4 feet all round. A copper tape should be carried down on opposite sides of the shaft to earth from the ring; and these conductors should be connected at the base, a test clamp being added to enable the continuity of the conductors and the state of the earth connections to be ascertained when necessary.

XXVII. Metallic continuity should be insured at the joints of all conductors. Solder should never be used where this can be done by closely fitting riveted or screwed joints. The solder is seldom properly sweated through the joint, and often consists of an imperfectly adhering mass of metal hiding up badly fitting and dirty surfaces. Solder tends to set up galvanic action, which after a time will destroy the connection; and, in the case of copper conductors, its use is objectionable, because it interposes an alloy of high resistance and low melting point in the joint. Soldering conductors in the vicinity of magazines and powder factories is attended with so many restrictions and precautions, that it is practically unsuited for War Department requirements.

For these reasons it is considered preferable to insure a perfect metallic contact between copper tapes by drawing the surfaces together by close riveting or by screw clamps, and to exclude damp from the joint by paint or other means. [Or they might be electrically welded. O. J. L.]

In riveting copper tapes, five rivets should be used, and the holes should be bored, not punched. The "arris" being removed, and the surfaces brightened with emery, the joint should be brought together with a hollow punch before riveting.

XXVIII. The connection between the lightning rod and the conductor is made by means of a slotted clamp similar in design to those employed for test or other joints. The lightning rod terminates at its lower extremity in a $\frac{3}{4}$-inch bolt, which is screwed into the clamp, thereby making firm contact with one or more tapes inside it. This joint admits of visual inspection.

XXIX. For the repair of old lightning conductors, however, it is sometimes necessary to use solder. For copper the solder usually employed consists of equal parts of tin and lead, which has a resistance nearly ten times that of copper. The surface of the joint should, therefore, be not less than $1\frac{1}{4}$ square inches.

Molten zinc should be used for soldering iron conductors. Being nearly twice as conductive as iron, the surface of the joint need not necessarily exceed that of the cross section of the conductor.

In both cases the joint should be put together previously by screws or rivets, and the soldered joint, especially in underground work, should be carefully protected from galvanic action by tarred tape.

XXX. Existing iron wire rope conductors may be connected to copper tapes in the following manner. Take a piece of sheet copper $4\frac{1}{2}'' \times 3'' \times \frac{1}{8}''$, and cut it down at one end to the size of the tape for a length of $1\frac{1}{2}''$, and rivet the tape to the sheet at this part; then bend the remainder of the sheet round the end of the rope, which has been previously frapped with fine wire, thus forming a tube which should be previously tinned inside. Sweat up the joint with zinc solder, and protect it by tarred tape bound tightly over it.

XXXI. Copper tape conductors may be connected to iron water mains by filing about a foot of the top of the pipe bright, binding the tape on to it by wire, and soldering with zinc; or a short length of iron bar, $2'' \times \frac{1}{2}''$, may be riveted and soldered to the copper tape, and then $\frac{1}{2}''$ screws studded together both through the bar and a flange of the pipe.

It is a good plan to coil several turns of the copper tape round the main, keeping it from actual contact by battens previous to fastening the end, so as to increase the earth connection should the joint ultimately fail.

The greatest care must in all cases be taken to protect the joint from galvanic action by layers of tarred rope, or by imbedding the main at the joint in cement.

XXXII. Metal surfaces on which the rays of the sun fall are exposed to a maximum range of temperature of about 144° Fahr. This range of temperature produces an expansion and contraction in copper conductors of about 1 inch in 60 feet, which should be provided for by forming small loops near the base of the lightning rods, and by allowing the conductor free play through the holdfast employed to fix it on the building. Conductors should never be screwed or nailed down. Vertical conductors on shafts, flagstaffs, and high walls should have small loops above every second or third holdfast, to take the weight of the conductor while still allowing for expansion and contraction. The holdfasts should in this case be about 4 feet apart.

XXXIII. Artisans, and especially painters employed on War Department Works, should be cautioned never on any pretence to disconnect, move, or tamper with any portion of a lightning conductor without first obtaining the written authority of the Royal Engineer Officer in charge.

XXXIV. Lightning conductor records should be kept in the office of the Commanding Royal Engineer of each district. In addition to the usual small scale plan, showing the details of the conductors, the position of the buildings protected and the earth connections should be marked on the lithographed plan. Copper conductors should be shown by red, and iron by blue lines. The lightning rods numbered 1, 2, 3, etc., conductors *a*, *b*, *c*, and earths E_1, E_2, E_3, etc., to correspond with a descriptive record containing as many of the following particulars as possible:

(1.) Date of erection or reconstruction of lightning rods.

(2.) Character of soil and substrata, and depth of permanent water level or wet soil.

(3.) Full particulars of lightning rods, conductors, and earth connections, nature of joints and connections, etc.

(4.) Details of all external or internal masses of metal entering into construction of building, and how connected to conductors.

(5.) Position of test joints, if any. Nearest earth available for testing, etc.

(6.) Quantity of powder, etc., kept in store.

(7.) Date of last inspection and *précis* of former tests, suggestions, etc., of inspecting Officer.

(8.) To whom notice of inspection should be sent, so that ladders may be ready for getting on the roofs, etc.

XXXV. Lightning conductors should be periodically inspected, once a year, and the Inspector should forward, on the accompanying form, a report to the Inspector-General of Fortifications, through the Commanding Royal Engineer of the district.

The Commanding Royal Engineer should take steps to have every portion of the system of lightning conductors visually examined by a competent mechanic, in order that all defects may be remedied, so far as the means at his disposal will admit, before the periodical inspection.

Annual Inspection of Lightning Conductors.

District.

Name of fort, battery, magazine, etc. . .

State of soil when inspected.

Date of inspection.

Lightning rod.		Conductor.		Earths.		
Number.	State of points and connections.	Letter.	Condition.	Letter.	Condition.	Tests.
1 2 3 etc.		*a* *b* *c* etc.		E_1 E_2 E_3 etc.		

XXXVI. In the case of lightning rods which require repair or reconstruction, the Inspector should submit, in addition, a special report, accompanied by tracings or hand-sketches, foolscap size, of the existing system of conductors, detailing fully the various defects either in design or construction, and his suggestions and proposals for their improvements.

XXXVII. The object of electrical tests of lightning conductors is to determine the resistance of the earth connections, and to localize the position of any defective joints or connections in the conductors. The resistance of the conductor itself is quite inappreciable (less than $\frac{1}{4}$ ohm per 1,000 yards for copper tapes $1'' \times \frac{1}{8}''$), and could not be detected by the portable testing apparatus employed. The resistance of the earths depends on the nature of the soil, and its state of moisture at the time of the test. As a rough guide, however, it may be mentioned that the joint resistance of two earths, such as those described in paragraph XIII., 30 yards apart in damp soil, will not exceed 2 ohms.

XXXVIII. The best system of testing lightning conductors is to balance the resistance of each of the earths of the building against the remainder of the system, from which the state of the earths may be inferred with sufficient accuracy for all practical purposes. With this object test joints should be added to all magazines and shafts.

When the system of conductors is permanently riveted up, the resistance of two small test-earths, placed not less than 20 yards apart in water or moist soil, or of any two convenient earths unconnected with the system of lightning rods, such as a gas and a water-pipe, is first ascertained. Then the sum of the resistance of each test-earth in succession, and the earths of the lightning conductor system, is ascertained, and the joint resistance of the latter calculated from the resulting three equations.

Thus:

Let L = conductivity resistance of the leads,

Then the resistance of leads and test-earths . . . . $= L + e_1 + e_2 = A$

Resistance of leads, one test-earth, and the earths of lightning conductors . . . . $= L + e_1 + E = B$

Resistance of leads, other test-earth and earths of lightning conductors . . . . $= L + e_2 + E = C$

$$\text{Then } E = \tfrac{1}{2}(B + C - A - L) \qquad (1)$$

$$e_1 = \tfrac{1}{2}(A + B - C - L) \qquad (2)$$

$$e_2 = \tfrac{1}{2}(A + C - B - L) \qquad (3)$$

Very short contacts should be made in order to avoid polarization of the test plates. If the resistances be very unequal, the results obtained from these equations may not be very accurate.

XXXIX. The resistance of the conductors from each lightning rod point to that part of the conductor from which the resistance of the earth connections was ascertained, is then found. As the resistance of the conductor should be practically *nil*, any resistance in excess of that of the leads must be due to defective joints, and the faults should be localized and marked for repair.

XL. Conductors made in short lengths and connected by joints, the metallic continuity of which has not been insured by soldering, or by a good system of riveting, if tested electrically, give results of little value.

Underground joints and earths, the nature of which is not known, should be dug up and examined when practicable.

APPENDIX III.

The rules for the Navy are very brief, and are as follows:

RULES OF THE ADMIRALTY.

Admiralty, S.W, 4th November, 1880.

(*Lightning Conductors—Mode of fitting.*)

The Superintendent at Yard is informed that the following instructions with regard to the fitting of Lightning Conductors in Her Majesty's ships should be adhered to in future:

(1.) *In Wooden Ships* with wooden masts, the copper strip down the mast is to be connected with a continuous strip of copper passing along the lower deck beams and down the side, and connected with the copper sheathing by a copper bolt leading through the bottom, and at the heel of the mast with a strip in contact with two of the bolts securing the mast step, which are also in contact with the metal sheathing of the bottom.

(2.) With iron masts the lower mast itself forms the conductor, and is to be connected in the same way with the sea.

(3.) *In Iron Ships* with iron masts the connection with the iron step, which is made bright before the mast is stepped, is considered sufficient.

(4.) In Iron Ships sheathed with wood, with wood or iron masts, the connection with the sea is to be made by copper bolts passing through the bottom and insulated from the metal on the bottom of the ship by vulcanite, as a preventive against galvanic action.

(5.) No conductor can be better or more direct than an iron mast in direct metallic contact with the outside of the Ship, so that where this can be attained it is not necessary to fit copper conductors to the lower masts or shrouds.

(6.) When a wooden mast is stepped in an Iron Ship, whether sheathed or not, the wire rigging forms an alternative route for the electricity of larger sectional area than the conductor on the mast, and this route being only broken by the lanyards of the rigging, it is quite conceivable, especially in wet weather, that a man might accidentally form part of the circuit. In this case it is considered necessary that a metallic connection between the wire shrouds and the chain plates should be fitted.

(7.) On completion at a Dockyard the Lightning Conductors of every Ship are to be tested for continuity, from the masthead to the sea, by the Yard Officers, by means of the Galvanometer supplied for the purpose; and the report of completion, Form No. 237, is to be understood as including the Lightning Conductors.

WM. HOUSTON STEWART,

Controller.

APPENDIX IV.

SOME SPECIAL CASES AND OTHER DETAILS.

A Remarkable Flash of Lightning.[1]

BY W. KOHLRAUSCH.

On May 15th last I was asked to inspect a stable on the farm of Mr. C. Jagau, near Hanover. On May 8th it had been struck by lightning, which set it on fire and killed a horse which was in it, and this, although the building had four conductors, in the immediate neighbourhood of one of which the lightning struck.

The accompanying illustration shows the positions of the buildings and the conductors. To the east lies the yard, which seems to be quite unoccupied. To the south also there are no buildings within the distance of 100m. It was at the request of Mr. Rudolf Siemsen, who had erected the conductors, that I went to see the place. The iron rods are 4m. high, and the 12-strand copper conductors have a cross section of 0·06in. The earth-plates are 100 by 50 c.m.; those at *a* and *b* hang right under water in two wells; that at *c* is buried 1·5m. deep in the ground. From the gilded tips of the rods down to the earth-plates, all joints were properly soldered. It might have been well to connect rods 2 and 3 along the roof, but with that exception no fault could be found with the arrangements. Those in the house had no idea that the stable had been struck until, shortly after, they saw smoke coming from the roof, and found that some loose hay lying in the loft was burning, which was easily extinguished; but it was found that a horse was lying dead just within the door. Distinct traces of the flash, in the form of numerous small splinterings, could be seen on the first two beams going to the

[1] From the "Elektrotechnische Zeitschrift."

north-west, under the ceiling of the stable behind the door *e*. Wood was splintered both on the inner and outer sides of the door itself, and here and there its nails and bars were melted at the ends and angles. The yard gate was also splintered. Finally, on the birch-tree *d*, at about 1·7m. from the ground, we found two burns, which were undoubtedly fresh, and which had gone through the bark to the better conducting interior of the tree. The most careful search has not brought to light any other traces of the lightning's course. We took down the conductor 1, but even with a magnifying glass could find no

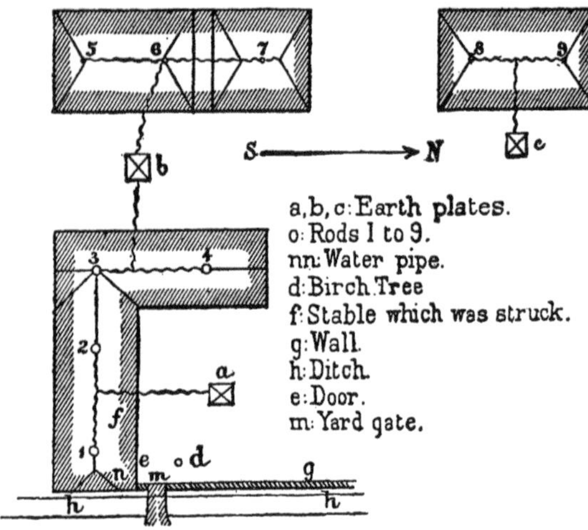

traces of its being touched by lightning. Rod 2 we examined through a telescope with the same result. The common conductor of rods 1 and 2 goes to earth at the well *a*. The water was about 1·5m. below the ground. The well is built in sandstone, and is about 3·5m. deep. The earth-plate *a* was quite under water.

Earths *a*, *b*, *c* were 10 ohms, 11 ohms, and 32 ohms respectively. There is thus nothing much to be said against the arrangement of the conductors. It would have been better had the resistances of the earths been smaller, and had rods 2 and 3 been connected; but I think one can hardly say that the buildings were insufficiently protected, and I can find no definite reason for the fact that the lightning injured them; and, moreover, injured them at a spot certainly not beyond the commonly supposed protective distance of the rod, and which was nowhere more than 5m. from either end of the rod. In this region the ironstone, which lies generally about a metre below the surface, in various sized beds, has a great deal to do with determining the course of a flash of lightning; but, on digging in the stable, near the birch, and round the neighbourhood where the lightning struck, we came to the water-level at 1·25m. down, but found no ironstone. The nearest we found was a small bed in the field, towards the east and about 25m. from the stable. The field lies about 0·8m., and the water in the ditch *h* about 1·3m. lower than the yard. In the stable or beneath it, was no amount of metal worth mentioning. No farm implements were kept there; the stable was empty, save for the horse, which happened to have been put there temporarily.

The following explanation is possible: The lightning struck the highest point in the neighbourhood, the birch *d*, passed through the twigs at *n*, which touched the roof of the building, and also through the tree, coming out lower down the tree, from whence it injured the yard gate *m*. All visible traces left by the flash fit in with this theory, for all the injuries to the building itself are just under the twigs *n*.

Notes on a Curious Lightning Flash.

TO THE EDITOR OF THE "ELECTRICIAN."

SIR,—Knowing that at the present time everybody's attention is turned towards thunderstorms, I send herewith what may be interesting to some of your readers.

On the afternoon of July 17th a storm broke over Surbiton of rather a severer character than has been known for some time past in this neighbourhood. The storm commenced between five and six o'clock, after a very sultry day, accompanied with a great darkness and much rain, the rain falling in dense sheets. At 6.40 I was fortunate enough to see one of those most appalling sights known commonly as globular lightning. The room I was standing in was situated about 30ft. from the ground, when all at once at the time mentioned it seemed to me as if a huge ball of reddish yellow fire was suspended in mid air, about the level of the window. (The colour was that of an electric arc as seen in a dense fog.) It did not last a moment, then it rapidly changed colour, and turned to an intense whiteness, more vivid than any other light I have seen, and burst with a terrific explosion, increasing in size many times over, and dozens of small splashes darted out of it on all sides; its brilliancy dazed me for several moments after. I have made several inquiries as to whether anybody else saw it; several people heard the explosion, but only a small boy saw anything of the real thing

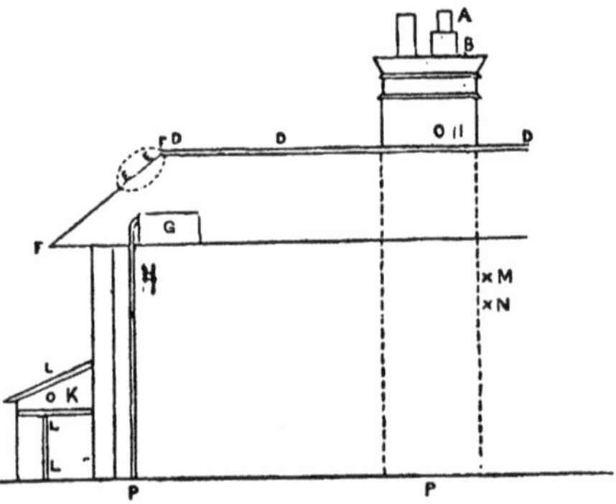

A, Zinc top; *B*, Earthenware (these were shattered to pieces); *C* Slits found in chimney; *D*, *D*, Lead edging; *E*, *E*, Portion of roof quite smashed up; *F*, *F*, Slate; *G*, Cold-water cistern; *H*, Water-pipe, indications of fusing between *H* and wall; *K*, Verandah of wood and glass; *L*, *L*, *L*, Iron ridging, gutter and pipe; *M*, Mantel-piece, broken; *N*, Iron register, broken and thrown into the room; *P*, *P*, Supposed earthing.

besides myself. He states that he saw a big red ball moving along towards the house; but it soon passed out of his sight, owing to his situation in the room; but he heard a terrific bang just after it passed from his view. As far as I am able to discover, it came and went leaving no trace at all of its existence. Another fireball that was also seen in the neighbourhood was said by those who saw it to have been like a red ball surrounded by several smaller ones. I also heard of another one, but up to the present moment I have had no satisfactory details of either of these last two. Below will be found the account of two houses struck simultaneously within a few hundred yards of this house. By the kind courtesy of Mr. Dunlop and Mr. Tiffin, I have been able to make a careful examination of the result of this said flash; and I enclose a rough sketch, which will explain itself. It appeared to have struck the chimney stack, which is common to both houses, and descended inside for a few feet. Here it evidently separated into two branches: one branch continued down the chimney to a fireplace of the first-floor bedroom; here it broke the masonry around the fireplace, also breaking the register and throwing the same into the room in fragments. After this we have no further clue of this branch, except that a large quantity of soot was found on the kitchen range, so I presume that it burst into the kitchen flue, and earthed itself through the pipes under the range, etc.

Now, going back to *B*, another portion of the current burst through the chimney, leaving two small slits in the same to reach the lead edging, by which it continued its path to *E*, where the lead stops, and a slate edging commences. A little way down under this slate-edging there is a water-tank, and over the tank there is a pipe from the hot-water cistern. The lightning leaving the lead, which it fused at the end, made for this water-pipe, fracturing a hole some 2ft. 6in. by 1ft. 6in. in the roof, splintering slates and rafters. The slate-edging here is very thick, being about the size of an ordinary man's wrist. A piece of the slate was driven right through the ceiling below, and another was driven with such force as to enter about 1in. into a beam. The hot-water system has been blocked since the house was struck.

We now come to what seems to me to be the most curious of everything: it will be seen by reference to the sketch that a verandah adjoins the house, partially formed of wood and partially of iron; at the spot indicated a round hole of about 2in. diameter has been pierced, as if a stone had been thrown from underneath, as I found glass outside the hole on top; it seemed as if the glass had been fused round the edges, but not to the extent one would expect. A very small quantity of glass was found inside. There was no further indication of the lightning going to earth to be found. I have been looking through some periodicals, and I came across one of Dr. Oliver Lodge's Papers read before the Institute of Electrical Engineers, in which he quotes "Note on the Lightning Flash at Antwerp Railway Station," by M. Melsens. It seems to me, on reading this, that this case bears a little resemblance, although the fusing of the glass in this instance is very slight. The fusing of the glass is on the inside of the glass.

If it was not for the fusing of the glass, which is very slight (which glass is still preserved), I would naturally think that this hole was caused by a piece of slate in falling, etc. I would also here note that the verandah door was wide open. That the house next door, which is only a few yards off, and certainly higher, is thoroughly well protected by lightning conductors. Also that, except for the fusing between the different pipes on descending the house, and also of the lead, there is no indication of scorching or charring.—Yours, etc.,

Surbiton, July 22, 1890. D. F. Adamson.

An Erratic Lightning Flash.

During a thunderstorm at Larnaca the entrance door of a house occupied by one of the Eastern Telegraph Company's *employés* was struck by lightning. The woodwork round the iron lock on the inside of the door and down to the top step of a flight of stone steps was completely splintered, and the lock was almost wrenched from the woodwork. Attached to the outside wall of the house was an iron stove-pipe, reaching considerably above the roof. No trace whatever could be found of this having been touched, and nothing inside or outside of the house showed any sign of an electrical discharge except the door. Within a stone's throw of the house in question are the buildings of the Anglo-Egyptian Bank, with a lightning rod of considerable altitude; also, quite close, are several high consular flagstaffs. None of these were touched. In the immediate vicinity are several iron telegraph poles, with lightning rods attached, and not far distant are several mosque minarets of great height, with metal caps. None of these were touched. It is difficult to understand, on the principle of the "protected area," why the flash struck the door almost on a level with the street instead of the iron stove-pipe, or the adjacent lightning rod, or the flagstaffs, telegraph poles, or minarets; or why it did not make direct earth in the muddy and flooded street within a few feet of the wooden door. Again, the houses adjoining and on the opposite side of the street—a narrow one—are built of sun-dried mud bricks, and these were saturated with rain, making capital earth; yet they escaped. The house of which the door was struck is built of trimmed blocks of stone.

There is an interesting account of damage by lightning in the Quarterly Journal of the Royal Meteorological Society by Mr. Alfred Hands.

Lightning in Electric Light Leads.

TO THE EDITOR OF THE "ELECTRICIAN."

SIR,—The enclosed letter from a correspondent in Portugal reached me shortly after the recent meeting of the Institution of Electrical Engineers, where I had been surmising that such occurrences might be not uncommon. Although the damage done was only slight, the account is perhaps of sufficient interest for insertion in your paper.—Yours, etc.,

Liverpool, April 26, 1890. OLIVER J. LODGE.

(COPY OF LETTER.)

Companhia de Luz Electrica, Oporto, Portugal,
April 19, 1890.

To Prof. Oliver J. Lodge, University College, Liverpool.

SIR,—Though personally unacquainted with you, I take the liberty of sending you a specimen of a lightning-injured wire, and a statement of the circumstances, which seem to me to go to confirm your views on the lightning-rod question.

The electric light system here is operated by means of both overhead and subterranean wires on the low-tension direct system, and the station stands well in the middle of the city and its district, in a roughly-pentagonal space bounded by five streets.

About 9 p.m. on the night of April 17th (Thursday last), a severe flash struck apparently into this space; but, dividing between many telephone wires, two or three lightning conductors, and tall houses very wet from four days' nearly continuous rain, did no structural damage. The report in the electric light works followed instantaneously after the flash. Nothing was observed on the lightning protectors of the lines, and no damage was done to any part of the apparatus or machinery.

But two faults were caused in installations within 100ms. of the works, the wire I enclose being one. In this case the shop is supplied from an overhead line. The damaged wire was inside a tubular arm on a combination (gas and electric) pendant, and the pendant gave a good earth through gas-pipes. The people in the shop say that the lamp gave two sharp cracks and then went out. The lamp, however, was not damaged. It merely went out because the discharge ruptured the wire. No other wire or lamp in the shop was damaged, and the faulty wire was at the extreme end of the shop, as far from the entrance of the leads as could be—perhaps 8ms. or so. The brass tube enclosing the wire showed no injuries externally. One end of the wire earthed on the pipe, and the fault was detected by the insulation test of the following morning before any complaint reached me. The overhead line runs very low, and is crossed by many telephone wires at up to 10 or 15ms. greater altitudes.

The other case occurred in a shop connected to the underground line. A spark passed between an electric bell wire and a twin wire leading to a single lamp, at a spot where the bell wire crossed the light conductor. All these wires, the P. and N. light conductors and the bell wire, were ruptured, the bell wire losing about 1cm. by volatization or fusion. The insulations showed no signs of overheating, and the fuse wire in the cut-out was not fused. Surrounding woodwork was slightly scorched and smoked at the point of rupture of the wires.

This case is rather curious. The bell wire and its circuit is wholly within the house. The electric lighting line is entirely underground, and no discharge could have got into it from an overhead line without passing through the works and running the gauntlet of two lightning protectors, one on each line terminal. A telephone wire runs near the bell circuit inside the house, and this wire got some discharge, the aerial line parting outside the house, and the instrument being rendered useless.

Does it not appear probable that the spark which passed between the bell and the light wire was due to an induced current set up in them, or perhaps to a resonance effect? It seems impossible that any direct lightning discharge can have got at either circuit, at least without leaving behind it unmistakable traces. And in the first case, though not clearly impossible, it seems to me improbable that any part of the actual flash reached the line, it being so well protected by the higher network of telephone wires. Also any lightning flash would pretty certainly have done much more damage, and would hardly have picked out the last pendant in the shop to burst through in. In this case the main gas-pipes run down one side of the shop, the electric light wires down the other, thus in plan—

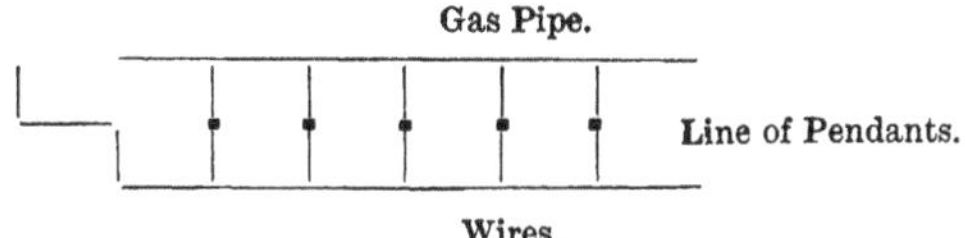

perhaps thus forming an incomplete resonant circuit, of which the fault formed the spark gap.

On the same night, probably a few minutes later, a flash struck near or actually on to a small theatre lighted from our circuit. This theatre is roofed and walled with galvanized iron, so that any entry of a flash would seem impossible. But a lamp in the manager's office burst, short-circuiting across the platinums inside at the moment of the flash. This is an accident that sometimes, but rarely, occurs to a lamp at normal pressure; nearly always, however, to new lamps, whereas this lamp was old. It happens that the lead and return of this particular lamp do not run together, but form a loop; in fact, the lead comes off one circuit and the return off another, the two circuits being in separate casings, parallel, about 3 or 4ms. apart. Might this have been another case of a resonant circuit? Anyway, something raised the potential difference between the two wires considerably at the moment.

No complaint has been received from any other installation on the same circuit, and its insulation was normal on the following morning. It is, I should say, an overhead circuit, about 350ms. in length, the theatre being at the far end from the works.

My apology for troubling you must be the interest you have shown in collecting information as to lightning effects; and the reiterated complaints of those who study the matter of the difficulty of getting such information untinctured with optical illusion effects, and so on. I could tell you of some wonderful things seen in shops, etc., lighted by us, on the occasion of this flash (which was most startling); but I did not notice anything of the kind, and rather doubt whether anyone who caught the full light of the flash could know for a minute after whether the lamps burnt blue or green; or, indeed, whether they were alight or out. If these slight remarks are of any value, you will please put them to any use you see fit. H. M. Sayers.

Lightning and Telephone Wires.[1]

The following article relating to telephone circuit protection, and recording experience with wire systems, Mr. A. R. Bennett permits me to reprint.

It has, I believe, been generally accepted that protection from lightning is unnecessary at both ends of a metallic circuit, provided it is run on poles fitted with earth-wires and has the exchange end guarded in the usual way. Experience has long shown that metallic circuits are frequently struck—indeed, it has been said that they are more liable to that accident than single wires, for translator coils have often been fused when the indicator coils on the single earthed lines running on the same poles and entering the same switch-room have all escaped—but, until quite recently, no further damage has ever resulted. The greater frequency of discharges from, or to, metallic circuits, may be accounted for by the fact that such circuits are almost invariably much longer than single wires, and consequently expose a greater area to the atmospheric influences; and it may almost be inferred that, being insulated from the earth at all points, induced currents, instead of dissipating gradually and quietly, as in single wires, sometimes become stored up until striking potential is attained. The charge nearly always escapes by breaking down the insulation between the primary and secondary coils of the translator and so reaching the earth. Therefore, when translators are used at both ends, as is always the case with metallic surfaces joining two exchanges, further protection, on the score of safety, seems superfluous,

[1] From the "Telephone," February 1, 1889.

although to avoid interruption of service through damage to the coils, it is the custom to lead both wires of the circuit through guards of the ordinary toothed pattern. But metallic circuits occasionally terminate in subscribers' offices or houses. This is the case with private lines erected with double wires, and also where a metallic loop is run from the exchange to a subscriber's station. Such loops, so far as my observation has extended, it has been the rule to leave unprotected—at both ends in the case of the private—and at the subscriber's end in that of the exchange line; a practice justified by some telephonists on the assumption—obviously an erroneous one —that as metallic circuits have no earth connections they are safe from lightning, and by others on the plea that the earth-wires on the poles, if well fixed, are a sufficient protection. The latter opinion I have hitherto shared, and an experience of some ten years without the occurrence of a single accident would seem to show that it is not an utterly unreasonable one. But ten into Eternity goes a good many times; a thing that does not happen once in ten years, may chance to occur several times in eleven, and the fallacy of the opinion has lately been demonstrated in a remarkable manner. A metallic circuit about two miles long, erected on poles sufficiently earth-wired, a soldered connection from the main earth being taken to the bolt of each insulator, terminated at one end in an exchange, where it was led through a toothed guard and the secondary of a translator, and at the other in a private house, where it was joined through an instrument of the usual magneto type. During a slight thunderstorm a discharge took place between the test-plates at the window and an adjacent gas-bracket. Fire was immediately afterwards found to have broken out in quite another part of the house, and examination showed that a length of the fusible composite gas-tubing in such common use had been melted and the gas ignited. The tubing was laid behind wainscoting, which had to be torn down to get at the fire, and could not possibly have been fused by any other means than lightning. Fortunately the damage done, owing to the prompt measures taken, was inconsiderable; but one cannot help speculating as to what would have been the consequences had the occurrence happened at night. The coils at the exchange were not fused, nor was any special disturbance noticed there. How, then, did a discharge of such violence occur at one end without indicating its presence at the other? Either one of the spans nearest the house must have been struck and the charge dissipated between the gas-pipes in the building and the earth-wires of the nearest poles; or the discharge may have been a back stroke, and jumped from the gas-bracket to the wires, there dissipating to the clouds. The marks left by the lightning were unmistakable and indicated that the discharge was in reality a back stroke, for the brass test-plates were burned and blackened, while the bracket was unharmed. But the cause is of less interest to telephonists than the effect. One thing *is* certain, and that is that overground metallic circuits protected at one end only are *not* safe. As a consequence, all such circuits under my care are now being guarded as indicated in the figure. A third plate is fixed between the two test-plates usually placed at the window or other point at which the outside and inside wires meet. The two line wires are attached to the outer plates, and from the middle one a high-conductivity copper wire is led to the instrument, stapled between the two line wires. At the instrument it is joined to the terminal of the discharger thereon provided, and thence continued to a soldered connection with the water main or other efficient earth. Use of the gas-pipes is strictly forbidden for the purpose. The wires should be led as far away from the gas-fittings as possible, and crossing of gas-pipes should be avoided. The plates at the window should preferably be mounted on ebonite to secure the insulation of the loop, which would suffer if they were screwed direct to damp wood so near the earth-plate. The plates may be toothed as shown, but this is scarcely necessary, since the earth is so close to the line wires throughout the whole of their inside course that jumping in any other direction need not be feared. This arrangement, when properly carried out and maintained, will render metallic circuits as safe as single wires, whether from direct or back strokes. The expression "as safe as single wires" will not, I am afraid, find favour with some telephonists; but, nevertheless, I regard it as incontestable that single wires properly earthed on the water-pipes, or main gas-pipes on the ground side of the meter and beyond the last red or white-lead coupling, with all joints soldered, offer such a ready passage to any ordinary discharge that there is little likelihood of the lightning deserting the line for a gas-bracket or other object having a more or less perfect connection with the ground. Under such circumstances, a very heavy discharge has been known to fuse a long span of iron wire into globules and pieces not exceeding half-an-inch in length, and to destroy the bell-coils of the magneto, and yet do no damage to the building; whilst moderate discharges generally fuse the bell-coils only. The reason is readily understood, for there is always the earth on the magneto, and the lightning, after fusing the coils, has only an inch or so to jump to the earth-wire, which is still its best path. Of course imperfect earths of any kind are dangerous, and notably those taken off gas-pipes. One of the difficulties of practical telephony is to get the very young men employed by the companies as fitters and inspectors to realize this sufficiently and work accordingly. When a gas-pipe is handy and water a long way off, there is a great temptation to use the former, and I have found it necessary to rule strictly that no gas earths be employed without the express sanction of the engineer in charge, who is made personally responsible for the efficiency of any that it may be found necessary to use. In some cases it is better to sink a special earth-plate, or to run a span to a more suitable building, than to use the gas. Iron wire, or small-gauge copper, should not be employed for earth-leads, for although adequate for telephonic purposes, they have not sufficient capacity to carry off a lightning stroke harmlessly. With such precautions as these, danger from lightning has been reduced to a minimum for single wires. The accident described suggests the reflection that a telephone exchange system, consisting entirely or chiefly of overground metallic circuits, would not, without special precautions, be nearly so safe as one on the single-wire plan. At present, the open wires strung in all directions over a town act as a protection to the houses beneath them, for the multiplicity of earth-connections afford paths of next-to-no resistance for the discharge of atmospheric electricity, so that an actual storm with visible lightning is requisite to produce any effect appreciable by our senses. But a well-insulated metallic circuit system similarly situated would act differently. During electrical storms, indeed, charges—as already remarked—would be stored up until relief were obtained by a series of jumps to the lightning guards or other convenient earths. The switch-rooms and subscribers' stations could be protected, by proper precautions, at small cost, but the buildings supporting standards and wires would require to be fitted with expensive conductors in order to be safe. The roofs of few buildings in towns have any proper earth-connections; the effect of the gutters, spouts, and inside gas and water-fittings being to make the building the path of least resistance, thereby inviting a stroke that cannot be disposed of without disruptive effects when it comes. The placing of a copper conductor from the summit of each standard to the water-main in the basement would be the least that could be done. Such conductors are sometimes fitted now to satisfy the requirements of way-leave grantors, but with single wires they are not needed. It would be well if architects could be got to discountenance the use of fusible gas-piping. Lightning does not require electrical fittings to attract it—the discharge I have alluded to might well have taken place had no telephone wires been present, only, in that case, it would have passed straight from the house to the clouds—and destructive adjustments of potential may take place with no louder sound than a sharp crack, and at times when no visible storm is raging. After the occurrence described, I am by no means disinclined to think that at least a proportion of the destructive fires which are continually taking place without apparent cause are due to lightning acting on fusible gas-pipes.

Lightning Conductors on the Melsens System.

The following account by Mr. J. W. Pearse of the Melsens system is taken from the "Electrician."

The two leading systems of lightning conductors were ably defined, as follows, by M. Mascart, Professor of Physics to the College of France, at the Paris Electrical Congress of 1881:

1. The Gay-Lussac system, based on the use of a small number of conductors of considerable sectional area and rods of great height; and

2. The Melsens system, which consists in surrounding the edifice to be protected with a kind of metallic cage, formed by many conductors of small sectional area, and provided with short but numerous rods or points.

The reason why lightning conductors have hitherto been mounted so high above the edifices they are intended to protect is probably because most of the formulæ for giving the zone of protection have been based upon the height; therefore, the greater the height the greater the supposed zone of protection. But it is a remarkable fact that the zone of protection admitted in the famous instructions of the French Academy, drawn up by Gay-Lussac in 1823, has, according to the opinion of scientific men, continued to decrease, until at the present time it has dwindled down to nearly one-fiftieth of what was then considered admissible. It will be seen by the annexed diagram (Fig. 14) of the various zones of protection admitted by the leading authorities, that taking as unity the capacity of the zone given by Gay-Lussac in 1823, that of the various authorities has successively decreased

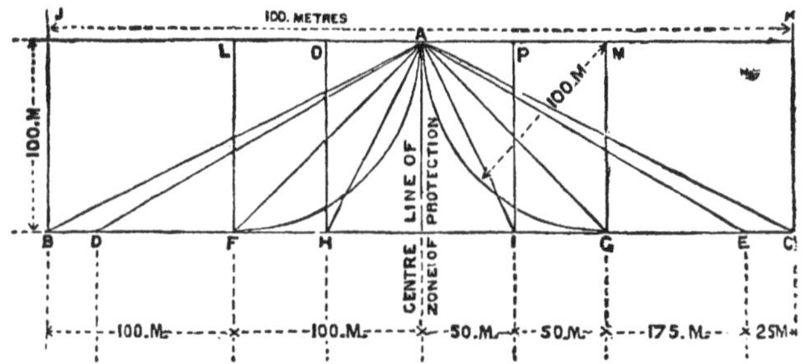

REFERENCE TO FIGURE.

1.	JBCK	Cylinder	Gay Lussac	1823
2.	BAC	Cone	De Fonveille	1874
3.	DAE	Cone	Paris Commission	1875
4.	LFGM	Cylinder	Chapman	1875
5.	FAG	Cone	Adams	1881
6.	OHIP	Cylinder	Hypothesis	
7.	FAG	Special Cone	Preece	1881
8.	HAI	Cone	Melsens	

Fig. 14. ZONES OF PROTECTION.

(approximately) to $\frac{1}{3}$, $\frac{1}{4}$, $\frac{1}{12}$, $\frac{1}{16}$, and $\frac{1}{42}$. M. Melsens is inclined to think that even $\frac{1}{48}$ of the original area cannot be implicitly relied upon. At any rate the transept of the church of Sainte-Croix, at Ixelles-lez-Bruxelles, was injured by lightning on the 3rd July, 1874, within a cone the radius of whose base is $1\frac{1}{4}$ the height of the lightning conductor, which was afterwards found to be in good condition. This is equal to one-fourth the zone originally supposed to be protected; and instances have occurred on board ship of lightning striking the deck, when the lightning conductor on the masthead remained intact.

M. Melsens has adopted as the motto of his system, "*Divide et impera*," and he carries it out by multiplying the terminals, the conductors, and the earth connections. In fact, his *paratonnerre* resembles a tree, with branches extending into the air, and roots ramifying in the soil. His terminals are very numerous, and assume the form of an *aigrette* or brush with five or seven points (Fig. 15), the central point being a little higher than the rest, which form with it an angle of 45°. The galvanized iron wires composing them are from six to eight millimetres thick; and the extremities, which may be of copper, are drawn out to sharp points, and should be tinned. The terminals are placed in great number on the more prominent parts of the building. It is now generally admitted that lightning does not strike buildings at a single point, but rather in a sheet. It is obvious that in such a case, or in the event of the globular form being assumed, the brush will constitute a much more effective protection than a single point. The conductors are also numerous, and consist each of a single wire of small sectional area (8mm. = 0·3 inch), so as to be easily laid, and readily follow the contour of a building. The wire should be employed in long continuous lengths, and certainly not in the form of a surveyor's chain, in which there is but slight contact between the links, and that a very poor one.

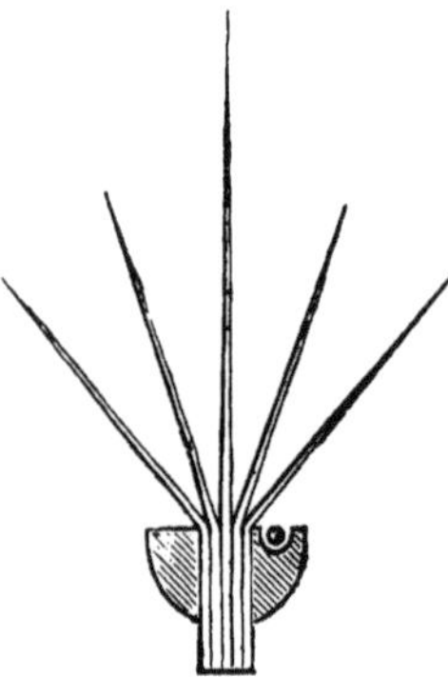

Fig. 15. AIGRETTE.

While copper may be advantageously employed for the terminals, on account of its resistance to oxidization, it is not adopted for the conductors, and this for the following reasons:—Its first cost is great; and its high value offers a temptation to theft, which would leave the building in a worse condition than if no lightning conductor had been provided.

But, more than this, although the conductibility of copper is six or seven times that of iron in the case of a continuous current of slight tension, M. Melsens' experiments show that, for long lengths of small sectional area, iron will conduct as well as copper an instantaneous discharge of high tension. He found that Ohm's law was not applicable in its entirety to sparks of a high tension furnished by charged Leyden jars, frictional electric machines, a Holtz machine, or a large Ruhmkorff coil, in which case the current passes with equal facility through conductors of the same sectional area composed of iron and copper. He has employed bundles or twisted ropes and *aigrettes* of the same shape, composed of different metals, but always obtained the same results. The current was sometimes divided between the two metals, although the portion in the iron seemed generally weaker than that in the copper. The current also passed alternately by one or the other; and a change of direction appeared to exercise no influence on the result of the experiment. There is this difference between the spark produced by the Holtz machine and that of the Ruhmkorff coil. In the former, a single spark passing simultaneously through two conductors, one iron and the other copper, of the same dimensions, generally gives two sparks issuing from their extremities, while in the latter the spark is not divided, but passes sometimes through one and sometimes through the other, even when the length, and consequently the resistance, of the iron is considerably increased. It is true that if Leyden jars, powerfully charged, be discharged through short iron and copper wires, say only a few centimetres long and $\frac{1}{3}$rd of a millimetre in diameter, the iron becomes red-hot and burns, whereas the copper remains intact. But if, on the other hand, copper and iron wires be taken several metres long and $\frac{1}{10}$th of a millimetre in diameter, or finer still, in this case the iron will stand, while the copper is partially melted in the form of a string of beads, or even completely pulverized, and projected to a distance in the state of dust composed of spherical metallic grains, and sometimes in the form of blackish dust, which is oxide of copper. Moreover, practical experience confirms the results ascertained by experiment, as ordinary telegraph line wires must frequently have been struck by lightning; but there is no record of their having been melted. With regard to the division of the current between several conductors, M. Melsens has found by experiment that it will take place between as many as 390 conductors, of conductivity varying from 1 to 6, and of diameters varying from 0·08mm. to 6·3mm., the direction of the current exerting no influence on the division. He concludes that the spark still becomes divided if, instead of striking the point where the several conductors meet, it strikes only one of them, provided, however, that this one be not too delicate. On placing any one of the 390 wires in contact with gas and water pipes, he could easily detect electrical manifestations in both of them. But the sectional area and conductivity of some of the wires experimented upon could not have allowed to pass more than the $\frac{1}{900}$ or even the $\frac{1}{2300}$ part of the single spark from a large Ruhmkorff coil. Indeed, subsequent experiments showed the remarkably small fraction of $\frac{1}{60000}$. The experiments proved, in addition, that, in the case of homogeneous wires of the same length, when deterioration takes place, it is the same for all; that is to say, it is divided equally among the conductors, or the mechanical energy is the same for all. Moreover, strong sparks from a Leyden jar battery, passing through fine wires stretched so as to be parallel, produced a series of irregular undulations, which were the same in all the wires.

The failure of lightning conductors is generally due to deficient communication with the earth. In order to transmit to water with perfect freedom—that is to say, without any other resistance than what is offered by a good conductor such as iron—the electricity which traverses it, or the lightning which strikes it, a conductor of a square centimetre (0·155 sq. in.) sectional area ought to terminate in a square iron plate, the side of which is 225 metres or 738 feet, entirely immersed. Again, to realize the same conditions in damp earth merely, the square plate ought to have a side of not less than 450 metres, or 1,476 feet. It is evident that these conditions cannot be carried out practically; but in order to reconcile practice as far as possible with theory, it is well to increase, by all the means at disposal, the surface of contact with water or damp soil; to increase the superficial area of the underground portion of the conductor in the pit; and, especially in towns where large buildings are chiefly situated, to connect the conductor with the immense ramifications of the gas and water pipes, and also with warming and ventilation tubes. It is true that M. H. Aerts, manager of the Brussels Gas Works, in a paper before the Society of Belgian Gas Engineers, objected to the connection of lightning conductors with gas-pipes, on the ground that joints were frequently formed of non-conducting substances. But M. Melsens contends that continuity is preserved by the approximate contact of iron with iron, and also by the damp soil surrounding large

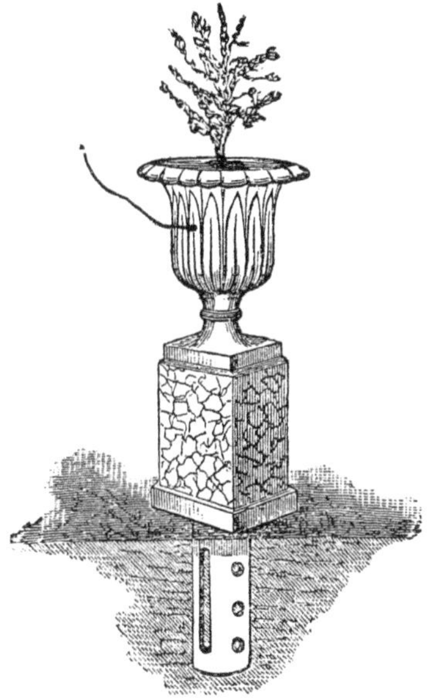

Fig. 16. EARTH CONNECTION.

pipes, and that there is practically nothing to fear on this account; at the same time, such gas-pipes should not constitute the only earth contact. It is highly important that the earth connection should be easily accessible to inspection, so as to make sure that, if water, it do not fail, and if earth, it be always damp. In the case of farms and isolated country houses, if the connection be not made with a well or pool, M. Melsens recommends that the conductor terminate in a hollow cylinder of cast iron perforated with holes, projecting above the ground, so as to be always in sight. The upper part may be made into the form of a vase (Fig. 16), so as to receive a plant that requires a great deal of water, when its thriving or otherwise will afford a good indication of the state of the ground below.

The conditions for protecting edifices from the lightning shock have changed considerably since Franklin's discovery. Large masses of iron are now employed in building which it is highly important to place in electrical communication with the conductor. And to this end it is also important that the lightning conductor be taken into consideration in the design of a building, just as are the arrangements for warming and ventilation. This condition has been realized in the case of the new athenæum now being built at Antwerp, in which provision has been made for connecting all the ironwork with the lightning conductor from the very foundations. On the other hand, the want of such foresight is strikingly exemplified in the new Palais de Justice, Brussels, in which there are no less than 9,615 tons of iron. Of this quantity it has been found impossible, except at too great expense, to connect 5,887 tons, or more than half, to the conductors on the Melsens system, connected with numerous brushes on the roof. If the lightning conductor had been taken into consideration in the design, it would have been possible, at comparatively slight expense, to make every particle of metal concur in the system of protection, and, by thus providing innumerable channels by which the electric fluid might reach the common reservoir, to have rendered the building absolutely invulnerable to the lightning shock. Fortunately, however, in the case of the dome (102½ metres = 336 feet high), the point most exposed to be struck, everything was provided for from the commencement; and all the iron in this part of the structure, weighing no less than 3,252 tons, is connected with the conductors and *aigrettes*. The following rule, which is rather more strict than that generally admitted, is laid down by M. Melsens on this subject:—"All parts of metal, if they be of any considerable size, should be placed in communication with the lightning conductor, in such a manner as to form closed metallic circuits, that is to say, by two points, or with connection to two leads at least." At the same time, if the greater portion of the ironwork be electrically connected, so as to realize Faraday's cage, some portion may be left unconnected. This fact has been proved by the following experiment, which M. Melsens lately communicated to the Belgian Academy of Science. On placing a rat, or other small animal in a cage composed entirely of metal, and leaving its body in immediate contact with the wires, a current of sufficient tension to cause instant death may be sent through the cage without affecting its occupant. If the tail be held forcibly outside the cage and in proximity with the opposite pole of the battery, a current sent through the cage will leave the wires where the tail projects, and only affect that organ, just as it will if sent through the tail and out by the cage.

M. Melsens holds, with Gay-Lussac, that lightning conductors for powder mills and magazines need not differ

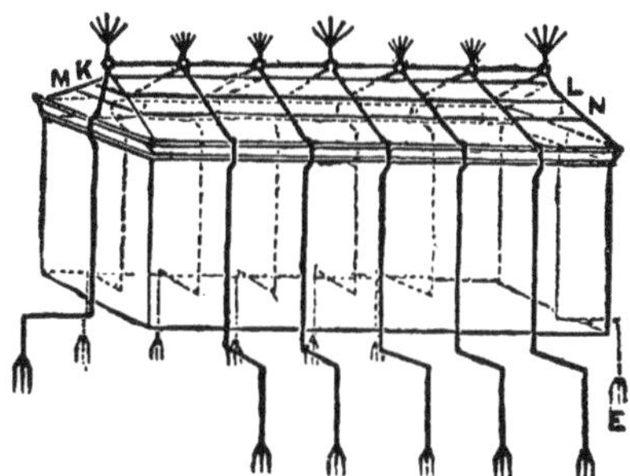

Fig. 17. POWDER MAGAZINE.

essentially from those for any other kind of building. He does not disapprove of the use of metal in their construction, provided it be all in electrical communication, and be sufficiently connected with water or damp earth. If—which is not likely—there be gas or water pipes in proximity, the conductors should be placed in connection with them. As all the electricity with which a body is charged remains on the outside, the interior, even if electrified by powerful apparatus, is exempt from electrical manifestations, as proved by the most sensitive instruments, and also by Faraday's experiments. M. Melsens therefore advised the director of the Wetteren Powder Mills, covering 24 hectares (nearly 60 acres) of ground, to provide all the buildings with lightning conductors having multiple terminals, leads, and earth connections, and to put all of them in intercommunication, thus constituting one single lightning conductor.

Fig. 17 shows a type of powder magazine most exposed to be struck by lightning. Instead of employing only two conductors of large sectional area, crossing one another over the roof, as recommended by Gay-Lussac, M. Melsens adopts several transverse conductors of small sectional area, crossed by one longitudinal, with *aigrettes* at the intersections, and further connected by belts, *K L* and *M N*, consisting of iron wire six to seven millimetres (about $\frac{1}{4}$ inch) in diameter. This arrangement, which realized the Romas metallic cage, affords a comparatively large number of earth contacts, which may be easily supplemented by a branch leading to a sheet of water or damp soil, either closely adjoining, though sufficiently far from the powder, or at some distance. There is nothing to prevent the magazine, thus protected, from being surrounded by poles surmounted by *aigrettes*, connected by wires crossed in all directions, thus realizing a double cage, and affording additional protection. If by any chance the lightning were not arrested by the outer network, it would be hardly likely to penetrate the inner also.

A similar application to telegraph lines, consisting in providing each post with an *aigrette*, connected with the earth, and also with a supplementary wire led above the rest, would probably destroy, or at any rate considerably weaken, the instantaneous or other currents in telegraph wires which so frequently interrupt the service. M. Melsens has so simplified the Marianini rhe-electrometer that it can be made at a cheap rate and easily applied. The indications of this instrument, when fitted both to telegraph wires and to the protecting wire mentioned above, would probably permit of ascertaining the direction of induced and other currents, and would, also, when placed in suitable positions, probably afford valuable information as to the meteorological phenomena which accompany storms.

The application of this rhe-electrometer to lightning conductors generally would also permit of ascertaining the direction of the electric fluid, because there is reason, supported by Faraday's opinion, to believe that the lightning discharge commences from the earth more frequently than is generally supposed. If this be the case, a lightning conductor provided with numerous points is in a far better position for transmitting the electric fluid to the clouds than one having only a single point.

The lightning certainly took an upward direction in a remarkable discharge which occurred on 10th July, 1865, at the Antwerp Railway Station, and which cannot be accounted for by any of the theories at present admitted. The electric fluid passed through a pane of glass 4mm., = 0·15 in., thick, in the

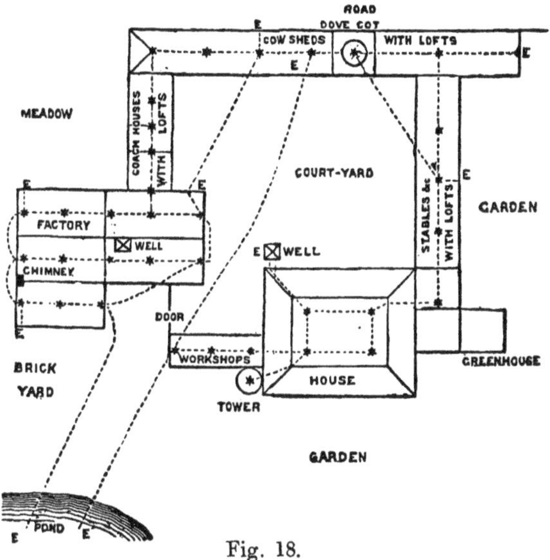

Fig. 18.

station roof, making a hole like that produced by a projectile in an upward direction, moving at the rate of 100ft. to 150ft. a second, the edges of the hole being melted. Strange to say, the electric fluid passed through a bad conductor, while only a few centimetres distant were lead and iron conductors in perfect metallic connection with all the iron of the station, exceeding 120 tons in weight. But the anomaly does not stop here, because. to the right and left of the lantern in which was situate the pierced pane, the zinc roof of 2,500 square metres (26,910 square feet) weighs at least 15 tons, while three high lightning conductors were in immediate contact with the zinc, the whole being connected to twenty-eight cast-iron columns, which also served to carry off the water from the roof.

The preventive as well as protective action of such a lightning conductor is shown by the observations, extending over

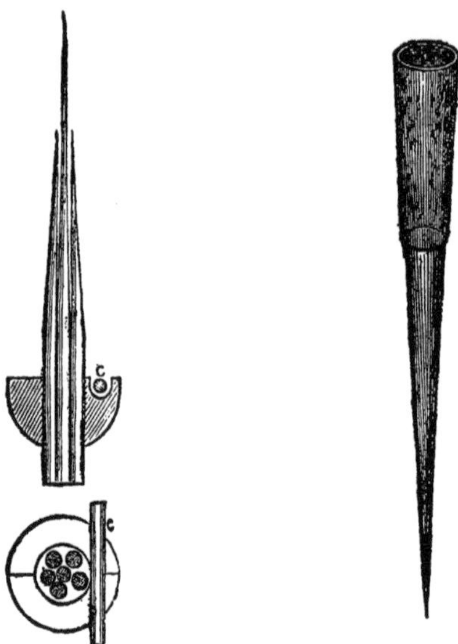

Fig. 19. AIGRETTE. Fig. 20. MODE OF FIXING AIGRETTE.

several years, of Dr. Mann, President of the Meteorological Society, who states that, at Pietermaritzburg, in Natal, after the erection, at his instigation, of a large number of lightning conductors provided with numerous points, lightning shocks, which had been frequent, then became very rare.

M. Melsens is of opinion that insufficient attention is paid to the protection of small houses and farms in the country by means of inexpensive lightning conductors. Fig. 18 shows the plan of a farm protected under his direction, and, as he believes, placed in perfect security from lightning. The roofs, in nine different horizontal planes, are about 300 metres (985 feet) long. The *aigrettes,* to the number of 36, having altogether 210 points, are shown by stars; and the dotted lines represent the conductors, which follow the ridges, or are suspended from one to another. There are 11 earth contacts, marked *E*, including two in the pond, and two others connected with the pump pipes of two wells. They are arranged so as to be always open to inspection; and testing with the galvanometer showed that there was scarcely any resistance to the current. The *aigrettes* are of the form shown at Fig. 19. They consist of iron wire of six or seven millimetres (about $\frac{1}{4}$ inch) in diameter, cut to lengths of 1, $\frac{3}{4}$, or $\frac{1}{2}$ metre, sharpened with a file, and tinned or galvanized. Five or six are bound together with wire and soldered, and then surrounded, by means of an iron mould, with a lump of zinc, leaving a groove for the conductor *C*. The *aigrette* is inserted in a hollow spike (Fig. 20) made out of gas-pipe and galvanized, which is driven into the timber of the roof. The branches are then bent out so as to form an angle of 45° with the central point, as shown at Fig. 15. An exact account was kept of the cost of this conductor; it is as follows:

	fr.	c.
600 metres of iron wire, 6mm.=$\frac{1}{4}$in. in diameter	91	30
Total cost (labour and materials) of 37 *aigrettes*, with galvanized points, at 1fr. 60c.	59	20
Old zinc for joints of earth contacts, 20 kilogrammes at 0fr. 35c.	7	00
Solder, 2 kilogrammes at 2fr. 50c.	5	00
Labour, including fixing on roof	150	00
120 hooks for fixing conductors	24	00
Pitch, tar, oakum, and drain-pipes	20	00
10 cast-iron pipes, 420 kilogrammes at 8c.	33	60
£15 12*s.* 1*d.* =	390	10

Taking an even sum, 400fr. or £16, this outlay represents 20c., or 2*d.* per square metre (1·196 square yard) of surface protected. The cost of lightning conductors on the Melsens system applied to three large public buildings in Brussels is, on an average, 66 centimes, or 6*d.* per square metre of surface protected. These are the Bourse (47c.), the new Palais de Justice (65c.) and the Hôpital St. Pierre (47c.) Compared with these figures, the protection on the old system of several large buildings cost on an average 4fr. 46c. (3*s.* 6*d.*) per square metre. Among them may be mentioned the king's palace at Brussels (3fr. 2c.), the king's stables (6fr. 23c.), the king's palace at Laeken (9fr. 68c.), and the hothouses at the Botanical Gardens (3fr. 99c.).

Sir William Thomson relates that when he urged the

Scotch manufacturers to protect their works, they contended that it was cheaper to insure them than to erect lightning conductors. Now, M. Melsens wishes to change all that, and especially to eliminate the empirical element, so that "everywhere, both in town and country, one may afford the luxury of protecting one's habitation by a lightning conductor, just as one now has the luxury of a fireplace to counteract the cold, and a chimney to carry off the products of combustion." The *aigrettes* or brushes, such as are described above, may be made by a country blacksmith at a cost of 2*s*. to 2*s*. 6*d*. each, including both labour and materials, while a highly-finished specimen in phosphor bronze may be produced for 12*s*.; at the same time it is far easier and cheaper to erect several small conductors than one large. M. Melsens has abstained from taking out any patents for his applications, but places his researches at the disposal of all. He also gives his advice freely to those constructors who consult him on the subject which he has so deeply studied. He contends that his lightning conductor, with its perfect and multiple earth contacts, but provided with numerous divergent points placed on the exterior of an edifice, is represented by Faraday's cage, and that these points have certainly not the property of provoking electrical manifestations in the interior of a metallic framework in perfect communication with the common reservoir. The sharp points adopted are but a return to Franklin's original model, conical points more or less obtuse having been employed subsequently. There are many thousands of these points in Belgium—the Brussels Hotel de Ville is furnished with no less than 510; and yet, when properly tinned or galvanized, they have not become blunt.

Tall Chimney Climbing.

The following is an extract from a pamphlet issued by Messrs. Richard Anderson and Co. (successors to Sanderson and Co.), 101, Leadenhall Street, E.C.

The plan which was adopted in the earlier days of big chimneys for gaining access to their tops for purposes of examination and repair was deserving of some admiration and praise on account of its ingenuity. It was managed by bringing into operation the ascending power of the kite. A kite was flown by a trained and skilful hand over the top of the chimney, until its string was placed obliquely across the orifice of the shaft, and the kite was then pulled down to the ground by a second string attached to the one which was used in managing and controlling the flight, leaving, in this way, the string looped over the top. The kite being then removed, a stout cord was attached in its place and drawn over the top of the chimney until the cord had taken the place of the string, rising from the ground, crossing over the mouth of the chimney, and descending to the ground at the other side. This process was then repeated, stouter and stouter cordage being used each time, and finally a strong iron chain, until at length a tackle was raised and fixed, from which an adventurous workman could be pulled up to complete the adjustments and attachments of more reliable machinery above. In instances in which hot vapours were issuing from the chimney during the raising of the tackle, the part of the kite-string which had, in the first instance, to be looped across the top of the shaft, was formed of a strand of metal wire. In most of the large chimneys of this earlier date, upwardly curved hooks of iron were left fixed at the rim in a position conveniently arranged for catching the kite-string.

Some practice and skill were required for the attainment of distinction in the art of kite-flying for getting at the top of tall chimneys. Mr. Solomon Sanderson, of Huddersfield, had some years ago acquired a high reputation for his successful practice of this craft. As recently as nine years ago it was no uncommon thing to come across him practising with his kites at Melton Moor, five miles from Huddersfield, where he seems also to have had great delight in repeating Franklin's renowned experiment of getting sparks from the string of a kite, when highly-charged storm-clouds were hovering above. But the tall chimney engineering owes a larger debt to this ingenious constructor, who died about three years ago, after having spent a long life in useful and successful work. He contrived the means which have now practically superseded the use of the kite, and which it is one object of this little sketch to bring under review.

One very great disadvantage of the kite-flying process was the delay that continually occurred in getting the tackle attached to the top of the chimney by its instrumentality. A contracting firm who had undertaken any particular work of reconstruction or repair, very naturally hesitated to send down a competent staff of workmen to any distant place until there was good assurance they could at once enter upon their task. A kite-flyer was therefore despatched as a preliminary measure, to establish a practical connection with the chimney-top. But when this avant-courier was once well away from the superintending eye, it very seldom indeed happened that a favourable wind could be secured. The public-houses of the place, which naturally became the refuge and resort of the kite-bearing artist and messenger, appear to have exerted some very curious meteorological influence upon the direction and force of the currents of the air. Weeks, and, in some special instances, months, slipped by before a favourable and manageable breeze would present itself for the raising of the kite. It was in these embarrassing circumstances that Solomon Sanderson determined to contrive some upward path that would be independent alike of the caprices of the wind and of the seductions of the drinking-shops. He signally succeeded in his design, and about fifteen years ago he introduced the ingenious method of getting at the tops of tall chimneys which is now almost universally followed.

Mr. Sanderson's method consists of pushing length after length of short segments of a ladder, as it were telescopically, up against the perpendicular face of the shaft of the chimney, and of climbing simultaneously upon the lengthening-out ladder as it goes—a most formidable-looking proceeding, it will be allowed, when it is a chimney of 250 or 300 feet that is so attacked, but one which has, nevertheless, been so perfected by the sagacity of the inventor and his successors, that it is now employed, in the hands of good climbers, with an almost complete immunity from dangerous risk.

The ladders which are used in this process were in the first instance in lengths of twelve feet, until it was shrewdly pointed out by some workman familiar with the conditions of railway transport how great an advantage would be derived from changing the standard measure of the section of the ladder to fifteen feet, because that corresponds with the adopted length of the railway truck. Fifteen feet ladders are as easily carried by railway as twelve feet. There is consequently a material saving in the carriage of ladders of the fifteen feet span, when large works are in hand.

It will, therefore, be understood that a number of ladders of fifteen feet length are in the first instance prepared, which are identical with each other in detail and form, and which are so fashioned that the bottom of any one ladder can be dropped into sockets provided at the top of any of the rest. The sides of each segment are pivots at the bottom and sockets at the top. There are also standards or pegs about eight inches long projecting out from one face of each segment, which serve the purpose of keeping it just so far off from the brickwork when it is fixed, and of, by this means, providing a secure foothold and handhold.

The first step in the erection of the ladder consists in placing one of the sections standing perpendicularly upon the ground, against the bottom of the chimney. A workman then drives an iron dog or holdfast firmly into the brickwork one foot up from the bottom of the ladder, and one foot down from its top. These holdfasts are of a hooked form, so that they can each be made to clamp one of the rungs of the ladder when they are driven home upon it into the brickwork. The segment of the ladder is as firmly attached to the shaft of the chimney, when this has been accomplished, as it would have been if it were originally an essential part of the structure.

When one section of the ladder has been attached in this way, a free ladder is sloped against it, and the climber then ascends upon this until he can reach about a foot above the top of the fixed segment. He there drives in a holdfast, and attaches to it a pulley and block, so that one end of the rope reeved into the pulley can be brought half down a second loose section of the ladder, placed perpendicularly and side by side with the first. The rope is there fastened at midway height, and by means of the block the second section of the ladder is hauled up by men standing upon the ground until it projects half-ladder height above the section No. 1. In that position it is temporarily lashed to the fixed section, rung to rung, so that the climber can mount to its top, and drive a holdfast into the brickwork a foot above its upper extremity. He then shifts the pulley and block to this upper holdfast and descends to the ground. Section 2, still attached to the rope at its middle part, is then hoisted up to its full height above Section 1. The climber, following its ascent, next inserts the bottom of its sides into the sockets at the top of section No. 1, mounts upon its steps as, still held by the pulley, it leans against the chimney, drives home two hooked holdfasts, clamping its rungs to the chimney, near the bottom and near the top; and this having been done the second section remains fixed in continuation of the first, and the ladder attached to the brickwork, and, affording a practicable way to the climber, has thus grown from fifteen to thirty feet of continuous height. The climber is then able to mount to its top, thirty feet up on the chimney, and, extending his arm about a foot higher upon the brickwork, drives in there the holdfast which becomes the *point d'appui* for the hauling up a third section of the ladder, first half its length and then full height, above the second segment, so that it can be in its turn pivoted into the sockets. The third section, in doing this, is handled in every essential particular like the first, pulled half-ladder high, temporarily lashed to the topmost rungs of the fixed ladder, then lifted to its full height, pivoted into the sockets of the fixed ladder there, and clamped firmly to the brickwork, and the fixed ladder has grown to a length of forty-five feet, by the junction of three segments of fifteen feet each. This process is afterwards repeated with other sections of the ladder again and again, half-lengths at a time, until a perpendicular path has been laid from the bottom to the top of the chimney. A chimney 355 feet high, it will be observed, requires seventeen sections of the ladder to reach to its top.

The essential points in this ingenious process which furnish a ready explanation of its success, thus are:—(1) the temporary lashing of each section of the ladder when it is half way up, so that the climber can get safely to the top, as it is held still attached to the pulley, and fix a fresh block above its upper extremity for the accomplishment of the second half of the hoist; (2) the joining of the sections by appropriate sockets as each one is placed in position upon the one beneath; and (3) the fixing of each section, when it is once lifted into its place, by the holdfasts driven into the brickwork of the chimney. The ladder virtually creeps up to the top of the chimney, joint above joint, and fixes its tenacious fangs into the brickwork as it goes. The process is so easily performed by practised hands that the highest chimneys are scaled in incredibly brief intervals of time. The chimney at the Abbey Mills pumping station, and which is some 230 feet high, was laddered completely from the ground to the summit in three hours and a half.

INDEX.

CHISWICK PRESS:—C. WHITTINGHAM AND CO., TOOKS COURT, CHANCERY LANE.

Milton Keynes UK
Ingram Content Group UK Ltd.
UKHW032320161024
449665UK00001B/21